Fundamentals of
Atmospheric Energetics

Fundamentals of Atmospheric Energetics

Aksel Wiin-Nielsen
Geophysical Institute
University of Copenhagen

Tsing-Chang Chen
Department of Geological and Atmospheric Sciences
Iowa State University

New York   Oxford
OXFORD UNIVERSITY PRESS
1993

Oxford University Press

Oxford   New York   Toronto
Delhi   Bombay   Calcutta   Madras   Karachi
Kuala Lumpur   Singapore   Hong Kong   Tokyo
Nairobi   Dar es Salaam   Cape Town
Melbourne   Auckland   Madrid

and associated companies in
Berlin   Ibadan

Copyright © 1993 by Oxford University Press, Inc.

Published by Oxford University Press, Inc.,
200 Madison Avenue, New York, New York 10016

Oxford is a registered trademark of Oxford University Press

All rights reserved. No part of this publication may be reproduced,
stored in a retrieval system, or transmitted, in any form or by any means,
electronic, mechanical, photocopying, recording, or otherwise,
without the prior permission of Oxford University Press.

Library of Congress Cataloging-in-Publication Data
Wiin-Nielsen, A.
Fundamentals of atmospheric energetics /
Aksel Wiin-Nielsen, Tsing-Chang Chen.
p. cm.   Includes bibliographical references and index.
ISBN 0-19-507127-1
1. Dynamic meteorology.   2. Atmospheric physics.
I. Chen, Tsing-Chang.   II. Title.
QC880.W49   1993
551.5—dc20   92-24612

1 3 5 7 9 8 6 4 2

Printed in the United States of America
on acid-free paper

PREFACE

Atmospheric energetics, as we know it today, has developed over the last few decades. Its advance has been fostered by a number of events that have made it possible to gain considerable insight into those physical processes responsible for the behavior of the atmosphere from an energetical point of view.

The most important development has been the establishment of an entirely new framework for studying the energy relations of the atmosphere. Atmospheric energetics has benefited not only from the introduction of new concepts such as the available potential energy, but also from research that has proceeded to the baroclinic and barotropic states, the nondivergent and divergent parts of the atmospheric flow. In conjunction with developments in both numerical weather prediction and simulations of the atmospheric general circulation, procedures have been developed to study atmospheric energetics by Fourier analysis and decomposition using other sets of orthogonal functions, in particular the spherical harmonic functions. Progress has been made in research on the nonlinear interactions among spectral components, and while some problems remain unsolved, our understanding of atmospheric energy processes has greatly increased the speculative ideas of earlier times.

In contrast to the time before the establishment of a reasonably dense global network of observations and the invention of electronic computers, most quantities appearing in the field of atmospheric energetics may now be estimated. However, calculations in this field depend typically on quantities such as the net heating, the vertical velocities, and the frictional forces. None of these is observed directly, and no general theory is available for their calculation. Uncertainty is therefore still a part of almost all the estimates.

The present book is written for the graduate student in the atmospheric sciences. The foundation of atmospheric energetics rests on a mathematical framework, which in the beginning may seem difficult and cumbersom to handle. Most papers on atmospheric energetics published today assume knowledge of this foundation. We have chosen to deal with the mathematical derivations in some detail, but it is indeed true that most results can be obtained if one is familiar with integrations by parts and masters the conversion of area integrals to line integrals along the boundary. Even so we are aware of the long deviations, which appear in the first part of the text. On the other hand, we have selected to stay with the framework employed in the great majority of studies in atmospheric energetics. We feel that the more advanced treatments of the concept of available potential energy are well represented in other books or review articles. For the same reason we have abstained from using vertical coordinates other than height and pressure except when it has been convenient for our purposes, as in the chapter on available potential energy.

The theory behind atmospheric energetics is developed for the global domain. Applications to limited, open domains require at least that the boundary processes are taken into account. This condition is the only difficulty, if one stays with those energy quantities that can be defined locally, such as internal, potential, and kinetic energy. The application of the concept of available potential energy to limited, open domains is difficult due to the requirement of a reference state, which in the general formulation is global in nature. A local reference state may be defined, but it is seldom independent of time, which makes comparisons between two events separated in time very difficult. Nevertheless, there is considerable interest in the energetics of more localized atmospheric events. We have tried to deal with the results of the studies of energetics in open domains in the latter part of the book.

The manuscript for the book was prepared in stages over the last few years. The first author wants to express his appreciation to the Department of Atmospheric, Oceanic and Space Sciences at the University of Michigan for the great assistance given during a visiting professorship in the winter term of 1988, where the theoretical chapters were typed for the first time. The final manuscript was prepared in part during a visit by the first author to the Department of Atmospheric Sciences at Iowa State University in January 1991. The contribution of the second author to the book was derived from his past research of atmospheric energetics. This research was made possible under the sponsorship of the National Science Foundation, the National Space and Aeronautics Administration, and the National Oceanic and Atmospheric Administration. Partial support for the illustrations was provided by Iowa State University.

Comments and suggestions were offered by the following professors: Dr. Phillip Smith of Purdue University, Dr. Joseph Tribbia of the National Center for Atmospheric Research, and Mrs. C. Wiin Christensen of Denmark. We also would like to thank Jason Lohman, Steve Heistand and Dr. Ming-Cheng Yen for their technical assistance in the final manuscript.

Copenhagen, Denmark   A.W.-N.
Ames, Iowa   T.-C.C.

CONTENTS

1. Introduction, 1
2. Some Elementary Considerations, 4
3. Basic Aspects of Atmospheric Energy, 15
4. Available Potential Energy, 31
    4.1 The Quasi-Geostrophic Case, 41
    4.2 An Elementary Derivation, 43
5. Baroclinic and Barotropic Flow, 46
6. Transports of Sensible Heat and Momentum, 55
7. Zonal and Eddy Energies, 70
8. Divergent and Nondivergent Flow, 85
9. Wavenumber representations, 95
10. Interaction Among Waves, 110
11. Energetics and Predictability, 123
    11.1 Nondivergent, Horizontal Flow, 124
    11.2 The Quasi-geostrophic Case, 126
12. Energetics of An Open Domain, 131
    12.1 Eulerian Energy Budget Analysis, 131
    12.2 Quasi-Lagrangian Energy Budget Analysis, 136
    12.3 The Kinetic Energy Budget of Baroclinic and Barotropic Flow in an Open Domain, 140
    12.4 The Kinetic Energy Budget Of Rotational and Divergent Flow in an Open Domain, 145
13. Energetics of Some Special Phenomena, 151
    13.1 Subtropical Jet Streams, 151
        13.1.1 Regional Ageostrophic Circulation Mechanism, 152
        13.1.2 Hemispheric Interaction Mechanism, 158
        13.1.3 Relation Between the Two Mechanisms, 161
    13.2 Spectral Energetics of Blocking, 162
        13.2.1 Spectral Energetics Analyses of Some Blocking Events, 164
    13.3 Energetics of Stationary Eddies, 177
        13.3.1 Energetics Scheme, 178
        13.3.2 The Three-Dimensional Structure of Stationary Eddies, 181
        13.3.3 Energetics of Stationary Eddies, 184

14. Quasi-Periodic Variation of Atmospheric Energetics, 194

  14.1 Annual Variation in the Northern Hemisphere, 195
     14.1.1 Annual Variation of the Lorenz Energy Cycle, 195
     14.1.2 Fourier Analysis of Energy Variables, 200
  14.2 Annual Variation of the Kinetic Energy Budget over North America, 204
  14.3 Vacillation of Atmospheric Energetics, 210
     14.3.1 Vacillation of Eddy Energy, 211
     14.3.2 Energy Vacillation of Long- and Short-Wave Regimes, 216

15. Energetics of the Tropics: Planetary Scale, 224

  15.1 Overview of Tropical Planetary-scale Circulation, 224
  15.2 Conventional Spectral Energetics, 231
  15.3 Low-Frequency Variation of Tropical Energetics, 237
  15.4 Spectral Energetics of Baroclinic and Barotropic Flows, 246
  15.5 Spectral Energetics of Tropical Divergent and Rotational Flows, 251
  15.6 Spectral Analysis of the Tropical Enstrophy, 257
  15.7 Kinetic Energy Budget of the Tropical Easterly Jet, 268
  15.8 Exchange of Kinetic Energy between Low and Middle Latitudes, 271

16. Energetics of the Tropics: Synoptic Scale, 275

  16.1 Equatorial Waves over the Western Pacific, 276
  16.2 African Waves, 283
  16.3 Monsoon Depression, 291

17. Energetics of the Southern Hemisphere, 297

  17.1 Comparison of the Annual Variations in the Atmospheric Energetics between the Southern and Northern Hemispheres, 299
     17.1.1 Sensible Heat and Momentum Transport, 299
     17.1.2 Annual Variation of Energetics, 304
  17.2 Spectral Energetics, 309
  17.3 Vacillation of the Southern Hemisphere Atmospheric Energetics, 315
  17.4 Jet Streams, 320
     17.4.1 Summer Australian Jet, 323
     17.4.2 Winter Jets, 325

Problems, 330

Exercises, 342

Answers, 351

Bibliography, 354

Author Index, 369

Subject Index, 371

Fundamentals of
Atmospheric Energetics

1

INTRODUCTION

Atmospheric energetics is a part of the general field of dynamic meteorology. Its purpose is to account for the amounts of energy of various kinds in the atmosphere and to determine, theoretically and observationally, the ways in which energy is generated, transformed, and dissipated in the atmosphere. The energy relations applicable to the global atmosphere are derived from the physical laws by suitable integrations over the spherical domain. The result therefore, describes the behavior of the atmosphere in this integrated sense. The general goal of the research is to describe the flow of energy through the atmospheric system from generation, through one or several transformations, to dissipation, which is a reconversion into heat.

It is often of interest to investigate the contributions from various atmospheric systems to the total amount of a given kind of energy or to a certain transformation. One may attempt to do this by dividing the atmospheric fields into components and treating each of them separately. The formal requirement is that a suitable equation governing the development of each component can be found. The general situation is that the various sub-components will interact with each other. It is thus necessary to account for these interactions, which are normally expressed in terms of transformation among the components. Such subdivisions of the fields are naturally artificial. They have nevertheless helped to settle certain questions posed by classical meteorology.

The field of atmospheric energetics is relatively new because its development had to await the establishment of a network of upper air stations permitting evaluation of the various processes that enter the theoretical framework. Studies based on observations started to appear around 1950, using the radiosonde stations established during and after World War II. This is not to say that there was no interest in atmospheric energy relations before this time. Notable among early investigations are those by the Austrian meteorologist, Margules, who investigated energy relations in relatively simple fluid systems consisting of two or more layers, each of constant density and separated by surfaces of discontinuity. Such systems, initially in an unbalanced state, will eventually approach a stable arrangement under the assumption of no mixing among the layers. Margules was able to compute the initial and final energy amounts and, selecting suitable arrangements he sought to show that kinetic energy of some atmospheric systems could indeed be obtained from transformations from other forms of energy.

The modern framework for atmospheric energetics was originally formulated by *Lorenz* (1955), who introduced the concept of available potential energy and provided a decomposition of the energy into that of the zonally averaged flow and that of eddies, the latter being defined as the difference between the total flow and the zonally averaged flow. Operating in a hydrostatic atmosphere with just two kinds of energy – that is, available potential energy and kinetic energy – and subdividing each of these into zonal and eddy kinetic energy, he produced a four-box diagram that in a schematic way could illustrate the generations, transformations, and dissipations. This framework was used almost immediately by *Phillips* (1956) to summarize the results of his now classical numerical experiment that simulates the general circulation of the atmosphere using a beta plane geometry and a two-level quasi-geostrophic model.

While Lorenz and Phillips (loc. cit.) could speculate on the amounts of energy in the various reservoirs as well as on the directions and magnitudes of the generations, transformations, and dissipations, it fell to other investigators to calculate all these quantities from samples of atmospheric data. Such data studies had been initiated a few years after World War II by a number of research groups concerned mainly with describing the general (or averaged) circulation of the atmosphere. They were interested primarily in providing zonally-averaged pictures of the distribution of temperatures and winds in the meridional plane, but also in studying the structure of atmospheric waves or eddies by calculating the meridional transport of heat, moisture, and momentum by the waves – simply because the zonally-averaged fields interact with the eddies, as one can see from the appropriate equations. As it turned out, when Lorenz formulated the energy framework, it was exactly those meridional transports that were the elements entering the energy transformations between the zonal and the eddy energies. During the following decade or so it was possible to make a first estimate of all quantities in Lorenz's energy diagram, and he could report on the contributions from a large number of investigations in his book: *"The nature and theory of the general circulation of the atmosphere"* Lorenz (1967) based on the first IMO lecture.

Some expansions of the original four-box diagram have been proposed. A further vertical subdivision in the vertical mean flow, the so-called barotropic component, and the deviation from this flow, the vertical shear flow – or the baroclinic component – has been proposed and used by *Wiin-Nielsen* (1962) in studies based on observations and, independently, by *Smagorinsky* (1963) in the analysis of his basic general circulation experiment.

A further breakdown of the eddies has been proposed by *Saltzman*(1957), who formulated the equations for the energetics by employing a Fourier analysis along the latitude circles while maintaining the finite difference representation in the meridional and vertical directions. The formulation in terms of spherical harmonic functions came as a by-product of the so-called

INTRODUCTION

spectral prediction models. However, the total atmospheric energetics has not yet been performed in the two-dimensional spectral domain.

The concept of available potential energy has been treated further by *Dutton and Johnson*(1967) and by *van Mieghem*(1956). *Lorenz*(1978) has also returned to the question of available potential energy in a moist atmosphere.

Atmospheric energetics is treated in a monograph by *van Mieghem*(1973), who uses a very general approach that permits a treatment of energetics on many atmospheric scales. He covers a broad spectrum of problems, but his treatment of large-scale atmospheric energetics is rather schematic due to the fact that his book necessarily must avoid many details. It appears therefore that there is a need for a textbook on the subject directed toward the graduate student of meteorology, giving a rather didactic treatment and, in any case, incorporating some later research results. The present text has been prepared with the purpose of covering this additional need.

2
SOME ELEMENTARY CONSIDERATIONS

Before we turn our attention to the atmosphere itself, it is useful to consider some simple cases to illustrate the general concepts to be used later. As a first example we consider a homogeneous fluid in a cylindrical container. Let the height of the fluid be $H = H(x, y, t)$ and let the horizontal area be S. The potential energy per unit mass is gz, as is known from physics. As a reference level for z we use the flat bottom of the container. The total potential energy per unit area is therefore

$$P = \frac{1}{S}\int_0^H \int_S gz\rho \, dS \, dz = \frac{1}{S}\int_S \frac{g\rho H^2}{2} \, dS. \tag{2.1}$$

We may write H as

$$H = \overline{H} + H' \tag{2.2}$$

where $\overline{H}$ is the averaged depth of the fluid, – that is,

$$\overline{H} = \frac{1}{S}\int_S H \, dS. \tag{2.3}$$

We find therefore

$$P = \frac{g\rho \overline{H}^2}{2} + \frac{1}{2}g\rho \frac{1}{S}\int_S H'^2 \, dS. \tag{2.4}$$

Equation (2.4) shows that P is a minimum provided $H' = H'(x, y, t)$ is equal to zero everywhere – that is, if the fluid has a constant depth, in which case

$$P_{\min} = \frac{g\rho}{2}\overline{H}^2. \tag{2.5}$$

We may define a quantity A as the excess potential energy over the minimum value – that is,

$$A = P - P_{\min} = \frac{g\rho}{2}\frac{1}{S}\int_S H'^2 \, dS. \tag{2.6}$$

It follows from the conservation of mass in the system that $\overline{H}$ is a constant independent of time. We may see this from the calculation of the total

SOME ELEMENTARY CONSIDERATIONS

mass, which is

$$M = \int_0^H \int_S \rho \, dz \, dS = \rho \int_S H \, ds = \rho \overline{H} S. \tag{2.7}$$

Since M is independent of time, it follows from (2.7) that so is $\overline{H}$. The same result may of course also be obtained from the continuity equation for the system which under hydrostatic conditions is

$$\frac{\partial H}{\partial t} + \vec{v} \cdot \nabla H + H \nabla \cdot \vec{v} = \frac{\partial H}{\partial t} + \nabla \cdot \left(H \vec{v} \right) = 0. \tag{2.8}$$

If (2.8) is averaged over the area S, we obtain immediately

$$\frac{\partial \overline{H}}{\partial t} = 0 \tag{2.9}$$

because the area integral of the divergence term can be written as a line integral along the boundary of the container

$$\frac{1}{S} \int_S \nabla \cdot \left(H \vec{v} \right) ds = \frac{1}{S} \int H v_n \, dl, \tag{2.10}$$

but v_n, the normal component to the vertical boundary, vanishes due to the kinematic boundary condition. Therefore we may write (2.8) in the form

$$\frac{\partial H'}{\partial t} + \overline{H} \nabla \cdot \vec{v} + \nabla \cdot \left(H' \vec{v'} \right) = 0. \tag{2.11}$$

Now we shall look at the energy equations for the system. It is then convenient to define the geopotential

$$\phi = gH. \tag{2.12}$$

Equation (2.8) becomes then

$$\frac{\partial \phi}{\partial t} + \vec{v} \cdot \nabla \phi = -\phi \nabla \cdot \vec{v} \tag{2.13}$$

while (2.1) becomes

$$P = \frac{1}{2} \frac{\rho}{g} \frac{1}{S} \int_S \phi^2 dS \qquad \frac{dP}{dt} = \frac{\rho}{g} \frac{1}{S} \int_S \phi \frac{\partial \phi}{\partial t} dS. \tag{2.14}$$

Combining (2.13) and (2.14) by multiplying (2.11) by ϕ and integrating, we get

$$\frac{dP}{dt} = \frac{\rho}{g} \frac{1}{S} \int_S \left[-\vec{v} \cdot \nabla \left(\frac{1}{2} \phi^2 \right) - \phi^2 \nabla \cdot \vec{v} \right] dS \tag{2.15}$$

$$= -\frac{1}{2}\frac{\rho}{g}\frac{1}{S}\int_S \phi^2 \nabla \cdot \vec{v}\ dS. \qquad (2.16)$$

Since $P_{\min}$ is independent of time, we find also

$$\frac{dA}{dt} = -\frac{1}{2}\frac{\rho}{g}\frac{1}{S}\int_S \phi^2 \nabla \cdot \vec{v}\ dS. \qquad (2.17)$$

We next shall consider the kinetic energy. From physics we know that the kinetic energy per unit mass is $k = \frac{1}{2}\vec{v}^2$. The total kinetic energy per unit area is therefore

$$K = \frac{1}{S}\int_0^H \int_S k\rho\ dS\ dz = \frac{1}{S}\int_S \rho k H\ dS \qquad (2.18)$$

or

$$K = \frac{\rho}{g}\frac{1}{S}\int_S k\phi\ dS. \qquad (2.19)$$

The equations of motion for our system are

$$\frac{\partial u}{\partial t} + \vec{v}\cdot \nabla u = -\frac{\partial \phi}{\partial x} + fv, \qquad (2.20)$$
$$\frac{\partial v}{\partial t} + \vec{v}\cdot \nabla v = -\frac{\partial \phi}{\partial y} - fu,$$

where f is the Coriolis parameter.

Multiplying the first eqution by u and the second by v we find after addition

$$\frac{\partial k}{\partial t} + \vec{v}\cdot \nabla k = -\vec{v}\cdot \nabla \phi. \qquad (2.21)$$

Multiplying this time (2.13) by k and (2.21) by ϕ and adding the resulting equations, we get

$$\frac{\partial (k\phi)}{\partial t} + \vec{v}\cdot \nabla(k\phi) = -k\phi\nabla\cdot\vec{v} - \vec{v}\cdot\nabla\left(\frac{\phi^2}{2}\right). \qquad (2.22)$$

It follows from the form of (2.19) and (2.22) that

$$\frac{dK}{dt} = \frac{\rho}{g}\frac{1}{S}\int_S \left[-\nabla\cdot\left(k\phi\vec{v}\right) - \nabla\cdot\left(\frac{\phi^2}{2}\vec{v}\right) + \frac{\phi^2 \nabla\cdot\vec{v}}{2}\right]dS \qquad (2.23)$$

or

$$\frac{dK}{dt} = \frac{1}{2}\frac{\rho}{g}\frac{1}{S}\int_S \phi^2 \nabla\cdot \vec{v}\ dS. \qquad (2.24)$$

SOME ELEMENTARY CONSIDERATIONS

When equations (2.17) and (2.24) are added, we obtain

$$\frac{d(A+K)}{dt} = 0. \tag{2.25}$$

Thus we have shown that the sum of A and K is constant. Since the total potential energy has a minimum $P_{\min}$, and since (2.25) holds, it is natural to call A the available potential energy, or that part of the potential energy that is available for conversion into kinetic energy. It is, as we see from (2.6), defined as an integral quantity and is proportional to the variance of depth of the fluid around its unchanging mean value.

From (2.24) we note also that the integral in the equation measures the rate of increase of the kinetic energy. Since the sum of A and K is constant, there is a corresponding decrease in A as is verified from (2.17). It is thus natural to talk about the integral as the transformation or the conversion between the two forms of energy. The notation is normally

$$C(A, K) = \frac{1}{2}\frac{\rho}{g}\frac{1}{S}\int_S \phi^2 \nabla \cdot \vec{v}\ dS, \tag{2.26}$$

which should be read as the conversion from A to K. It follows that

$$C(K, A) = -C(A, K). \tag{2.27}$$

We may express (2.26) in a different way by noting that the homogeneous fluid is also incompressible. The continuity equation is therefore

$$\frac{\partial w}{\partial z} = -\nabla \cdot \vec{v} \tag{2.28}$$

or

$$w = -z\nabla \cdot \vec{v}. \tag{2.29}$$

In particular, the vertical velocity at the surface of the fluid is

$$w_H = -H\nabla \cdot \vec{v} \qquad gw_H = -\phi \nabla \cdot \vec{v}. \tag{2.30}$$

Therefore we also may write (2.26) as follows:

$$C(A, K) = -\frac{\rho}{2}\frac{1}{S}\int_S w_H \phi\ dS = -C(K, A). \tag{2.31}$$

Thus A will tend to increase if w_H and ϕ are positively correlated, or in other words if the fluid goes up where it already is high and down when it already is low. It is immediately understandable that this process will increase the variance of the height field and thus increase A.

We may use the example to obtain some idea of the ratio between A and P. For this purpose we consider the example where the homogeneous fluid is set into a solid rotation – that is, where the velocity is Ωr when Ω is the angular velocity and r the distance from the axis of rotation. In our example we shall assume that the container is a circular cylinder with a flat bottom and a radius of R. The shape of the surface can in this simple case be calculated from the equation

$$g \frac{dH}{dr} = \Omega^2 r \qquad (2.32)$$

or

$$H = H_0 + \frac{1}{2}\frac{\Omega^2}{g}r^2. \qquad (2.33)$$

The integration constant H_0, which is the depth in the middle of the fluid, can be related to the mean depth $\overline{H}$ by mass or volume continuity. For the volume we have

$$V = \int_0^H \int_0^R \int_0^{2\pi} r\, d\theta\, dr\, dz = 2\pi \int_0^R Hr\, dr \qquad (2.34)$$

$$V = \pi R^2 \left(H_0 + \frac{1}{4}\frac{\Omega^2}{g} R^2 \right), \qquad (2.35)$$

but with a flat surface where $H = \overline{H}$ everywhere, we have

$$V = \pi R^2 \overline{H}. \qquad (2.36)$$

Therefore,

$$H_0 = \overline{H} - \frac{1}{4}\frac{\Omega^2}{g} R^2. \qquad (2.37)$$

The minimum value of P is

$$P_{\min} = \frac{\pi g \rho}{2} R^2 \overline{H}^2, \qquad (2.38)$$

while the total potential energy is

$$P = \int_0^H \int_0^R \int_0^{2\pi} gz\rho r\, d\theta\, dr\, dz = 2\pi g\rho \int_0^R H^2 \frac{r}{2}\, dr, \qquad (2.39)$$

which is evaluated to be

$$P = \frac{\pi g \rho}{2} \overline{H}^2 R^2 + \frac{1}{96}\frac{\Omega^4}{g^2} \pi g \rho R^6. \qquad (2.40)$$

SOME ELEMENTARY CONSIDERATIONS

We have thus

$$A = \frac{1}{96}\frac{\Omega^4}{g}\pi\rho R^6 \tag{2.41}$$

or

$$\frac{1}{S}A = \frac{1}{96}\frac{\Omega^4}{g}\rho R^4. \tag{2.42}$$

We have finally

$$\frac{A}{P_{\min}} = \frac{1}{48}\frac{\Omega^4}{g^2}\frac{R^4}{\overline{H}^2}. \tag{2.43}$$

We may evaluate (2.43) using $\overline{H} = 10^4$ m, $\Omega = 7.29 \times 10^{-5}$ s^{-1}, $g = 9.8$ m s^{-2} and finally $R = 2 \times 10^6$ m. One finds then that $A/P_{\min} \sim 10^{-3}$ indicating that only a small fraction of the total potential energy can be converted into kinetic energy. No particular significance can be attached to the numerical value because it is totally dependent on the ratio $R^2/\overline{H}$. However, the above values seem to be of the correct order of magnitude.

Examples similar to those given above have been constructed by several authors; see for example V.H.Ryd (1923), who calculates the pressure field from a given wind distribution by using the gradient wind equation. We may briefly consider his example. He starts from the gradient wind equation

$$g\frac{dh}{dr} = fV + \frac{V^2}{r} \tag{2.44}$$

and notes that this equation is most easily handled if one starts by specifying the wind $V = V(r)$ in the case of a circular stationary vortex. Denoting the nondimensional radius by $x = r/r_*$ where r_* is to be specified later, he selects

$$V = V_* x e^{1-x} \tag{2.45}$$

as the wind distribution. It is easily seen that V has a maximum V_* for $x = 1$. Therefore the distance of the maximum wind is r_*. Introducing x and V in (2.44), we find

$$\frac{dh}{dx} = \frac{fr_* V_* e}{g} x e^{-x} + \frac{V_*^2 e^2}{g} x e^{-2x}, \tag{2.46}$$

which may be integrated directly to give

$$h = h_0 + A\left[1 - (1+x)e^{-x}\right] + \frac{B}{4}\left[1 - (1+2x)e^{-2x}\right] \tag{2.47}$$

in which

$$A = g^{-1} f r_* V_* e \qquad B = g^{-1} V_*^2 e^2. \tag{2.48}$$

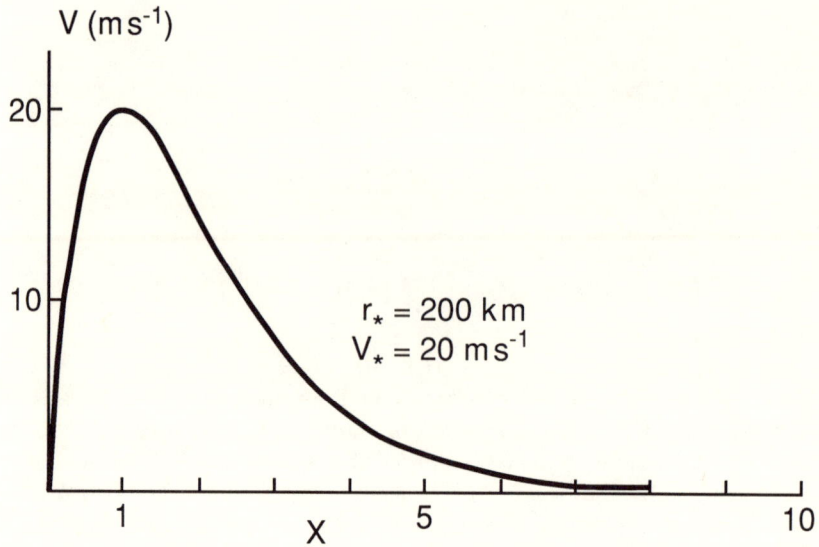

Fig. 2.1: The tangential wind as given by Eq. (2.45) as a function of $x = r/r_*$ for $r^* = 200$ km and $V_A = 20$ m s^{-1}.

Fig. 2.1 shows $V = V(x)$ for $V_* = 20$ m s^{-1}, while Fig. 2.2 shows $h - h_0$ for the same values and $g = 9.8$ m s^{-2}, $f_0 = 10^{-4}$ s^{-1}, $r_* = 2 \times 10^5$ m. The expression (2.45) is not based on any physical consideration, but a maximum wind some distance from the center is characteristic for atmospheric low pressure systems. Given (2.45) and (2.47), it is a straightforward matter to calculate the available potential energy and the kinetic energy. Let x_m denote the maximum distance of x. We shall then repeatedly meet integrals of the form

$$\int_0^{x_m} x^n e^{ax} \, dx.$$

As illustrated by Fig. 2.1 and as known from integral calculus, for a sufficiently large x_m we may approximate such an integral by

$$\int_0^\infty x^n e^{-ax} \, dz = \frac{n!}{a^{n+1}}, \qquad (2.49)$$

which will be used through all the calculations. We find first the area mean depth by calculating the integral

$$\overline{h} = \frac{1}{\pi x_m^2} \int_0^{2\pi} \int_0^{x_m} h(x) x \, d\theta \, dx = \frac{2}{x_m^2} \int_0^\infty hx \, dx, \qquad (2.50)$$

SOME ELEMENTARY CONSIDERATIONS

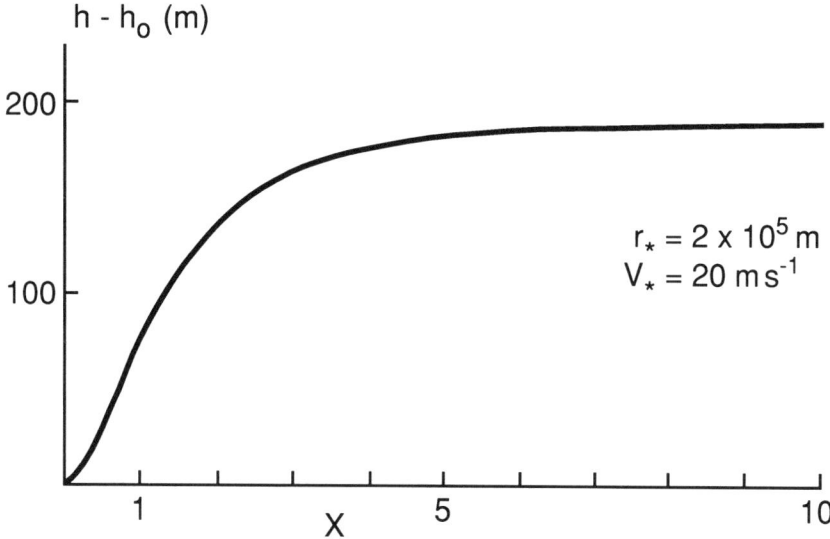

Fig. 2.2: The deviation of the height h from the central height h_0 in meters as a function of x. Parameters as in Fig 2.1.

which gives the result

$$\overline{h} = \left(h_0 + A + \frac{B}{4}\right) - x_m^{-2}\left(6A + \frac{3B}{8}\right), \tag{2.51}$$

and consequently

$$h' = h - \overline{h} \tag{2.52}$$

$$h' = x_m^{-2}\left(6A + \frac{3}{8}B\right) - A(1+x)e^{-x} - \frac{1}{4}B(1+2x)e^{-2x}. \tag{2.53}$$

The height deviation $h - \overline{h}$ as a function of x is displayed in Fig. 2.2. The potential energy corresponding to the flat surface (2.51) is easily calculated to be

$$\overline{P} = \frac{1}{\pi x_m^2} \int_0^{2\pi} \int_0^{x_m} \int_0^h (gz)\rho \, dz \, dx \, x \, d\theta = \frac{g\rho \overline{h}^2}{2} \tag{2.54}$$

or

$$\overline{P} = \frac{g\rho}{2}\left[h_0 + A + \frac{B}{4} - x_m^{-2}\left(6A + \frac{3B}{8}\right)\right]^2. \tag{2.55}$$

For the available potential energy we find in a similar way

$$APE = \frac{g\rho}{x_m^2}\left[\frac{9A^2}{8} + \frac{9B^2}{512} + \frac{13AB}{54} - \frac{1}{2x_m^2}\left(6A + \frac{3B}{8}\right)^2\right]. \tag{2.56}$$

12   FUNDAMENTALS OF ATMOSPHERIC ENERGETICS

As an example we may use $f_0 = 10^{-4}$ s^{-1}, $V_* = 60$ m s^{-1}, $g = 9.8$ m s^{-2}, $r_* = 2 \times 10^6$ m, and $x_m = 10$. We find then

$$\overline{P} = 3.46 \times 10^5 \text{ kJ m}^{-2} \qquad (\overline{h} = 8400 \text{ m})$$

while

$$A = 3.3285 \times 10^3, \qquad B = 2.7143 \times 10^3,$$

and consequently

$$APE = 1231.45 \text{ kJ m}^{-2}.$$

We note as before that APE is small compared to $\overline{P}$ because $APE/\overline{P} \approx 0.003$. The kinetic energy may of course also be calculated for our example because

$$K = \frac{1}{\pi x_m^2} \int_0^{2\pi} \int_0^{x_m} \int_0^h \frac{\rho}{2} V^2 \, dz \, dx \, (x \, d\theta) \qquad (2.57)$$

$$K = \frac{\rho}{x_m^2} \int_0^{x_m} V^2 h x \, dx, \qquad (2.58)$$

which using the previous results becomes

$$K = \frac{V_*^2 e^2 \rho}{x_m^2} \left(\frac{3}{8} h_0 + \frac{131}{648} A + \frac{39}{512} B \right). \qquad (2.59)$$

Using the same parameters as before we calculate K to be

$$K = 881.16 \text{ kJ m}^{-2} \qquad (2.60)$$

and therefore $K/APE = 0.7$.

It is seen by calculating several examples that the amounts of the various forms of energy and their ratio depend very strongly on the selection of parameters. The two most important parameters are naturally V_* and r_*. The purpose of the example, however, is not to model the atmosphere, but only to illustrate the concept of available potential energy in a homogeneous, rotating fluid. Note also that both examples consider balanced motion.

Before leaving these simple considerations we shall also briefly mention the early considerations of Margules (1904) who was interested in the mechanism by which the kinetic energy in cyclones is generated. Among several other possibilities he was interested in the kinetic energy that can be generated by a rearrangement of matter in the atmosphere. He considers what he called "closed systems" where the air is thought to be confined by a mantle of solid material. Margules uses several schemes of which the most frequently quoted is as follows. He imagines two homogeneous fluids with densities ρ_1 and ρ_2 such that $\rho_1 > \rho_2$. They are arranged in a container separated by a vertical wall (see Fig. 2.3).

SOME ELEMENTARY CONSIDERATIONS

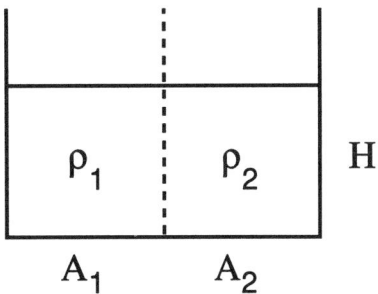

Fig. 2.3: The initial arrangement of two homogeneous fluids separated by a vertical wall.

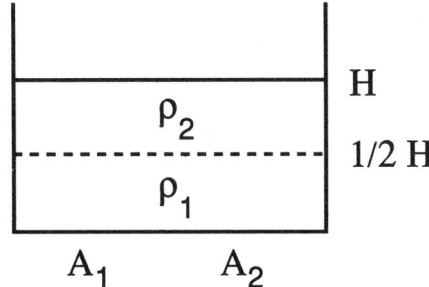

Fig. 2.4: The final state of the two homogeneous fluids starting from the initial arrangement as shown in Fig. 2.3.

The horizontal areas of the two fluids are A_1 and A_2 and the common depth is H. It is assumed that the fluids are at rest. Consequently, they have only potential energy. The total potential energy in the above arrangement is easily computed with the result that

$$P_i = \frac{g}{2} H^2 \left(\rho_1 A_1 + \rho_2 A_2 \right), \qquad K_i = 0. \tag{2.61}$$

The vertical wall is then removed, and since $\rho_1 > \rho_2$ the fluids will eventually rearrange themselves as indicated in Fig. 2.4.

The potential energy in the final stable state is also easily calculated to be

$$P_F = \frac{g\rho_1}{8} H^2 (A_1 + A_2) + \frac{g\rho_2}{2}(H^2 - \frac{H^2}{4})(A_1 + A_2). \tag{2.62}$$

We assume now for convenience that $A_1 = A_2 = A/2$. The difference $P_i - P_f$ must be equal to the total kinetic energy that is generated under the assumption of energy conservation, i.e. no dissipation. We find

$$K_F = P_i - P_F = \frac{gH^2A}{8}(\rho_1 - \rho_2). \tag{2.63}$$

The total mass of the two fluids is

$$M = \frac{HA}{2}(\rho_1 + \rho_2), \tag{2.64}$$

and the kinetic energy per unit mass is therefore

$$\frac{K_F}{M} = \frac{gH}{4}\frac{\rho_1 - \rho_2}{\rho_1 + \rho_2} = \frac{gH}{8}\frac{\Delta\rho}{\overline{\rho}}, \tag{2.65}$$

where

$$\Delta\rho = (\rho_1 - \rho_2) \qquad \overline{\rho} = \frac{(\rho_1 + \rho_2)}{2}. \tag{2.66}$$

With good approximation one can write (2.65) in the form

$$\frac{K_F}{M} = \frac{gH}{8}\frac{\Delta T}{\overline{T}}, \tag{2.67}$$

where ΔT and $\overline{T}$ are the temperature difference $\Delta T = T_2 - T_1$ and the mean temperature $\overline{T} = \frac{1}{2}(T_2 + T_1)$, respectively. T_1 and T_2 are temperatures in the two fluids at the same level. Equation (2.67) expresses the kinetic energy per unit mass, which may also be written

$$\frac{\overline{V}^2}{2} = \frac{gH}{8}\frac{\Delta T}{\overline{T}} \tag{2.68}$$

or

$$\overline{V} = \frac{1}{2}\left(gH\frac{\Delta T}{\overline{T}}\right)^{1/2} \tag{2.69}$$

For $\Delta T = 10$ K, $\overline{T} = 283$ K, $g = 9.8$ m s^{-2}, we find $\overline{V} = 13$ m s^{-2} (H = 2000 m), $\overline{V} = 16$ m s^{-1} (H = 3000 m) and $\overline{V} = 23$ m s^{-1} (H = 6000 m). Margules used calculations of this kind to make it plausible that the kinetic energy in the atmosphere was generated by a rearrangement of mass, or, in other words, by release of potential energy. The weakness in the argument is of course that his examples are quite difficult to adapt to atmospheric conditions. It is, however, noteworthy that both Margules (1904) and later Ryd (1927) realized that not all of the potential energy can be converted into kinetic energy, and both of them speak of the energy available for conversion into kinetic energy.

3

BASIC ASPECTS OF ATMOSPHERIC ENERGY

There are three forms of energy in the atmosphere: internal, potential, and kinetic. As known from physics the amounts of energy per unit mass are

$$i = c_v T,$$

$$\phi = gz,$$

$$k = \frac{\vec{v} \cdot \vec{v}}{2}, \tag{3.1}$$

of which the first is the internal energy, the second the potential energy and the third the kinetic energy. In this chapter we shall be concerned with the global aspects of the energy relations. For a given form of energy e per unit mass, where e can be either i, ϕ, or k, we calculate the total amounts by integrating over the total mass of the atmosphere – that is,

$$E = \int_v e\rho \, dV. \tag{3.2}$$

Without going into the details of the computations it will be understood that atmospheric data of a global nature can be used to calculate the total amounts of each of the three forms of energy. Such calculations have for example been made by Oort (1971) based on five years of data. He gets the following annual mean values for the Northern Hemisphere for the layer from 100 kPa to 7.5 kPa:

Potential energy: $P = 567.5 \times 10^3$ kJ m^{-2}

Internal energy: $I = 1674.8 \times 10^3$ kJ m^{-2}

Kinetic energy: $K = 1153.4$ kJ m^{-2}

The amounts are expressed as energy per unit area. Since the area in this case is a hemisphere, they have been divided by $2\pi a^2$, where a is the radius of the earth. Assuming that the Southern Hemisphere contains the same average energy amounts, we would find that the total amount of energy of all three kinds is about 8.8×10^{20} kJ. It is thus seen that the atmospheric energy content is very large, and that the kinetic energy is a small fraction

(1/2000) of the total energy.

It may also be useful for certain purposes to express the energy amounts per unit mass. In this case where the calculation was made for the layer from 100 kPa to 7.5 kPa, the mass is 9.4 t m^{-2} and we find then

$$P = 6.037 \times 10^4 \text{ kJ t}^{-1},$$

$$I = 1.7817 \times 10^5 \text{ kJ t}^{-1},$$

$$K = 122.7 \text{ kJ t}^{-1}.$$

Since P per unit mass is gz, we may say that the above amount corresponds to a mean height of about 6160 m, while the second ($c_v T$) corresponds to a mean temperature of 248 K. The kinetic energy per unit mass is $\frac{1}{2}V^2$, and we may then calculate that K corresponds to a mean wind speed of 15.7 m s^{-1}.

While the amounts of energy are interesting in themselves, we shall primarily in the following be interested in the relations between the various energy forms, and in the rate of change per unit time which they undergo in the atmosphere. The early studies of atmospheric energy in the beginning of the twentieth century by Margules (1904), Ryd (1923, 1927), Hesselberg (1914) and others were concerned with energy of the storms – that is, the cyclones. How was the kinetic energy of the middle-latitude depressions created? Was it mainly the work of the pressure forces and therefore conversion from potential energy? When Margules (loc. cit.) found by estimates that the work of the pressure forces did not produce enough energy to account for the energy observed in the storms, he turned to the rearrangement of mass as the most likely energy source for the kinetic energy. Ryd (loc. cit.) disagreed in several ways with Margules and came to the conclusion that "kinetic energy will be generated when alterations in the distribution of temperatures render the potential energy available" (Ryd, 1927, p 67).

Characteristically these authors realize that only a small amount of the internal and potential energies can be converted into kinetic energy. On the other hand, none of them considers the global aspects of energy and energy transformation, presumably because global aspects did not come to mind with the geographically limited networks of meteorological stations. Furthermore, the network contained no systematic upper air observations.

In the following discussion we shall consider these global aspects and hope in later chapters to answer the questions posed by the early investigators. To ease these considerations we state a couple of mathematical theorems that will be used repeatedly.

Let S be an arbitrary scalar quantity. We shall then show that

$$\int_v \rho \frac{dS}{dt} \, dV = \int \frac{\partial(\rho S)}{\partial t} \, dV. \tag{3.3}$$

BASIC ASPECTS OF ATMOSPHERIC ENERGY

The proof is as follows:

$$\rho \frac{dS}{dt} = \rho \left(\frac{\partial S}{\partial t} + \vec{v} \cdot \nabla S \right) = \frac{\partial(\rho S)}{\partial t} - S \frac{\partial \rho}{\partial t} + \rho \vec{v} \cdot \nabla S. \tag{3.4}$$

Substituting from the continuity equation for $\partial \rho / \partial t$ we get

$$\rho \frac{dS}{dt} = \frac{\partial(\rho S)}{\partial t} + S \nabla \cdot \left(\rho \vec{v} \right) + \rho \vec{v} \cdot \nabla S \tag{3.5}$$

$$\rho \frac{dS}{dt} = \frac{\partial(\rho S)}{\partial t} + \nabla \cdot \left(\rho S \vec{v} \right). \tag{3.6}$$

Equation (3.3) follows from (3.6) by integration noting that the volume integral of the second term vanishes because it reduces to two surface integrals: at the surface of the earth one that vanishes because the normal wind component is zero, and at the top of the atmosphere, the other is zero because ρw goes to zero at the outer limit of the atmosphere.

Next we shall consider the global aspects of each of the three forms of energy. We have

$$\frac{d\phi}{dt} = g \frac{dz}{dt} = gw, \tag{3.7}$$

and therefore

$$\frac{dP}{dt} = \int_v g \rho w \, dV. \tag{3.8}$$

From the thermodynamic equation we obtain

$$c_v \frac{dT}{dt} = H - p \frac{d\alpha}{dt} = H - p\alpha \nabla \cdot \vec{v} = H - RT \nabla \cdot \vec{v} \tag{3.9}$$

and consequently

$$\frac{dI}{dt} = \int_v H\rho \, dV - \int_v p \nabla \cdot \vec{v} \, dV. \tag{3.10}$$

Thus the two processes will influence the amount of internal energy, which tends to increase if the atmosphere is heated where the density is high, and cooled where the density is low. This is often formulated in the form that the heat sources in the atmosphere are located at low altitudes where the density is high. Due to the second term, it also will tend to decrease if on the average the pressure is high where there is divergence and low where there is convergence. We note, however, that the second integral also may be written in the form

$$-\int_v p \nabla \cdot \vec{v} \, dV = -\int_v \nabla \cdot \left(p \vec{v} \right) dV + \int_v \vec{v} \cdot \nabla p \, dV, \tag{3.11}$$

and since the first integral on the right-hand side of (3.11) vanishes, we might as well say that the work of the pressure force will change the internal energy.

Next, we turn our attention to the kinetic energy. We note first that

$$\frac{dk}{dt} = \vec{v} \cdot \frac{d\vec{v}}{dt}, \tag{3.12}$$

and since

$$\frac{d\vec{v}}{dt} = -\frac{1}{\rho}\nabla p - 2\vec{\Omega} \times \vec{v} + \vec{g} + \frac{1}{\rho}\vec{F}, \tag{3.13}$$

we find that

$$\frac{dk}{dt} = -\frac{1}{\rho}\vec{v} \cdot \nabla p - gw + \frac{1}{\rho}\vec{v} \cdot \vec{F} \tag{3.14}$$

from which we obtain

$$\frac{dK}{dt} = -\int_v \vec{v} \cdot \nabla p \, dV - \int_v g\rho w \, dV + \int_v \vec{v} \cdot \vec{F} \, dV \tag{3.15}$$

or

$$\frac{dK}{dt} = \int_v p\nabla \cdot \vec{v} \, dV - \int_v g\rho w \, dV + \int_v \vec{v} \cdot \vec{F} \, dV. \tag{3.16}$$

We compare the three equations (3.8), (3.10), and (3.16), and we note that the second integral in (3.10) appears with opposite sign in (3.16). Using the earlier notation we may write

$$C(I, K) = \int_v p\nabla \cdot \vec{v} \, dV. \tag{3.17}$$

Similarly, the integral in (3.10) is found again in (3.16) with the opposite sign. We may thus write

$$C(P, K) = -\int_v g\rho w \, dV. \tag{3.18}$$

The first integral in (3.10) may be called the generation of internal energy while the last integral in (3.16) with the opposite sign is the dissipation of kinetic energy – that is,

$$G(I) = \int_v H\rho \, dV,$$

$$D(K) = -\int_v \vec{v} \cdot \vec{F} \, dV. \tag{3.19}$$

In symbolic form we may write

$$\frac{dP}{dt} = -C(P, K),$$

BASIC ASPECTS OF ATMOSPHERIC ENERGY

$$\frac{dI}{dt} = G(I) - C(I,K),$$

$$\frac{dK}{dt} = C(P,K) + C(I,K) - D(K). \tag{3.20}$$

If we consider a long-term average, it may be assumed that we are very close to steady-state conditions in which there is no change in the energy amounts – that is, all three time-derivatives in (3.20) are zero. In this case we find that

$$C(P,K) = 0,$$

$$G(I) = C(I,K) = D(K). \tag{3.21}$$

In the formulation given here, we have taken H to be the total heating per unit mass including the heating that results from the dissipation due to frictional forces. We may write

$$H = H_{NF} + H_F \tag{3.22}$$

in which H_{NF} is due to all heating processes except friction, while H_F measures this heating – that is,

$$\rho H_F = - \vec{v} \cdot \vec{F}. \tag{3.23}$$

We have then

$$G(I) = \int_v H_{NF}\rho \; dV + \int_v H_F\rho \; dV \tag{3.24}$$

$$G(I) = \int_v H_{NF}\rho \; dV - \int_v \vec{v} \cdot \vec{F} \; dV, \tag{3.25}$$

but since (3.21) has to be satisfied, we find for steady-state conditions that

$$G_{NF} = \int_v H_{NF}\rho \; dV \equiv 0. \tag{3.26}$$

The general energy relations expressed in (3.20) are illustrated in Fig. 3.1 where the dashed lines indicate the effect of the heating due to frictional forces. On the other hand, Fig. 3.2 shows the steady-state energy diagram.

We next obtain an expression for the local change of the total energy. Adding the three equations (3.7), (3.9), and (3.14), we obtain

$$\frac{d}{dt}(\phi + c_v T + k) = H - \frac{1}{\rho}\left(p\nabla \cdot \vec{v} + \vec{v} \cdot \nabla p\right) + \frac{1}{\rho}\vec{v} \cdot \vec{F}. \tag{3.27}$$

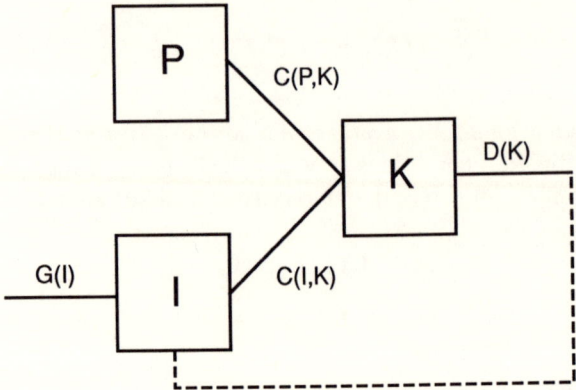

Fig. 3.1: The general energy diagram for internal, potential, and kinetic energy. Generation, conversions, and dissipation are indicated by connecting lines. The dashed line indicates the heat due to the frictional dissipation.

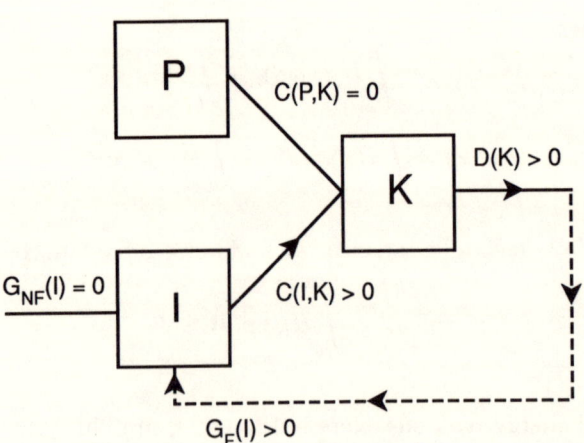

Fig. 3.2: The steady-state energy diagram corresponding to Fig. 3.1. Note that the generations of internal energy by the nonfrictional heating and the energy conversion from K to P vanish.

BASIC ASPECTS OF ATMOSPHERIC ENERGY

Denoting $S = \phi + c_v T + k$ and using (3.6), we obtain from (3.27)

$$\frac{\partial(\rho S)}{\partial t} = -\nabla \cdot \left(\rho S \vec{v}\right) - \nabla \cdot \left(p \vec{v}\right) + \rho h + \vec{v} \cdot \vec{F}. \qquad (3.28)$$

Noting from the gas equation that $p = RT\rho$ and $c_p = c_v + R$, we may write (3.28) in the final form

$$\frac{\partial}{\partial t}\left[(\phi + i + k)\rho\right] = -\nabla \cdot \left[(\phi + c_p T + k)\rho \vec{v}\right] + \rho H + \vec{v} \cdot \vec{F}. \qquad (3.29)$$

Equation (3.29) expresses the local change in total energy per unit volume. Thus the change may be due to three processes: the heating, the dissipation, and the three-dimensional convergence of the energy transport.

Using (3.22) and (3.23) to introduce the nonfrictional heating and introducing the transport vector,

$$\vec{w} = (c_p T + \phi + k)\rho \vec{v}. \qquad (3.30)$$

We may write (3.29) in the short form

$$\frac{\partial}{\partial t}\left[(c_v T + gz + k)\rho\right] = -\nabla \cdot \vec{w} + \rho H_{NF}. \qquad (3.31)$$

Equation (3.31) is quite general. It has been used mostly to study the heat budgets for certain well-defined volumes of the atmosphere. A typical example is a zonal ring which is defined as a volume bounded by vertical walls along two latitude circles φ_1 and φ_2, by the earth's surface, and by the top of the atmosphere. Denoting this volume by V and using Stokes' theorem, we find that

$$\int_v \frac{\partial(\rho S)}{\partial t} dV = \int_{\sigma_1} W_{\varphi_1} d\sigma_1 - \int_{\sigma_2} W_{\varphi_2} d\sigma_2 + \int_v H_{NF}\rho \, dV. \qquad (3.32)$$

The first two integrals are surface integrals over the two walls at latitudes φ_1 and φ_2, respectively. The general area element is

$$d\sigma = a \cos\phi \, d\lambda \, dz \qquad (3.33)$$

and W_ϕ is as follows:

$$W_\varphi = (c_p T + \phi + k)\rho v(\varphi), \qquad (3.34)$$

where v is the meridional velocity.

Together the first two integrals in (3.32) measure the net-influx in the meridional direction. Each of them can be computed from atmospheric data. Written in detail and using the hydrostatic equation we get the following form:

$$\begin{aligned}\int_\sigma W_\varphi \, d\sigma &= \int_0^\infty \int_0^{2\pi} (c_p T + gz + k) \, v\rho a \cos\varphi \, d\lambda \, dz \\ &= \frac{1}{g} \int_0^{p_0} \int_0^{2\pi} (c_p T + gz + k) \, va \cos\varphi \, d\lambda \, dp \\ &= \frac{2\pi a \cos\phi}{g} \int_0^{p_0} [(c_p T + gz + k) v]_z \, dp \end{aligned} \qquad (3.35)$$

in which the subscript z has been used to denote a zonal average – that is,

$$b_z = \frac{1}{2\pi} \int_0^{2\pi} b \, d\lambda. \qquad (3.36)$$

Equation (3.32) together with the following equations form the basis for the studies of the heat budget in zonal rings. We note that the heating H_{NF} may be studied by itself considering the various processes that contribute to the heating. On the other hand, the transport processes may be studied using atmospheric data. In the long-term average where the left-hand side of (3.32) vanishes, there should be a balance between the processes. The heating, especially in this case, can be computed from the transports. A treatment of H_{NF} and the transport processes has been summarized by Wiin-Nielsen (1973) for the steady-state case.

We shall next turn to the implications of the hydrostatic assumption for atmospheric energetics. The hydrostatic assumption is considered to be so well justified that is is incorporated in the evaluation of radiosonde observations. It is thus necessary to investigate how this fact will change the diagrams given in Figs. 3.1 and 3.2.

The major effect of the hydrostatic assumption is that the internal and potential energy become proportional to each other. This is true for each vertical column. We recall that for a vertical column we have

$$P = \int_0^\infty gz\rho \, dz = \int_0^{p_s} z \, dp. \qquad (3.37)$$

Using

$$z \, dp = d(pz) - p \, dz \qquad (3.38)$$

and the fact that $p = 0$ at the top and $z = 0$ at the bottom, we find

$$P = \int_0^\infty p \, dz = \frac{R}{c_v} \int_0^\infty c_v T\rho \, dz = \frac{R}{c_v} I, \qquad (3.39)$$

which says that the ratio P/I for each vertical column and therefore also for the total atmosphere is equal to $R/c_v = 0.4$. For the amounts given at the

BASIC ASPECTS OF ATMOSPHERIC ENERGY

beginning of this chapter and based on observations, we find $P/I = 0.34$. The difference is due to the uncertainty in the observations and also to the fact that these calculations were based on the layer 100 to 7.5 cb.

From *Dynamic Meteorology* (Wiin-Nielsen, 1973) we know that another implication of the hydrostatic assumption is that the vertical velocity no longer is a free variable, but a quantity that may be obtained from the condition that the atmosphere should be in hydrostatic equilibrium at all places and at all times. The diagrams in Figs. 3.1 and 3.2 are thus still applicable with the modification that w should be replaced by W_R, computed from Richardson's equation.

Let us consider this equation. It is obtained by deriving two pressure tendency equations. One of them is obtained from a combination of the thermodynamic equation and the continuity equation as follows. From the continuity equation

$$\frac{d\rho}{dt} + \rho \nabla \cdot \vec{v} = 0, \tag{3.40}$$

we obtain using the gas equation to eliminate the density:

$$\frac{dp}{dt} - \frac{p}{T}\frac{dT}{dt} + p\nabla \cdot \vec{v} = 0. \tag{3.41}$$

But

$$\frac{dT}{dt} = \frac{1}{c_v}\left(H - p\frac{d\alpha}{dt}\right) = \frac{1}{c_v}\left(H - RT\nabla \cdot \vec{v}\right). \tag{3.42}$$

Therefore, by elimination of dT/dt from (3.41) we get

$$\frac{\partial p}{\partial t} = -\vec{v} \cdot \nabla p - \frac{c_p}{c_v}p\nabla \cdot \vec{v} + \frac{R}{c_v}\rho H. \tag{3.43}$$

On the other hand, the hydrostatic equation gives by differentiation with respect to time

$$\frac{\partial}{\partial z}\left(\frac{\partial p}{\partial t}\right) = -g\frac{\partial \rho}{\partial t} = g\nabla \cdot \left(\rho \vec{v}\right). \tag{3.44}$$

Integrating this equation from the top of the atmosphere to an arbitrary height z, we obtain

$$\frac{\partial p}{\partial t} = g\rho W_R - \int_z^\infty g\nabla_2 \cdot \left(\rho \vec{v}\right) dz. \tag{3.45}$$

The final step is to equate the two expressions for the pressure tendencies in (3.43) and (3.45). In doing so we observe that each term, where possible, is divided in the horizontal and vertical parts. Richardson's equation is then

$$\frac{c_p}{c_v}p\frac{\partial W_R}{\partial z} = \int_z^\infty g\nabla_2 \cdot \left(\rho \vec{v}\right) dz - \vec{v}_2 \cdot \nabla p - \frac{c_p}{c_v}p\nabla_2 \vec{v} + \frac{R}{c_v}\rho H, \tag{3.46}$$

which also may be written in the form

$$p\frac{\partial W_R}{\partial z} = \frac{c_v}{c_p}\left(\int_z^\infty g\nabla_2\cdot\left(\rho\vec{v}\right)\,dz - \nabla_2\cdot\left(p\vec{v}_2\right) - \frac{R}{c_v}p\nabla_2\cdot\vec{v} + \frac{R}{c_v}\rho H\right). \tag{3.47}$$

W_R is obtained finally by integrating (3.47) from the ground to an arbitrary height. Let us now consider (3.18) in the hydrostatic case:

$$C(K,P) = \int_v g\rho W_R\,dV = -\int_v W_R\frac{\partial p}{\partial z}\,dV = \int_v p\frac{\partial W_R}{\partial z}\,dV, \tag{3.48}$$

where for simplicity we have taken the earth without topography – that is, $W_R = 0$ at $z = 0$. Using (3.47) we find that

$$C(K,P) = \frac{c_v}{c_p}\left[-\frac{R}{c_v}\int_v p\nabla_2\cdot\vec{v}\,dV + \frac{R}{c_v}\int_v \rho H\,dV\right]. \tag{3.49}$$

On the other hand, from (3.17) we get

$$C(I,K) = \int_v p\nabla\cdot\vec{v}\,dV = \int_v p\nabla_2\cdot\vec{v}\,dV + \int_v p\frac{\partial W_R}{\partial z}\,dV. \tag{3.50}$$

Comparing (3.48) and (3.50) we see that of the total amount converted from I to K, a fraction, equal to the second term in (3.50), is immediately converted to the potential energy to keep I and P in the correct proportion. Thus (3.50) may be written in the form

$$\int_v p\nabla_2\cdot\vec{v}\,dV = C(I,K) - C(K,P), \tag{3.51}$$

which inserted in (3.49) finally gives after rearrangement

$$C(K,P) = \frac{R}{c_v}\left[G(I) - C(I,K)\right]. \tag{3.52}$$

Comparing with (3.20) we see that (3.52) expresses the relation necessary to satisfy the equation

$$\frac{dP}{dt} = \frac{R}{c_v}\frac{dI}{dt}. \tag{3.53}$$

Equation (3.53) of course could have been written immediately using (3.20) from which (3.52) would follow. The analysis given here employing Richardson's equation shows more, however, because the conversion to the potential energy goes through the kinetic energy, or, to be exact, through the kinetic energy contained in the vertical motion.

We may gain further insight into these matters by dividing the kinetic energy in the general nonhydrostatic case into the kinetic energy of the horizontal motion and that of the vertical motion. To clarify matters we shall use component equations and spherical geometry. The three equations of motion are then

$$\frac{du}{dt} = -\frac{1}{\rho a \cos\varphi}\frac{\partial p}{\partial \lambda} + fv - ew + \frac{uv}{a}\tan\varphi - \frac{uw}{a} - \frac{1}{\rho}F_x$$

$$\frac{dv}{dt} = -\frac{1}{\rho a}\frac{\partial p}{\partial \varphi} - fu - \frac{u^2}{a}\tan\varphi - \frac{vw}{a} + \frac{1}{\rho}F_y$$

$$\frac{dw}{dt} = -\frac{1}{\rho}\frac{\partial p}{\partial z} - g + eu + \frac{u^2}{a} + \frac{v^2}{a} + \frac{1}{\rho}F_z. \quad (3.54)$$

The radius of the earth is denoted by a, longitude by λ, latitude by φ; $f = 2\Omega\sin\varphi$, and $e = 2\Omega\cos\varphi$.

Following the usual procedure we find

$$\frac{dK_H}{dt} = -\int_v \vec{v_2}\cdot\nabla p\, dV - \int_v euw\rho\, dV - \int_v \frac{u^2+v^2}{a}w\rho\, dV + \int_v \vec{v_2}\cdot\vec{F_2}\, dV \quad (3.55)$$

and

$$\frac{dK_w}{dt} = -\int_v w\frac{\partial p}{\partial z}\, dV - \int_v\int_v gw\rho\, dV + \int_v euw\rho\, dV \quad (3.56)$$

$$+ \int_v \frac{u^2+v^2}{a}w\rho\, dV + \int_v wF_z\, dV.$$

It is seen immediately that

$$C(K_H, K_w) = \int_v euw\rho\, dV + \int_v \frac{u^2+v^2}{a}w\rho\, dV. \quad (3.57)$$

The first integral in (3.55) may be written in the form

$$-\int_v \vec{v_2}\cdot\nabla p\, dV = \int_v p\nabla_2\cdot\vec{v}\, dV \quad (3.58)$$

while the first integral in (3.56) is

$$-\int_v w\cdot\frac{\partial p}{\partial z}\, dV = \int_v p\frac{\partial w}{\partial z}\, dV. \quad (3.59)$$

Comparing with (3.10) which we write in the form

$$\frac{dI}{dt} = \int_v H\rho\, dV - \int_v p\nabla_2\cdot\vec{v}\, dV - \int_v p\frac{\partial w}{\partial z}\, dV, \quad (3.60)$$

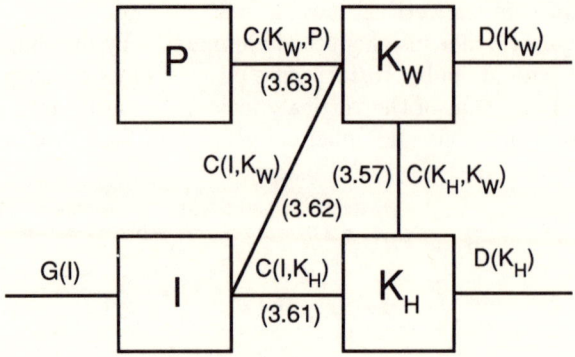

Fig. 3.3: The general energy diagram, where the kinetic energy has been divided into the kinetic energies of the horizontal motion and the vertical motion. The numbers in parentheses indicate the relevant equations.

we see that
$$C(I, K_H) = \int_v p \nabla_2 \cdot \vec{v} \, dV, \tag{3.61}$$
while
$$C(I, K_w) = \int_v p \frac{\partial w}{\partial z} \, dV. \tag{3.62}$$

Finally, comparing (3.56) with (3.8) we see that
$$C(K_w, P) = \int_v g\rho w \, dV. \tag{3.63}$$

On the basis of these formulas we may construct the energy diagram given in Fig. 3.3 in which the numbers in parentheses are the equations defining the conversions.

When the hydrostatic approximation is made – that is,
$$\frac{\partial p}{\partial z} = -g\rho, \tag{3.64}$$

we note that this means that $D(K_w) = 0$, $C(K_H, K_w) = 0$, and
$$C(I, K_w) = \int_v p \frac{\partial w_R}{\partial z} \, dz = -\int_v w_R \frac{\partial p}{\partial z} \, dz = \int_v g\rho w + R \, dz = C(K_w, P). \tag{3.65}$$

Equation (3.65) shows that K_w is catalytic in the sense that it participates in the energy transformations, but remains unchanged. The simplified energy diagram is given in Fig. 3.4. We can conclude from the formulas and

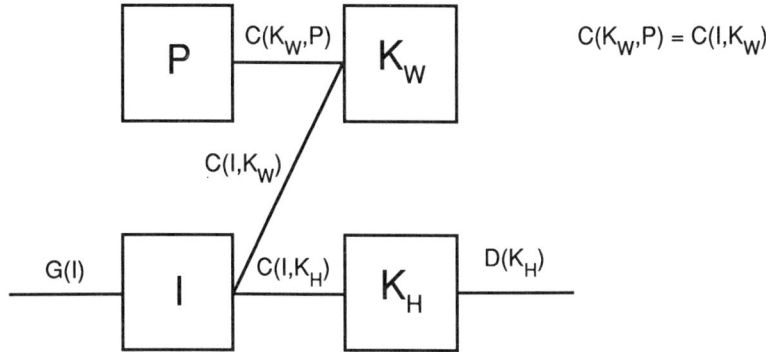

Fig. 3.4: The hydrostatic energy diagram corresponding to Fig. 3.3. Note that $D(K_W) = C(K_H, K_W) = 0$ and $C(I, K_W) = C(K_W, P)$.

the figure that in a hydrostatic system there is a well-defined vertical velocity that can be computed by integration of (3.47), but the total amount of kinetic energy in the vertical velocity remains unchanged during an integration.

In view of the proportionality between I and P and the very special role played by the vertical velocity in a hydrostatic system, it is more convenient to combine I and P into a single energy reservoir $E = I + P$ and to include only the kinetic energy of the horizontal motion in K. We note that

$$E = I + P = \int_v c_v T \rho \, dV + \int_v RT\rho \, dV = \int_v c_p T\rho \, dV. \qquad (3.66)$$

E, called total potential energy, may thus per unit mass be defined as

$$e = c_p T. \qquad (3.67)$$

In this new simplified hydrostatic system we have two energy reservoirs only, E and K, where $K = K_H$. It is also well known that changes of the vertical coordinate are both convenient and easy in hydrostatic systems. We therefore may take this opportunity to derive the energy expressions in a system with pressure as the vertical coordinate. For this purpose we use the thermodynamic equation in the form

$$c_p \frac{dT}{dt} - \alpha \frac{dp}{dt} = H. \qquad (3.68)$$

Any integral of the form

$$\int_v (\ ) \rho \, dV \qquad (3.69)$$

will be written as
$$\frac{1}{g}\int_0^{p_0}\int_S (\ )\, dp\, dS, \qquad (3.70)$$

where we have used the hydrostatic equation, where dS is the area element and where p_0 is taken as constant. We find therefore

$$\frac{dE}{dt} = \frac{1}{g}\int_0^{p_0} c_p \frac{dT}{dt}\, dS\, dp = \frac{1}{g}\int_0^{p_0}\int_S H\, dS\, dp + \frac{1}{g}\int_0^{p_0}\int_S \alpha\omega\, dS\, dp. \qquad (3.71)$$

From the equation of motion in the hydrostatic case we find

$$\begin{aligned}\frac{dK}{dt} &= \frac{1}{g}\int_0^{p_0}\int_S \frac{dk}{dt}\, dS\, dp \qquad (3.72)\\ &= -\frac{1}{g}\int_0^{p_0}\int_S \vec{v}_2\cdot\nabla\phi\, dS\, dp + \frac{1}{g}\int_0^{p_0}\int_S \alpha\, \vec{v}\cdot\vec{F}\, dS\, dp\end{aligned}$$

when ϕ is the geopotential. When the first integral is integrated by parts over the global domain using the continuity equation

$$\nabla_2\cdot\vec{v} + \frac{\partial\omega}{\partial p} = 0, \qquad (3.73)$$

we obtain

$$\frac{1}{g}\int_0^{p_0}\int_S \phi\nabla_2\cdot\vec{v}\, dS\, dp$$
$$= -\frac{1}{g}\int_0^{p_0}\int_S \phi\frac{\partial\omega}{\partial p}\, dS\, dp$$
$$= \frac{1}{g}\int_0^{p_0}\int_S \omega\frac{\partial\phi}{\partial p}\, dS\, dp \qquad (3.74)$$

Finally using the hydrostatic equation in the form

$$\frac{\partial\phi}{\partial p} = -\alpha, \qquad (3.75)$$

we find

$$\frac{dK}{dt} = -\frac{1}{g}\int_0^{p_0}\int_S \alpha\omega\, dS\, dp + \frac{1}{g}\int_0^{p_0}\int_S \alpha\, \vec{v}\cdot\vec{F}\, dS\, dp. \qquad (3.76)$$

Comparing (3.71) and (3.76) we may write

$$G(E) = \frac{1}{g}\int_0^{p_0}\int_S H\, dS\, dp, \qquad (3.77)$$
$$C(E, K) = -\frac{1}{g}\int_0^{p_0}\int_S \alpha\omega\, dS\, dp,$$
$$D(K) = -\frac{1}{g}\int_0^{p_0}\int_S \alpha\, \vec{v}\cdot\vec{F}\, dS\, dp.$$

BASIC ASPECTS OF ATMOSPHERIC ENERGY

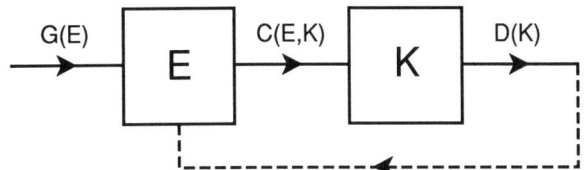

Fig. 3.5: The simple energy diagram, where E is the so-called total potential energy and K the kinetic energy of the horizontal motion.

The corresponding energy diagram is shown in Fig. 3.5

It is of some interest to give an example of a calculation of $C(E, K)$. We clearly need a method of calculating the vertical velocity ω. For this purpose we shall use a quasi-nondivergent two-level model. Furthermore, we shall select a particularly simple sinusoidal wave pattern. Our example will be concerned with a typical baroclinic wave where the temperature field is lagging behind the height field.

The two-level quasi-nondivergent model is well known; see for example Wiin-Nielsen (1973). The ω-equation for this model is

$$\nabla^2 \omega - \lambda^2 \omega = \frac{P}{f_0}\lambda^2 \left[-\vec{v}_* \cdot \nabla \psi_T - \vec{v}_T \cdot \nabla \psi_* - \beta v_T + \nabla^2 \left(\vec{v}_* \cdot \nabla \psi_T \right) \right]. \tag{3.78}$$

The baroclinic wave is specified by the two streamfunctions

$$\psi_* = -U_* y + A_* \cos kx \tag{3.79}$$
$$\psi_T = -U_T y + A_T \cos (kx + \gamma).$$

Equation (3.79) is used to evaluate each term on the right-hand side of (3.78). We find then

$$\nabla^2 \omega - \lambda^2 \omega = \frac{P}{f_0}\lambda^2 k^2 \left[\frac{\beta}{k^2}(kA_T) \sin \lambda \cos kx \right.$$
$$\left. + \left(\frac{\beta}{k^2}(kA_T) \cos \gamma - 2U_T k A_* \right) \sin kx \right], \tag{3.80}$$

where $P = 50$ kPa and $\lambda^2 = 2f_0^2/\sigma P^2$. Here σ is the static stability and f_0 a standard value of the Coriolis parameter. A solution to (3.80) is

$$\omega = \Omega_c \cos kx + \Omega_S \sin kx, \tag{3.81}$$

where Ω_c and Ω_S are found by insertion of (3.81) into (3.80). The result is

$$\Omega_c = -\frac{\lambda^2 k^2}{k^2 + \lambda^2} \frac{P}{f_0} \frac{\beta}{k^2} (kA_T) \sin \gamma \tag{3.82}$$

30   FUNDAMENTALS OF ATMOSPHERIC ENERGETICS

$$\Omega_S = -\frac{\lambda^2 k^2}{k^2 + \lambda^2} \frac{P}{f_0} \left(\frac{\beta}{k^2} (kA_T) \cos\gamma - 2U_T (kA_*) \right).$$

With this solution for ω we next may calculate the energy conversion $C(E, K)$ carried out by the baroclinic wave. We had

$$C(E, K) = -\frac{1}{g} \int_0^{p_0} \int_S \omega\alpha \, dS \, dp. \tag{3.83}$$

In this case we must apply (3.83) at the 50 cb level. In addition, we have

$$\alpha = -\frac{\partial \phi}{\partial p} = -f_0 \frac{\partial \psi}{\partial p} = \frac{2f_0}{P} \psi_T. \tag{3.84}$$

The meridional direction may be neglected as we are interested in the wave only. We get therefore

$$\begin{aligned} C(E, K) &= -\frac{1}{Lg} \int_0^{p_0} \int_0^L \frac{2f_0}{P} \omega\psi_T \, dx \, dp \\ &= -\frac{4f_0}{Lg} \int_0^L \omega\psi_T \, dx \end{aligned} \tag{3.85}$$

The expressions for the wave part of ψ_T from (3.78) and the solution for ω from (3.81) are used in (3.85), the integrations are carried out, and after reduction we obtain

$$C(E, K) = \frac{2P}{g} \cdot \frac{\lambda^2 \cdot k}{\lambda^2 + k^2} (2U_T) (kA_*) (kA_T) \sin\gamma. \tag{3.86}$$

In this expression we use $2P = 100$ kPa, $g = 9.8$ m s^{-2}, $\lambda^2 = 2.5 \times 10^{-12}$ m^{-2}, $U_T = 10$ m s^{-1}, $kA_T = v_{T,max} = 5$ m s^{-1}, $kA_* = v_{*,max} = 10$ m s^{-1} and $\gamma = 20°$. With these MTS units we get the result in the following units: kJ m^{-2} s^{-1}. Multiplying the outcome by 10^3 converts the result to J m^{-2} s^{-1} = Wm^{-2}. From (3.86) it is first of all seen that the conversion is positive when $U_T > 0$ and $\gamma > 0$. These conditions are fulfilled in general in the middle latitudes where $U = U(z)$ increases up through the atmosphere to the tropopause, and where the waves slope toward the west with height. The conversion is furthermore proportional to U_T and the two wave amplitudes. It will thus increase by a factor 8 if each of these are doubled. The dependence on wave number is found in the ratio

$$\frac{k}{\lambda^2 + k^2},$$

which has a maximum if $k = \lambda$,

$$C(E, K) = 2.75 \text{ W m}^{-2}.$$

4

AVAILABLE POTENTIAL ENERGY

The concept of available potential energy was introduced for simple fluid systems in Chapter 2. Here we define available potential energy for the atmosphere. Contrary to internal, potential, and kinetic energies defined for a parcel of air, the available potential energy is an integral concept that ideally applies to the total global atmosphere.

The basic idea as before, is that not all of the total potential energy E can be converted into kinetic energy, or, in other words, a fraction of E is unavailable for conversion. We consider as Lorenz (1955) a stably stratified, hydrostatic atmosphere. In such an atmosphere the potential temperature will increase upwards. At a point where the potential temperature is θ, the pressure is equal to the weight of a column with unit cross section of air and with potential temperatures larger than θ. Since we want to consider adiabatic rearrangements, it is convenient to use θ as the vertical coordinate, in other words isentropic surfaces. In an arbitrary observed state of the atmosphere the isentropic surfaces will deviate from surfaces parallel to the spherical earth – that is, locally horizontal surfaces. As a matter of fact some isentropic surfaces will be entirely in the atmosphere if θ for these surfaces is sufficiently large, while others with smaller values of θ will intersect the ground.

It is convenient to define the pressure on those parts of an isentropic surface that are below ground. Suppose that the isentropic surface characterized by the potential temperature θ at a point (λ, φ) is below ground. The pressure at such a point is then defined by the surface pressure p_0 at the geographical point (λ, φ) – that is,

$$p(\lambda, \varphi, \theta) = p_0(\lambda, \varphi) \text{ when } \theta < \theta_0, \tag{4.1}$$

where θ_0 is the potential temperature in the surface point (λ, φ). We may even extend this definition to isentropic surfaces that are entirely below ground, because according to (4.1) the pressure on these surfaces, point by point, is equal to the surface pressure. Therefore we may also talk about an isentropic surface with $\theta = 0° K$.

With these definitions we may define the average pressure on any isentropic surface as

$$\overline{p}(\theta) = S^{-1} \int_S p(\lambda, \varphi, \theta) \, dS \qquad dS = a^2 \cos\varphi \, d\lambda \, d\varphi. \tag{4.2}$$

We define next a so-called *reference state* which is a state of the atmosphere in which the isentropic surfaces are concentric spheres and where the pressure at each point of a given isentropic surface is equal to the mean pressure $\bar{p}(\theta)$ for the corresponding isentropic surface in the real atmosphere. Said in a different way: we rearrange the isentropic surfaces by adiabatic processes in such a way that each surface becomes a sphere. The available potential energy A is now defined as the difference between the total potential energy E in the observed state and the corrresponding value $\bar{E}$ for the reference state:

$$A = E - \bar{E}. \tag{4.3}$$

We recall the definition of the potential temperature

$$T = \theta \left(\frac{p}{p_{00}}\right)^\kappa, \qquad p_{00} = 100 \text{ kPa}, \qquad \kappa = \frac{R}{c_p}. \tag{4.4}$$

The first task is to express E using θ as the vertical coordinate. We have

$$E = \frac{1}{g} \int_0^{p_0} \int_S c_p T \, dS \, dp = \frac{c_p}{g} \int_0^{p_0} \int_S \theta \left(\frac{p}{p_{00}}\right)^\kappa dS \, dp \tag{4.5}$$

$$= \frac{c_p}{g p_{00}^\kappa} \int_0^{p_0} \int_S \theta p^\kappa \, dS \, dp = \frac{c_p}{g p_{00}^\kappa (1+\kappa)} \int_0^{p_0} \int_S \theta \, dp^{1+\kappa} \, dS.$$

At this point we make use of the identity

$$\theta \, dp^{1+\kappa} = d\left(\theta p^{1+\kappa}\right) - p^{1+\kappa} \, d\theta. \tag{4.6}$$

In the integration with respect to θ we use as explained above the limits $\theta = 0$ and $\theta = \infty$, and we obtain

$$E = \frac{c_p}{g p_{00}^\kappa (1+\kappa)} \int_0^\infty \int_S p^{1+\kappa} \, dS \, d\theta. \tag{4.7}$$

For the reference state we get

$$\bar{E} = \frac{c_p}{g p_{00}^\kappa (1+\kappa)} \int_0^\infty \int_S \bar{p}^{1+\kappa} \, dS \, d\theta, \tag{4.8}$$

and finally

$$A = \frac{c_p}{g p_{00}^\kappa (1+\kappa)} \int_0^\infty \int_S \left(p^{1+\kappa} - \bar{p}^{1+\kappa}\right) dS \, d\theta. \tag{4.9}$$

Equation (4.9) is the so-called exact expression for the available potential energy. Although well defined, it has seldom been used in calculations

AVAILABLE POTENTIAL ENERGY

involving data, presumably because the data are normally given in isobaric surfaces and not in isentropic coordinates. Lorenz (1955) derived an approximate expression in pressure coordinates which has been used in observational studies. We shall now derive such an expression.

The starting point is to write the pressure p as

$$p = \bar{p} + p', \tag{4.10}$$

where p' at a given point on an isentropic surface is the deviation from the global mean pressure. The integrand in (4.9) may then be approximated as follows:

$$p^{1+\kappa} - \bar{p}^{1+\kappa} = \bar{p}^{1+\kappa}\left[\left(1 + \frac{p'}{\bar{p}}\right)^{1+\kappa} - 1\right] \tag{4.11}$$

$$= \bar{p}^{1+\kappa}\left[(1+\kappa)\frac{p'}{\bar{p}} + \frac{\kappa}{2}(1+\kappa)\left(\frac{p'^2}{\bar{p}}\right) + \cdots\right].$$

Since the integration carried out in (4.9) involves an integration over the global area S, we find that the term containing $p'/\bar{p}$ will integrate to zero. The approximate result is

$$A \approx \frac{1}{2}\frac{c_p}{gp_{00}^\kappa(1+\kappa)}\int_0^\infty \int_S \bar{p}^{1+\kappa}\kappa(1+\kappa)\left(\frac{p'}{\bar{p}}\right)^2 dS\, d\theta. \tag{4.12}$$

We note next that p' may be written as follows:

$$p' = \frac{\partial p}{\partial \theta}\theta' \tag{4.13}$$

and

$$d\theta = \frac{\partial \theta}{\partial p} dp. \tag{4.14}$$

The last expression in (4.14) brings us back to the pressure as vertical coordinate, and θ' may then be considered as the deviation of the potential temperature from its global area average on the isobaric surface. A further approximation is that in the following we shall evaluate the derivative using the reference state – that is,

$$\frac{\partial \theta}{\partial p} = \frac{\partial \bar{\theta}}{\partial p}. \tag{4.15}$$

With $\kappa = R/c_p$, (4.12) becomes

$$A = \frac{1}{2}\frac{R}{g}\frac{1}{p_{00}^\kappa}\int_0^{p_0}\int_S p^{\kappa-1}\frac{\theta'^2}{(\partial\bar{\theta}/\partial\bar{p})^2}\left(-\frac{\partial\bar{\theta}}{\partial p}\right) dS\, d\bar{p}. \tag{4.16}$$

Considering then an arbitrary isobaric surface and denoting the averaged specific volume by $\bar{\alpha}$ and the averaged temperature by $\bar{T}$, we find from the gas equation where p is a constant that

$$\frac{\alpha'}{\bar{\alpha}} = \frac{T'}{\bar{T}}. \tag{4.17}$$

On the other hand, from the definition of potential temperature θ we find by a similar process

$$\frac{T'}{\bar{T}} = \frac{\theta'}{\bar{\theta}}. \tag{4.18}$$

We have therefore

$$\theta' = \frac{\bar{\theta}}{\bar{\alpha}}\alpha' \tag{4.19}$$

and thus

$$A = \frac{1}{2}\frac{R}{g}\frac{1}{p_{00}^\kappa}\int_0^{p_0}\int_S \bar{p}^{\kappa-1}\frac{\bar{\theta}^2}{\bar{\alpha}^2}\frac{\alpha'^2}{\partial\bar{\theta}/\partial\bar{p}}\,dS\,d\bar{p}. \tag{4.20}$$

We define the static stability measure

$$\bar{\sigma} = -\bar{\alpha}\frac{\partial \ln \bar{\theta}}{\partial \bar{p}} \tag{4.21}$$

and noting that

$$\bar{\theta} = \frac{\bar{\alpha}\,\bar{p}^{1-\kappa}}{Rp_{00}^\kappa} \tag{4.22}$$

we finally may write (4.20) in the form

$$A = \frac{1}{g}\int_0^{p_0}\int_S \frac{1}{2\bar{\sigma}}\alpha'^2\,dS\,dp, \tag{4.23}$$

which is the formulation most often used in data studies. Since $\alpha = -\partial\varphi/\partial p$, we see that it is well suited for data giving the geopotential on isobaric surfaces. One could of course equally well have used (4.18) to introduce T', in which case we would have obtained the approximate formula given by Lorenz (1955).

At first sight it seems unnecessary to introduce the concept of available potential energy because the classical energy forms seem quite sufficient in describing the energetics of the atmosphere. Apart from the fact that (4.9) shows that there is an upper limit to the amount of the total potential energy that can be converted into kinetic energy, the available potential energy can also be used in an independent way to estimate the efficiency of

AVAILABLE POTENTIAL ENERGY

the atmosphere as a heat engine. From the preceding discussion it is seen that a measure of the efficiency would be either $D(K)$ or $C(E,K)$. $G(E)$ cannot be used for this purpose because the generation of E by nonfrictional processes is zero, and the generation of E by frictional processes is of course equal to $D(K)$. As we shall see later in detail, $G(A)$ is different.

Qualitatively we may see this by the following argument. If a resting atmosphere is heated in a certain region, we increase the total potential energy, and at the same time we make some of the total potential energy available for conversion into kinetic energy. On the other hand, if we cool the atmosphere in a limited region, the total potential energy decreases; but the available potential energy increases, and some of it is converted into kinetic energy.

In the following discussion we shall formalize these considerations and find an expression for the generation of available potential energy by heating. For this purpose it is necessary to recall some equations formulated in the isentropic coordinate system – that is, a system with potential temperature as the vertical coordinate. Let us start with the thermodynamic equation. It is well known that this equation may be written in the form

$$\dot{\theta} = \frac{d\theta}{dt} = \frac{1}{c_p}\left(\frac{p_{00}}{p}\right)^{\kappa} H, \tag{4.24}$$

where $\dot{\theta}$ is the "vertical velocity" in the θ-system.

Next, we turn our attention to the continuity equation. The mass element is

$$\delta m = \rho\, \delta x\, \delta y\, \delta z, \tag{4.25}$$

but using the hydrostatic equation we find

$$\delta m = -\frac{1}{g}\frac{\partial p}{\partial z}\, \delta x\, \delta y\, \delta z = -\frac{1}{g}\frac{\partial p}{\partial \theta}\, \delta x\, \delta y\, \delta \theta. \tag{4.26}$$

Expressing in the usual way that the change of mass in a stationary volume is equal to the net-inflow, it is easily found that

$$\frac{\partial}{\partial t}\left(\frac{\partial p}{\partial \theta}\right) = -\nabla_3 \cdot \left(\frac{\partial p}{\partial \theta}\vec{v}\right), \tag{4.27}$$

which when expanded becomes

$$\frac{\partial}{\partial t}\left(\frac{\partial p}{\partial \theta}\right) + \nabla_2 \cdot \left(\frac{\partial p}{\partial \theta}\vec{v}\right) + \frac{\partial}{\partial \theta}\left(\frac{\partial p}{\partial \theta}\dot{\theta}\right) = 0. \tag{4.28}$$

This equation is integrated from an arbitrary level θ to the top of the atmosphere. Defining

$$\vec{B} = \int_{\theta}^{\infty} \frac{\partial p}{\partial \theta}\vec{v}\, d\theta, \tag{4.29}$$

we find
$$\frac{\partial p}{\partial t} = \nabla_2 \cdot \vec{B} - \frac{\partial p}{\partial \theta} \dot{\theta}. \tag{4.30}$$

Going back to (4.9) we find by differentiation with respect to time that
$$\frac{dA}{dt} = \frac{c_p}{gp_{00}^\kappa} \int_0^\infty \int_S \left(p^\kappa \frac{\partial p}{\partial t} - \overline{p}^\kappa \frac{\partial \overline{p}}{\partial t} \right) dS\, d\theta. \tag{4.31}$$

At this point we consider only the contribution from the heating which is related to the last term in (4.30), and find therefore
$$G(A) = \frac{c_p}{gp_{00}^\kappa} \int_0^\infty \int_S \left[-\frac{\partial p}{\partial \theta} p^\kappa \dot{\theta} + \overline{p}^\kappa \overline{\left(\frac{\partial p}{\partial \theta} \dot{\theta} \right)} \right] dS\, d\theta. \tag{4.32}$$

or
$$G(A) = \frac{c_p}{gp_{00}^\kappa} \int_0^\infty \int_S \left[-\frac{\partial p}{\partial \theta} p^\kappa \dot{\theta} + \overline{p}^\kappa \frac{\partial p}{\partial \theta} \dot{\theta} \right] dS\, d\theta. \tag{4.33}$$

The step from (4.32) to (4.33) is explained by the identity
$$\int_S \overline{p}^\kappa \overline{\left(\frac{\partial p}{\partial \theta} \dot{\theta} \right)} dS = \overline{p}^\kappa \overline{\left(\frac{\partial p}{\partial \theta} \dot{\theta} \right)} \cdot S = \int_S \overline{p}^\kappa \frac{\partial p}{\partial \theta} \dot{\theta}\, dS. \tag{4.34}$$

From (4.24) we introduce $\dot{\theta}$ and obtain finally
$$G(A) = \frac{1}{g} \int_0^\infty \int_S \frac{\partial p}{\partial \theta} \cdot \left(\frac{\overline{p}^\kappa}{p^\kappa} - 1 \right) H\, d\theta\, dS \tag{4.35}$$

or recalling (4.26)
$$G(A) = \int_M H \left(1 - \frac{\overline{p}^\kappa}{p^\kappa} \right) dm. \tag{4.36}$$

$G(A)$ will be discussed later in terms of the atmosphere but the theoretical development will be continued by finding the contribution from the first term (4.30) to the rate of change of the available potential energy. We expect naturally that it is the term containing the vector B that will result in a conversion between A and K, but due to the fact that the equations are expressed in isentropic coordinates it is far from obvious that the final result will agree with the result found for the conversion between E and K, derived earlier – that is,
$$C(E_g) = -\frac{1}{g} \int_0^{p_0} \int_S \omega \alpha\, dS\, dp. \tag{4.37}$$

AVAILABLE POTENTIAL ENERGY

The first step is to recall the form of the hydrostatic equation in isentropic coordinates. Using the Montgomery potential

$$\psi = gz + c_p T, \qquad (4.38)$$

we seek an expression for $\partial \psi / \partial \theta$ – that is,

$$\frac{\partial \psi}{\partial \theta} = g \frac{\partial z}{\partial \theta} + c_p \frac{\partial T}{\partial \theta}. \qquad (4.39)$$

Equation (4.39) may be developed as follows:

$$\frac{\partial \psi}{\partial \theta} = \left(\frac{\partial \phi}{\partial p} + c_p \frac{\partial T}{\partial p} \right) \cdot \frac{1}{\partial \theta / \partial p} = \frac{\alpha}{\partial \theta / \partial p} \frac{\gamma - \gamma_d}{\gamma_d}. \qquad (4.40)$$

On the other hand,

$$\frac{\partial \theta}{\partial p} = \frac{\alpha}{g} \cdot \frac{\theta}{T} (\gamma - \gamma_a). \qquad (4.41)$$

We obtain therefore

$$\frac{\partial \psi}{\partial \theta} = c_p \frac{T}{\theta} = c_p \left(\frac{p}{p_{00}} \right)^\kappa. \qquad (4.42)$$

The next step to recall is the transformation formula from isentropic to isobaric coordinates for the gradient. Textbooks on atmospheric dynamics show that

$$\nabla_\theta \cdot \vec{B} = \left(\frac{\partial B x}{\partial x} \right)_\theta + \left(\frac{\partial B y}{\partial y} \right)_\theta = \nabla_p \cdot \vec{B} - \frac{\partial \vec{B}}{\partial \theta} \cdot \nabla_p \theta. \qquad (4.43)$$

Recalling the definition of $\vec{B}$ in (4.29) we see that

$$\frac{\partial \vec{B}}{\partial \theta} = -\frac{\partial p}{\partial \theta} \vec{v} \qquad (4.44)$$

and further that

$$\vec{B} = \int_0^\infty \frac{\partial p}{\partial \theta} \vec{v} \, d\theta = \int_p^0 \vec{v} \, dp. \qquad (4.45)$$

It is then immediately seen that

$$\nabla_p \cdot \vec{B} = -\int_0^p \nabla_p \cdot \vec{v} \, dp = \int_0^p \frac{\partial \omega}{\partial p} dp = \omega. \qquad (4.46)$$

Going back to (4.31) we find that

$$C(A, K) = -\frac{c_p}{gp_{00}^\kappa} \int_0^\infty \int_S p^\kappa \nabla_\theta \cdot B \, dS \, d\theta \qquad (4.47)$$

$$= -\frac{1}{g} \int_0^\infty \int_S \frac{\partial \psi}{\partial \theta} \left(\omega + \frac{\partial p}{\partial \theta} \vec{v} \cdot \nabla_p \theta \right) dS \, d\theta,$$

where we have used (4.42) and (4.44). Changing the vertical integration from θ to p we find

$$C(A, K) = \frac{1}{g} \int_0^{p_0} \int_S \left(\omega \frac{\partial \psi}{\partial p} + \frac{\partial \psi}{\partial \theta} \vec{v} \cdot \nabla_p \theta \right) dS \, dp. \qquad (4.48)$$

We note that

$$\frac{\partial \psi}{\partial p} = \frac{\partial \phi}{\partial p} + c_p \frac{\partial T}{\partial p} = -\alpha + c_p \frac{\partial T}{\partial p}, \qquad (4.49)$$

and therefore

$$C(A, K) = -\frac{1}{g} \int_0^{p_0} \int_S \omega \alpha \, dS \, dp$$

$$+ \frac{1}{g} \int_0^{p_0} \int_S \left(c_p \omega \frac{\partial T}{\partial p} + c_p \frac{T}{\theta} \vec{v} \cdot \nabla_p \theta \right) dS \, dp. \qquad (4.50)$$

It is thus seen that we will have obtained the desired result if we can show that the last integral in (4.50) vanishes. This is, however, straightforward, because

$$\int_0^{p_0} \int_S \left(c_p \omega \frac{\partial T}{\partial p} + c_p \left(\frac{p}{p_{00}} \right)^\kappa \vec{v} \cdot \nabla_p \left[T \left(\frac{p}{p_{00}} \right)^{-\kappa} \right] \right) dS \, dp$$

$$= c_p \int_0^{p_0} \int_S \left(\omega \frac{\partial T}{\partial p} + \vec{v} \cdot \nabla_p T \right) dS \, dp$$

$$= c_p \int_0^{p_0} \int_S \left[\frac{\partial T \omega}{\partial p} + \nabla_p \cdot (T \vec{v}) - T \left(\frac{\partial \omega}{\partial p} + \nabla_p \cdot \vec{v} \right) \right] dS \, dp$$

$$= 0 \qquad (4.51)$$

where in the last step we have used the continuity equation in pressure coordinates and the boundary condition $\omega = 0$ and $p = p_0$.

For the sake of completeness it ought to be mentioned that we obtain the same result if we start from the equations of motion in isentropic coordinates. We know that the pressure force in this coordinate system is $-\nabla_\theta \psi$. The energy conversion is therefore

$$C(A, K) = -\frac{1}{g} \int_0^{p_0} \int_S \vec{v} \cdot \nabla_\theta \psi \, dS \, dp, \qquad (4.52)$$

but we find also that

$$\begin{aligned}\nabla_\theta \psi &= \nabla_p \psi - \frac{\partial \psi}{\partial \theta} \nabla_p \theta = \nabla_p \phi + c_p \nabla_p T - c_p \frac{T}{\theta} \nabla_p \theta \\ &= \nabla_p \phi + c_p \nabla_p T - c_p \nabla_p T = \nabla_p \phi, \end{aligned} \quad (4.53)$$

and thus

$$\begin{aligned} C(A,K) &= -\frac{1}{g} \int_0^{p_0} \int_S \vec{v} \cdot \nabla \phi \, dS \, dp \\ &= -\frac{1}{g} \int_0^{p_0} \int_S \omega \alpha \, dS \, dp. \end{aligned} \quad (4.54)$$

With these final derivations we have completed the introduction of available potential energy, the generation of this energy form and its conversion to kinetic energy. As is seen in the definition it is most convenient to use the exact formulation in the hydrostatic system in terms of isentropic coordinates. On the other hand, the use of these coordinates, which are suitable for the theory, gives practical difficulties in the formulation of procedures for the evaluation of the various integrals. The reason is that meteorological data normally are arranged with respect to isobaric surfaces and not isentropic surfaces. Thus there is a need to obtain the approximate expression in (4.37), which will be used later in connection with data studies.

At this point we may summarize the results obtained in the present chapter. Available potential energy is generated in the atmosphere by heating. The effect of the heating in the generation process depends on the pressure distribution as expressed in the formula

$$G(A) = \int_M HN \, dm, \quad (4.55)$$

where N, the so-called efficiency factor, is

$$N = 1 - \frac{\bar{p}^\kappa}{p^\kappa}, \quad (4.56)$$

p being the pressure on the isentropic surface, and $\bar{p}$ its global mean value. To get a first idea we may consider a typical meridional cross section. It is well known that the isentropic surfaces slope upward from the equator toward the North Pole. This means that the pressure is larger than the mean pressure in the low latitudes ($p > \bar{p}$) while $p < \bar{p}$ in the high latitudes. We therefore should expect that N is generally positive in the low latitudes and negative in the high latitudes. With respect to the heating we know that $H > 0$ in the low latitudes and $H < 0$ in the high latitudes. Therefore there is a positive correlation between H and N leading to a positive value

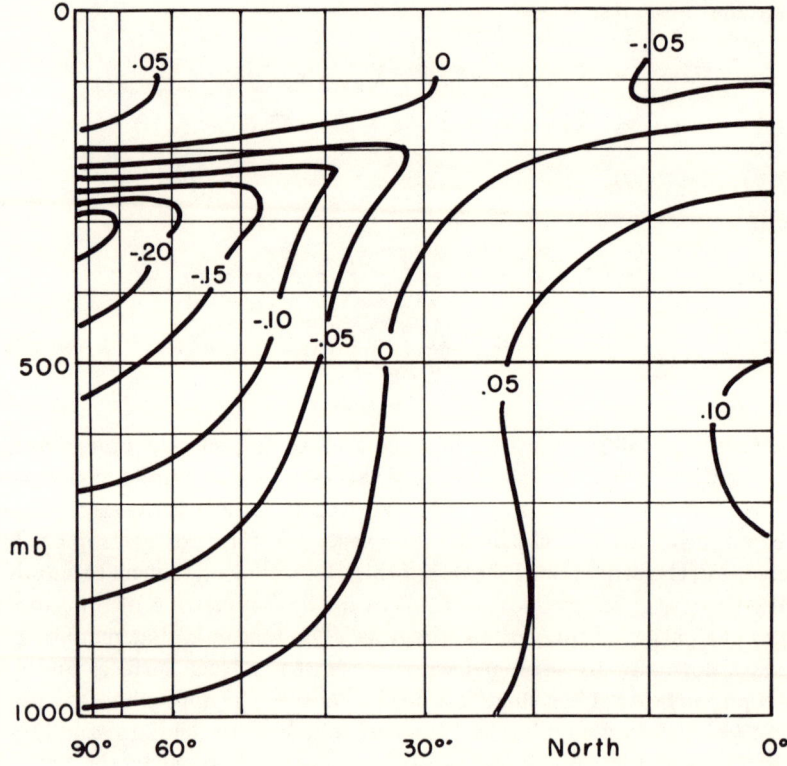

Fig. 4.1: The efficiency factor as computed by Lorenz (1967) and given by Eq. (4.56).

of $G(A)$ in our example. Lorenz (1967) has calculated N from a reasonable distribution of the isentropic surfaces. His results are shown in Fig. 4.1. They confirm the arguments made above. It is more difficult to determine the contribution from the heating in the longitudinal direction.

We have gone to some trouble to show that the energy conversion $C(A, K)$, derived from the so-called exact formula for an isentropic coordinate, reduces to the same expression as derived earlier for $C(E, K)$ – that is,

$$C(A, K) = C(E, K) = - \int_M \omega \alpha \, dm. \tag{4.57}$$

It could have been argued in advance that this would necessarily be so because no energy conversion can be obtained from the reference state where the isentropic surfaces are spherical or parallel to the earth. However, our derivations are at least a test of the consistency of the theory and a confirmation of the correctness of the more artificial aspects of the formulation such as the continuation of the isentropic surfaces below the ground.

AVAILABLE POTENTIAL ENERGY

4.1 The Quasi-Geostrophic Case

We shall finally make the point that the approximate formula derived in (4.37) is the correct form of the available potential energy for use in quasi-nondivergent models. One could not know this in advance because (4.37) was derived by series expansion using only the terms up to and including the quadratic terms in p'. Considering quasi-nondivergent models it is known that some of the main principles derived from scale-theory are that the nondivergent wind is used for advection purposes, that the static stability is a function of pressure only, and that the vertical velocity is determined from the so-called $\omega-$ equation which is the equation expressing that hydrostatic and nondivergent relations have to be satisfied everywhere and at all times.

To show this condition we start from the thermodynamic equation in the form [see (4.24)]

$$\frac{d \ln \theta}{dt} = \frac{1}{c_p T} H \tag{4.58}$$

or

$$\frac{\partial \ln \theta}{\partial t} + \vec{v}_\psi \cdot \nabla \ln \theta + \omega \frac{\partial \ln \theta}{\partial p} = \frac{1}{c_p T} H. \tag{4.59}$$

Using the definition of potential temperature and the fact that (4.59) is expressed in the pressure system, we see that

$$\frac{\partial \ln \theta}{\partial t} = \frac{\partial \ln \alpha}{\partial t} \qquad \nabla \ln \theta = \nabla \ln \alpha, \tag{4.60}$$

and thus

$$\frac{\partial \alpha}{\partial t} + \vec{v}_\psi \cdot \nabla \alpha - \bar{\sigma} \omega = \frac{R}{c_p} \frac{1}{p} H \qquad \bar{\sigma} = -\bar{\alpha} \frac{\partial \ln \bar{\theta}}{\partial p}. \tag{4.61}$$

We note that $\vec{v}_\psi$ is nondivergent and that the global average of ω is zero because

$$\omega = -\int_0^p \nabla \cdot \vec{v} \, dp. \tag{4.62}$$

Taking a global average over an isobaric surface, we find therefore

$$\frac{\partial \bar{\alpha}}{\partial t} = \frac{R}{c_p} \frac{1}{p} \bar{H}, \tag{4.63}$$

and by subtraction denoting $\alpha' = \alpha - \bar{\alpha}$ and $H' = H - \bar{H}$,

$$\frac{\partial \alpha'}{\partial t} + \vec{v}_\psi \cdot \nabla \alpha' - \bar{\sigma} \omega = \frac{R}{c_p} \frac{1}{p} H'. \tag{4.64}$$

Using (4.23) we find that

$$\frac{dA}{dt} = \frac{1}{g}\int_0^{p_0}\int_S \omega\alpha' \, dS \, dp + \frac{1}{g}\int_0^{p_0}\int_S \frac{R}{c_p\overline{\sigma}p} H'\alpha' \, dS \, dp, \quad (4.65)$$

which may also be written in the form

$$\frac{dA}{dt} = -C(A,K) + G(A) \quad (4.66)$$

when

$$C(A,K) = -\frac{1}{g}\int_0^{p_0}\int_S \omega\alpha' \, dS \, dp = -\frac{1}{g}\int_0^{p_0}\int_S \omega\alpha \, dS \, dp \quad (4.67)$$

and

$$\begin{aligned} G(A) &= \frac{1}{g}\int_0^{p_0}\int_S \frac{R}{c_p\overline{\sigma}p} H'\alpha' \, dS \, dp \\ &= \frac{1}{g}\int_0^{p_0}\int_S \frac{R^2}{c_p\overline{\sigma}p^2} H'T' \, dS \, dp. \end{aligned} \quad (4.68)$$

The expression (4.67) has the same form as in the more general cases considered previously, but the vertical velocity in the formula is the one calculated from the quasi-nondivergent model, and it therefore is not necessarily the same as the vertical velocity from a model based on the primitive equations.

The new expression for $G(A)$ shows that in a quasi-nondivergent model it is the correlation between the heating and the temperature in isobaric surfaces that determines the size and sign of $G(A)$. If we once again use a typical meridional cross section as an example, we see, as before, that $H' > 0$ in low latitudes and $H' < 0$ in high latitudes. With the normal tropospheric decrease of temperature from the equator to the North Pole it is seen that T' over large regions will have the same sign as H' leading to a positive value of $G(A)$.

In the various formulations given here it is always true that the generation, the conversion, and the dissipation depend on quantities such as the heating and the vertical velocity, which are difficult to obtain or calculate from normal atmospheric data. It is therefore understandable that a considerable effort of a computational nature goes into the determination of these quantities from those atmospheric data that are available. Fig. 4.2 shows the typical energy diagram concerning A and K.

AVAILABLE POTENTIAL ENERGY

Fig. 4.2: The most simple energy diagram for the available potential and kinetic energies.

4.2 An Elementary Derivation

It may help in the understanding of the concept of available potential energy to give an approximate derivation of the quasi-nondivergent expression for this quantity by considering the classical parcel method. The force on a particle of unit mass is

$$-\alpha \frac{\partial p}{\partial z} - g. \tag{4.69}$$

As usual in the parcel method we distinguish between the properties of the parcel and the properties of the environment, but with the assumption that the parcel adjusts its pressure instantaneously to agree with the pressure of the environment. Equation (4.69) may then be written

$$-\alpha \frac{\partial \overline{p}}{\partial z} - g = +g\frac{\alpha}{\overline{\alpha}} - g = g\frac{\alpha - \overline{\alpha}}{\overline{\alpha}}, \tag{4.70}$$

where we have used the hydrostatic assumption for the environment. From the definition of potential temperature it is known that for constant pressure we have

$$g\frac{\alpha - \overline{\alpha}}{\overline{\alpha}} = g\frac{\theta - \overline{\theta}}{\overline{\theta}}. \tag{4.71}$$

Let the parcel have the potential temperature θ_0 initially. Assuming that adiabatic conditions prevail during the displacement, we note that $\theta = \theta_0$ for the parcel. For the environment we may write

$$\overline{\theta} = \theta_0 + \frac{\partial \overline{\theta}}{\partial z} \cdot z. \tag{4.72}$$

The expression (4.71) is then with good approximation

$$F = -g\frac{1}{\overline{\theta}}\frac{\partial \overline{\theta}}{\partial z} \cdot z = -g\frac{\partial \ln \overline{\theta}}{\partial z} \cdot z. \tag{4.73}$$

The work carried out during the displacement from 0 to z is

$$W = \int_0^z -g\frac{\partial \ln \overline{\theta}}{\partial z} \cdot z\, dz \simeq -\frac{g}{2}\frac{\partial \ln \overline{\theta}}{\partial z} \cdot z^2. \qquad (4.74)$$

Since the potential temperature is conserved during the displacement, we have $\theta = \theta_0$, and thus from (4.72)

$$z = \frac{\overline{\theta} - \theta}{\partial \overline{\theta}/\partial z} = -\frac{\theta'}{\partial \overline{\theta}/\partial z}, \qquad (4.75)$$

and therefore

$$W = -\frac{g}{2}\frac{\partial \ln \overline{\theta}}{\partial z}\frac{\theta'^2}{(\partial \overline{\theta}/\partial z)^2} = \frac{g^2}{2\overline{\alpha}}\frac{\partial \ln \overline{\theta}}{\partial p} \cdot \frac{\theta'^2}{(g^2/\overline{\alpha}^2)(\partial \overline{\theta}/\partial p)^2} \qquad (4.76)$$

or

$$W = \frac{1}{2}\overline{\alpha} \cdot \frac{1}{\partial \ln \overline{\theta}/\partial p}\frac{\theta'^2}{\overline{\theta}^2} = \frac{1}{2\overline{\alpha}(\partial \ln \overline{\theta}/\partial p)} \cdot \alpha'^2. \qquad (4.77)$$

Finally we get

$$W = -\frac{1}{2}\frac{1}{\overline{\sigma}}\alpha'^2. \qquad (4.78)$$

However, the work carried out by displacing the particle is the opposite of the potential energy gained in the process. Therefore,

$$\Delta P = \frac{1}{2}\frac{1}{\overline{\sigma}}\alpha'^2. \qquad (4.79)$$

It is thus seen from (4.79) that in the normal atmosphere in which the isobaric surfaces are sloping, we can bring ΔP to zero for a parcel if we move it adiabatically from its postion to a height where the pressure in a reference atmosphere, consisting of spherical isobaric surfaces, is equal to the pressure from which the parcel started.

We conclude this chapter by giving some estimates of the available potential and kinetic energy based on data studies. Based on atmospheric data for one year (Feb. 1963 - Jan. 1964, incl.) for the Northern Hemisphere, using the levels 100, 85, 70, 50, 30, 20, and 10 kPa and the quasi-nondivergent formulas, the amounts of A and K were computed. The reason for the period is that January 1963 was very unusual. The calculations were carried out once a day for every day of the year. The annual mean values are $A = 3978$ kJ m^{-2} and $K = 1849$ kJ m^{-2} implying $K/A \approx 0.46$. Calculations have also been made for other years albeit using a lower vertical resolution. The derived data may also be used to estimate the annual

AVAILABLE POTENTIAL ENERGY

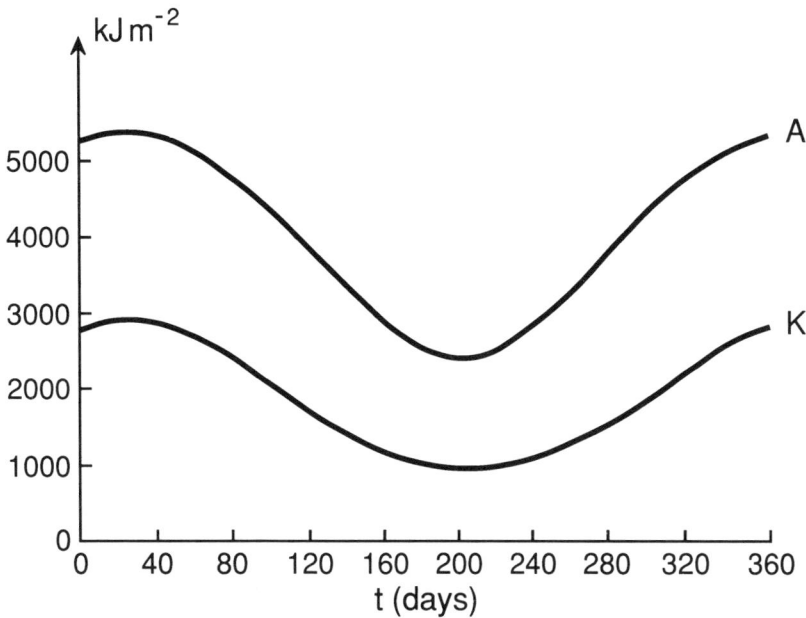

Fig. 4.3: The annual variation of the available potential and the kinetic energies computed from the mean value and the first two Fourier components. Energy amounts in kJ m^{-2} and time in days.

variation of A and K (Wiin-Nielsen, 1967). Figure 4.3 shows the annual variation computed from the annual mean value and the first two Fourier components. We get therefore a very smoothed distribution. It is, however, seen that the largest amounts are obtained in late January with the smallest values in late July. There is also a phase difference of about four days between the maxima in A and K with K attaining its largest value at the later time, although it is difficult to see this small difference in the figure.

5

BAROCLINIC AND BAROTROPIC FLOW

The atmospheric flow is baroclinic. Nevertheless, the barotropic approximation to atmospheric conditions has played a major role both in atmospheric dynamics and in numerical weather prediction. The reason such a simple approximation has been so important is that the vertically averaged flow is nearly barotropic, and this flow can to a good degree of approximation be identified with the flow at a mid-tropospheric level, normally 50 kPa. On the other hand, the vertical mean flow is not disassociated from the remaining part because it experiences frictional dissipation, and it follows therefore that the vertical mean flow must be supplied with energy from some source, which turns out to be the baroclinic flow in the atmosphere. In this chapter we explore the energetics of the vertical mean flow (the barotropic flow) and the deviation from this flow – the shear flow (the baroclinic flow).

The vertical mean flow or the vertical mean of any quantity is defined by the following operator:

$$(\ )_M = \frac{1}{p_0} \int_0^{p_0} (\ ) \, dp. \tag{5.1}$$

The vertical shear flow is defined as the difference between the total flow and the vertical mean flow – that is,

$$(\ )_s = (\ ) - (\ )_M. \tag{5.2}$$

Starting with the continuity equation

$$\nabla \cdot \vec{v} + \frac{\partial \omega}{\partial p} = 0 \tag{5.3}$$

and applying (5.1), we find with $\omega = 0$ at $p = 0$ and $p = p_0$ that

$$\nabla \cdot \vec{v}_M = 0. \tag{5.4}$$

The vertical mean flow is thus approximately nondivergent. For the total horizontal flow we have earlier derived the energy equation:

$$\frac{dK}{dt} = -\frac{1}{g} \int_0^{p_0} \int_S \vec{v} \cdot \nabla \phi \, dS \, dp + \frac{1}{g} \int_0^{p_0} \int_S \vec{v} \cdot \vec{F} \, dS \, dp \tag{5.5}$$

where $\vec{F}$ is the frictional force per unit mass, and where we note that the first integral was converted as follows:

$$-\frac{1}{g}\int_0^{p_0}\int_S \vec{v}\cdot\nabla\phi\, dS\, dp = -\frac{1}{g}\int_0^S \omega\alpha\, dS\, dp. \tag{5.6}$$

The total kinetic energy is equal to the sum of K_M and K_S because

$$K = \frac{1}{g}\int_0^{p_0}\int_S \frac{1}{2}\left(\vec{v}_M+\vec{v}_S\right)\left(\vec{v}_M+\vec{v}_S\right)dS\, dp \tag{5.7}$$

$$= \frac{1}{g}\int_0^{p_0}\int_S \frac{1}{2}\vec{v}_M^2\, dS\, dp + \frac{1}{g}\int_0^{p_0}\int_S \frac{1}{2}\vec{v}_S^2\, dS\, dp,$$

and therefore

$$\frac{dK_S}{dt} = \frac{dK}{dt} - \frac{dK_M}{dt}. \tag{5.8}$$

We next shall derive an expression for dK_M/dt. We start from the well-known equations

$$\frac{\partial u}{\partial t}+\frac{\partial(uu)}{\partial x}+\frac{\partial(uv)}{\partial y}+\frac{\partial(uw)}{\partial p} = -\frac{\partial\phi}{\partial x}+fv+F_x, \tag{5.9}$$

$$\frac{\partial v}{\partial t}+\frac{\partial(uv)}{\partial x}+\frac{\partial(vv)}{\partial y}+\frac{\partial(vw)}{\partial p} = -\frac{\partial\phi}{\partial y}-fu+F_y.$$

The following identities are easily verified:

$$\frac{\partial(uu)}{\partial x}+\frac{\partial(uv)}{\partial y} = \frac{\partial k}{\partial x}+u\nabla\cdot\vec{v}-v\zeta, \tag{5.10}$$

$$\frac{\partial(uv)}{\partial x}+\frac{\partial(vv)}{\partial y} = \frac{\partial k}{\partial y} = \frac{\partial k}{\partial y}+v\nabla\cdot\vec{v}+u\zeta.$$

Introducing (5.10) into (5.9) and taking the vertical average, we obtain

$$\frac{\partial u_M}{\partial t}+\frac{\partial(k)_M}{\partial x}+\left(u\nabla\cdot\vec{v}\right)_M-(v\zeta)_M = -\frac{\partial\phi_M}{\partial x}+fv_M+F_{x,M}, \tag{5.11}$$

$$\frac{\partial v_M}{\partial t}+\frac{\partial(k)_M}{\partial y}+\left(v\nabla\cdot\vec{v}\right)_M+(u\zeta)_M = -\frac{\partial\phi_M}{\partial y}-fu_M+F_{y,M}.$$

We note that in general

$$(\beta\gamma)_M = [(\beta_M+\beta_S)(\gamma_M+\gamma_S)]_M = \beta_M\gamma_M+(\beta_S\gamma_S)_M. \tag{5.12}$$

Defining

$$k_M = \frac{1}{2}\left(u_M^2+v_M^2\right), \tag{5.13}$$

we obtain from (5.12) through multiplication by u_M and v_M, respectively,

$$\frac{\partial k_M}{\partial t} + \vec{v}_M \cdot \nabla(k)_M + \vec{v}_M \cdot \left[(\nabla \cdot \vec{v})\vec{v}\right]_M - \vec{k} \times \vec{v}_M \cdot \left(\zeta \vec{v}\right)_M$$
$$= -\vec{v}_M \cdot \nabla \phi_M + \vec{v}_M \cdot \vec{F}_M. \qquad (5.14)$$

Integrating (5.14) using (5.12) and the fact that $\nabla \cdot \vec{v}_M = 0$, we obtain

$$\frac{dK_M}{dt} = -\frac{1}{g}\int_0^{p_0}\int_S \left(\vec{v}_S \cdot \vec{v}_M\right)\nabla \cdot \vec{v}_S \, dS \, dp$$
$$-\frac{1}{g}\int_0^{p_0}\int_S \vec{k}\cdot\left(\vec{v}_S \times \vec{v}_M\right)\zeta_S \, dS \, dp$$
$$+\frac{1}{g}\int_0^{p_0}\int_S \vec{v}_M \cdot \vec{F}_M \, dS \, dp. \qquad (5.15)$$

In (5.14) and (5.15) $\vec{k}$ is the vertical unit vector. The first two integrals of (5.15) are the conversion from K_S to K_M, while the last integral is the frictional dissipation – that is,

$$\frac{dK_M}{dt} = C(K_S, K_M) - D(K_M). \qquad (5.16)$$

Subtracting (5.15) from (5.5) we find

$$\frac{dK_S}{dt} = -C(K_S, K_M) + C(A, K_S) - D(K_S) \qquad (5.17)$$

where $C(A, K_S)$ is given in (5.6), and $D(K_S)$ is

$$D(K_S) = -\frac{1}{g}\int_0^{p_0}\int_S \vec{v}_S \cdot \vec{F}_S \, dS \, dp. \qquad (5.18)$$

The energy reservoir K_M is characterized by both a single dissipation that supposedly is positive and a transformation $C(K_S, K_M)$ that consequently also must be positive in the long-term average. On the other hand, K_S, the baroclinic energy, receives energy through conversion from A, gives energy to K_M through $C(K_S, K_M)$, and supposedly has a positive dissipation. If this is so, we have shown that $C(K_S, K_M)$ can give a lower limit to the intensity of the general circulation. We should also stress that the conversion from A goes into the baroclinic component of the flow, as would be expected.

The two integrals in $C(K_S, K_M)$ are quite different although both of them come from the horizontal advection of momentum. The first integral is of course identically zero in a quasi-nondivergent model because the advecting velocity by assumption is nondivergent in such a model. It is therefore natural to define the integrals as follows:

$$C_D(K_S, K_M) = \frac{1}{g} \int_0^{p_0} \int_S \left(\vec{v}_S \cdot \vec{v}_M\right) \left(-\nabla \cdot \vec{v}_S\right) dS\, dp, \qquad (5.19)$$

$$C_{ND}(K_S, K_M) = \frac{1}{g} \int_0^{p_0} \int_S \vec{k} \cdot \left(\vec{v}_M \times \vec{v}_S\right) \zeta_S\, dS\, dp. \qquad (5.20)$$

At a later stage we shall report on the results of evaluating the two integrals (5.19) and (5.20) from atmospheric data. We may discuss, however, the conditions under which the integrals have positive or negative values. Starting with (5.20) we observe that if the flow was totally zonal ($v_M = v_S = 0$), $C_{ND}(K_S, K_M)$ would be zero. The contribution must then come from the waves. Evaluation of the integrand shows that

$$\zeta_S \vec{k} \cdot \left(\vec{v}_M \times \vec{v}_S\right) = \zeta_S (u_M v_S - u_S v_M) = \zeta_S J(\psi_M, \psi_S) \qquad (5.21)$$

if we assume that the winds are nondivergent. We may think of ψ_M as the streamfunction at some mid-tropospheric level. $\psi_S = \psi - \psi_M$ then corresponds to the thermal streamfunction if the level is higher in the atmosphere than that of ψ_M, and to the negative value of the thermal streamfunction if the level is below that of ψ_M. On the other hand, $\zeta_S = \nabla^2 \psi_S$ will be the thermal vorticity or its negative value again depending on whether the level is above or below the level of ψ_M. We may thus consider the Jacobian in (5.21) as an expression of the advection of the thermal streamfunction in the field of ψ_M. It is easily seen that $J(\psi_M, \psi_S) < 0$ if we have warm air advection and $J(\psi_M, \psi_S) > 0$ in cold advection.

As an example we may consider a typical baroclinic wave (see Fig. 5.1). We know that such a wave normally has a westward slope with height, or equivalently, that the thermal field is lagging behind the streamfunction (or height) field. It is thus seen that warm air advection will take place around the ridge when the vorticity is negative, while the trough, where the vorticity is positive, will experience cold air advection. The contribution from such a wave system to the integral $C_{ND}(K_S, K_M)$ will thus be positive. Since the atmosphere on average is dominated by baroclinic waves, we should expect $C_{ND}(K_S, K_M)$ to be positive.

If neglecting the divergent part of the horizontal velocity in quasi-nondivergent models is well justified, we should expect $C_D(K_S, K_M)$ to be small. Contrary to $C_{ND}(K_S, K_M)$, it is, however, possible that $C_D(K_S, K_M)$ has a non-zero value in an axially symmetric, zonal circulation. In this case we find

$$C_D(K_S, K_M) = \frac{1}{g} \int_0^{p_0} \int_S u_S u_M \left(-\nabla \cdot \vec{v}_S\right) dS\, dp. \qquad (5.22)$$

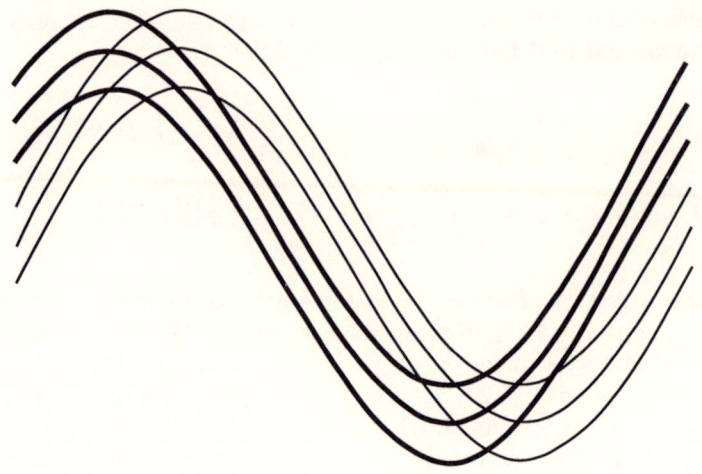

Fig. 5.1: The typical relationship between the temperature (thin lines) and geopotential (thick lines) fields in a baroclinic wave.

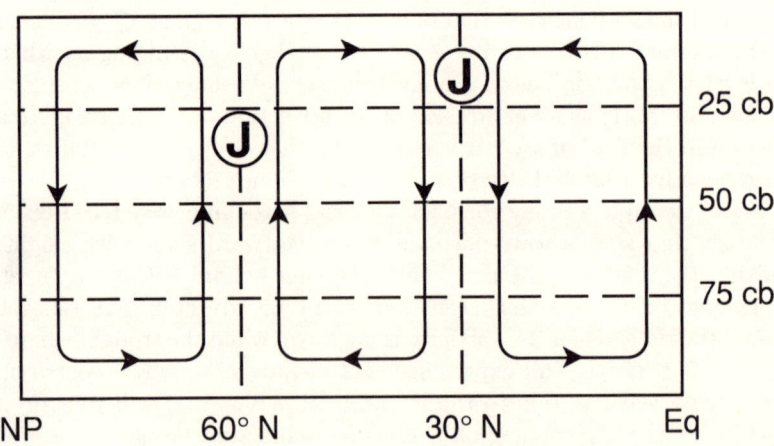

Fig. 5.2: A schematic meridional cross section showing the meridional cells and the positions of the polar and the subtropical jet streams. The equatorial and the polar cells are thermally direct, energy producing cells of the Hadley type, while the middle cell is a thermally indirect, energy consuming cell of the Ferrel type.

The question is naturally how the product $u_S u_M$ will correlate with the convergence $-\nabla \cdot \vec{v}$. For this purpose we consider a meridional cross section, shown schematically in Fig. 5.2. It contains the tropical Hadley cell and the mid-latitude Ferrel cell; the polar cell and the locations of the polar and subtropical jets are also shown. The subtropical jet is located in a region of convergence. In the same region we have $u_M > 0$ and $u_S > 0$ giving a positive contribution. At low levels and at $30°$ N there is divergence, but at this location we have $u_M > 0$ and $u_S < 0$ again giving a positive contribution. In the region of the polar jet the conditions are reversed with respect to the convergence, but are the same with respect to u_M and u_S. We should therefore expect a negative contribution from the region around $60°$ N, but it should be numerically smaller due to the smaller strength of the zonal winds. A small positive value of $C_D(K_S, K_M)$ considering only a meridional cross section may therefore be expected.

It may be of interest to calculate $C_{ND}(K_S, K_M)$ using the same example of a simple sinusoidal, baroclinic wave as was given at the end of Chapter 3. Evaluation of the two factors in the integrand gives

$$\vec{k} \cdot \left(\vec{v}_M \times \vec{v}_T \right) = u_M v_T - u_T v_M \tag{5.23}$$
$$= -kA_T U_M (\cos\gamma \sin kx + \sin\gamma \cos kx) + kA_* U_T \sin kx$$

and

$$\zeta_T = -k^2 A_T \cos\gamma \cos kx + k^2 A_T \sin\gamma \sin kx. \tag{5.24}$$

When Eq. (5.23) is multiplied by (5.24) and the integrations are performed, the result is

$$C_{ND}(K_S, K_M) = \frac{Pk}{g} (kA_T)(kA_*) U_T \sin\gamma. \tag{5.25}$$

In evaluations of (5.25) we have used $P = 50$ kPa, $g = 9.8$ m s^{-2}, $kA_T = kA_* = 10$ m s^{-1}, $U_T = 10$ m s^{-1} and $k = \frac{m}{(a \cos\phi_0)}$. For $\gamma = 20°$ and $m = 1$ we find $C_{ND}(K_S, K_M) = 0.3876$ W m^{-2}. For $m \approx 7$ we find $C_{ND}(K_S, K_M) = 2.71$ W m^{-2}.

The expression (5.25) verifies our earlier result that $C_{ND}(K_S, K_M) > 0$ if the temperature field is lagging behind the height field.

One can also calculate $C_D(K_S, K_M)$ in (5.22) using the same simple baroclinic wave and using the solution of the ω-equation that was obtained in Chapter 3 to estimate $\nabla \cdot \vec{v}$. The result, however, is zero. We must therefore conclude that waves more complicated than sinusoidal waves are necessary to produce a nonzero value of $C_D(K_S, K_M)$.

We shall finally make some comments on the frictional dissipation of K_M and K_S. In the previous discussion it was assumed that both of these were positive. It may help in the following discussion to consider Fig. 5.3, which gives the diagram for A, K_S, and K_M. We consider first the

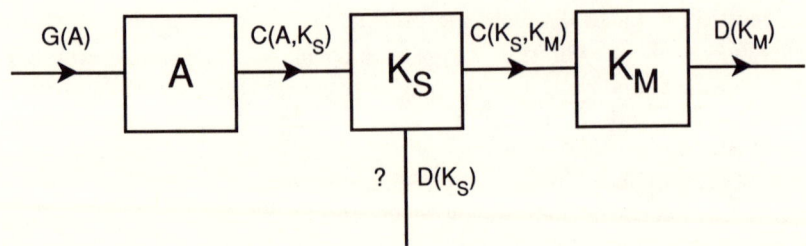

Fig. 5.3: The energy diagram for the available potential energy, the shear flow kinetic energy, and the mean flow kinetic energy.

dissipation of K_M, which is

$$D(K_M) = -\frac{1}{gS} \int_0^{p_0} \int_S \vec{v}_M \cdot \vec{F}_M \, dS \, dp. \qquad (5.26)$$

The force of friction in the coordinate system with pressure as the vertical coordinate is $+g\partial \vec{\tau}/\partial p$, where $\vec{\tau}$ is the stress. From this formulation it follows that

$$\vec{F}_M = +\frac{g}{p_0} \vec{\tau}_0, \qquad (5.27)$$

where $\vec{\tau}_0$ is the surface stress. Inserting (5.27) into (5.26) we find

$$D(K_M) = -\frac{1}{S} \int_S \vec{v}_M \cdot \vec{\tau}_0 \, dS. \qquad (5.28)$$

One of the simpler formulations lead to an approximation in which $\vec{\tau}_0$ is proportional to the surface wind $\vec{v}_0$. If so, we find

$$D(K_M) = \frac{1}{S} \int_S k_* \, \vec{v}_M \cdot \vec{v}_0 \, dS, \qquad (5.29)$$

where $\vec{\tau}_0 = -k_* \, \vec{v}_0$, $k_* > 0$. The expression (5.29) will in general be positive because the angle between $\vec{v}_M$ and $\vec{v}_0$ is less than 90° on average. It can thus be expected that $D(K_M)$ is positive. The situation is less certain with respect to $D(K_S)$. To illustrate this point we may once again use a quasi-nondivergent two-level model, as for example the one formulated by Charney (1959). In this model the stress is needed at the surfaces 100 and 50 kPa. The formulation at 100 kPa is as above for $\vec{\tau}_0$. At 50 kPa Charney makes the assumption that the stress is proportional to the vertical wind shear, which in turn leads to the fact that the curl of the stress is proportional to the vertical change of vorticity. Writing only those terms that relate to the rate of change of vorticity and to the friction we find in

BAROCLINIC AND BAROTROPIC FLOW

Charney's model that

$$\frac{\partial \zeta_1}{\partial t} \cdots = \cdots - 2k_i \zeta_T, \qquad (5.30)$$

$$\frac{\partial \zeta_3}{\partial t} \cdots = \cdots + 2k_i \zeta_T - k_0 \cdot \zeta_4.$$

In (5.30) we have used the normal numbering of the surfaces – that is, 1 for 25 kPa, 2 for 50 kPa, 3 for 75 kPa and 4 for 100 kPa. $k_i = \frac{1}{2} \times 10^{-6}$ s^{-1} is the value used for the internal coefficient of friction while $k_0 = 4 \times 10^{-6}$ s^{-1} is the value for the surface.

Adding and subtracting the two equations in (5.30) and dividing by 2, we find

$$\frac{\partial \zeta_*}{\partial t} \cdots = \cdots - \frac{1}{2} k_0 \zeta_4, \qquad (5.31)$$

$$\frac{\partial \zeta_T}{\partial t} \cdots = \cdots - 2k_i \zeta_T + \frac{1}{2} k_0 \zeta_4,$$

where the first equation will be identified with the mean flow and the second with the shear flow. We note here that the rate of change of the kinetic energy of a nondivergent flow may be calculated as follows:

$$\frac{dK}{dt} = \frac{1}{gS} \int_0^{p_0} \int_S \nabla \psi \cdot \nabla \left(\frac{\partial \psi}{\partial t} \right) \, dS \, dp = -\frac{1}{gS} \int_0^{p_0} \int_S \psi \frac{\partial \zeta}{\partial t} \, dS \, dp. \qquad (5.32)$$

Applied to the two-level model we find

$$\frac{dK_*}{dt} = \cdots \frac{1}{2} k_0 \frac{p_0}{g} \frac{1}{S} \int_S \psi_* \zeta_4 \, dS = \cdots - \frac{1}{2} k_0 \frac{p_0}{g} \frac{1}{S} \int_S \vec{v}_* \cdot \vec{v}_4 \, dS \qquad (5.33)$$

in agreement with (5.29). For the shear flow we find

$$\frac{dK_T}{dt} = \cdots - 2k_i \frac{p_0}{g} \frac{1}{S} \int_S \vec{v}_T \cdot \vec{v}_T \, dS + \frac{1}{2} k_0 \frac{p_0}{g} \frac{1}{S} \int_S \vec{v}_T \cdot \vec{v}_4 \, dS. \qquad (5.34)$$

It is seen from (5.34) that the dissipation is

$$D(K_T) = 2k_i \frac{p_0}{g} \frac{1}{S} \int_S \vec{v}_T \cdot \vec{v}_T \, dS - \frac{1}{2} k_0 \frac{p_0}{g} \frac{1}{S} \int_S \vec{v}_T \cdot \vec{v}_4 \, dS. \qquad (5.35)$$

The sign of $D(K_T)$ will thus depend on the relative magnitude of the two terms in (5.35) of which both of the two integrals probably are positive on average over the global domain. If we make use of the numerical values quoted above (that is, $k = 8k_i$) we may write (5.35) in the form

$$D(K_T) = 2k_i \frac{p_0}{g} \frac{1}{S} \int_S \vec{v}_T \cdot \left(5 \vec{v}_T - 2 \vec{v}_* \right) \, dS. \qquad (5.36)$$

While the specific integral in (5.36) probably is positive on average, it should be pointed out that the numerical coefficients in (5.36) depend on the relative magnitudes of k_i and k. The sign of $D(K_T)$ (see Fig. 5.3) is important for the interpretation of results concerning $C(K_S, K_M)$. If $D(K_T) > 0$, we have for the long-term average that

$$C(K_S, K_M) < C(A, K_S), \qquad (5.37)$$

while the opposite inequality holds if $D(K_T) < 0$. No calculation of $D(K_T)$ using observed data has so far been made.

It is naturally of interest to compute values of K_M and K_S from data. This has been done by Wiin-Nielsen and Drake (1965,1966) based on data from January, April, July, and October 1962 and from January 1963. The averaged values, representing an estimate of the annual means, are $K_M = 1796$ kJ m^{-2} and $K_S = 678$ kJ m^{-2} with a ratio $K_S/K_M \approx 0.38$. The annual variation of K_M and K_S is very similar to that of the total kinetic energy – that is, a maximum in January and a minimum in July.

The same data were used to calculate the energy conversion $C(K_S, K_M)$. Based on the four months it was found that the nondivergent part $C_{ND}(K_S, K_M)$ generally gives the larger contribution to the total conversion, as can be seen in Table 5.1.

The numbers in Table 5.1 represent as far as we know a rather nor-

Table 5.1:

		$C_{ND}(K_S, K_M)$	$C_D(K_S, K_M)$	$C(K_S, K_M)$
January	1962	4.65	0.03	4.68
April	1962	2.88	0.17	3.05
July	1962	1.24	0.06	1.30
October	1962	2.96	0.27	3.23
Unit:	W m^{-2}			

mal situation. On the other hand, there are periods where the atmosphere operates in a somewhat different mode characterized by very large anomalies on a large scale. Such a period happened in January 1963, which was dominated by large-scale blocking in both the Atlantic and Pacific sectors. For this month we find: $C_{ND}(K_S, K_M) = 4.16$ W m^{-2} and $C_D(K_S, K_M) = 1.46$ W m^{-2} with a very large contribution from the mean-meridional circulation (see Chapter 7 for a further discussion of this period).

6

TRANSPORTS OF SENSIBLE HEAT AND MOMENTUM

Most of the older theories, described by Lorenz (1967), try to describe the general circulation in terms of the maintenance of the zonal flow such as the low latitude easterlies, the mid-latitude westerlies, and the polar flow. Since the days of Hadley there have been several proposals for arrangements of various meridional circulations which should explain the observed zonal circulation. In the early part of the twentieth century Jeffries (1926) and Defant (1921) suggested the possibility that the atmospheric waves – that is, the deviations from the zonal flow – could play a major part in the maintenance of the zonal current.

Jeffries pointed out that there was a systematic transport of momentum in atmospheric waves, while Defant considered the influence of atmospheric waves (the highs and lows) as a mixing process on a global scale. The former ideas were taken up again after the establishment of an upper air network of sufficient density to warrant global calculations based on observed data. These investigations were aimed more at describing the maintenance of the zonal flows and the zonal temperature field than at atmospheric energetics, but since the results are necessary ingredients in energy investigations (to be described later), we shall consider them here.

We consider first the thermodynamic equation

$$c_p \frac{dT}{dt} - \alpha\omega = H, \tag{6.1}$$

or, expanded in pressure coordinates

$$\frac{\partial T}{\partial t} + \vec{v}\cdot\nabla T + \omega\frac{\partial T}{\partial p} - \frac{1}{c_p}\alpha\omega = \frac{1}{c_p}H, \tag{6.2}$$

or through the use of the continuity equation

$$\frac{\partial T}{\partial t} + \nabla\cdot\left(T\vec{v}\right) + \frac{\partial(T\omega)}{\partial p} - \frac{R}{c_p}\frac{1}{p}(T\omega) = \frac{1}{c_p}H. \tag{6.3}$$

Here and in the following we shall use the zonal average, which is defined through the integral

$$(\ )_z = \frac{1}{2\pi}\int_0^{2\pi}(\ )\,d\lambda, \tag{6.4}$$

where the subscript z denotes the zonal average. We note that the zonal average of a derivative with respect to longitude is zero because

$$\left(\frac{\partial a}{\partial \lambda}\right)_z = \frac{1}{2\pi}\int_0^{2\pi}\frac{\partial a}{\partial \lambda}\,d\lambda = \frac{1}{2\pi}[a(2\pi) - a(0)] = 0. \tag{6.5}$$

The deviation from the zonal average is called the eddy quantity – that is, $a_E = a - a_z$. It follows immediately that the zonal average of an eddy quantity is zero because

$$(a_E)_z = a_z - a_z = 0. \tag{6.6}$$

We note finally that the zonal average of a product may be written as follows:

$$(ab)_z = (a_z b_z + a_z b_E + a_E b_z + a_E b_E)_z = a_z b_z + (a_E b_E)_z. \tag{6.7}$$

After these preparations we may take the zonal average of the thermodynamic equation (6.3), with the following result:

$$\frac{\partial T_z}{\partial t} = -\frac{1}{a\cos\varphi}\frac{\partial (Tv)_z \cos\varphi}{\partial \varphi} - \frac{\partial (T\omega)_z}{\partial p} + \frac{R}{c_p}\frac{1}{p}(T\omega)_z + \frac{1}{c_p}H_z. \tag{6.8}$$

We may say that to determine the rate of change of the zonally averaged temperature, we need to calculate three quantities:

1. the meridional transport $(Tv)_z$,

2. the vertical transport $(T\omega)_z$, and

3. the heating H_z.

To simplify matters we may take the vertical average of (6.8), giving

$$\frac{\partial T_{z,M}}{\partial t} = -\frac{\partial (Tv)_{z,M}\cos\varphi}{a\cos\varphi\,\partial\varphi} + \frac{1}{c_p}(\alpha\omega)_{z,M} + \frac{1}{c_p}H_{z,M}, \tag{6.9}$$

using the boundary conditions $\omega = 0$ at $p = 0$ and $p = p_0$.

If (6.9) was integrated with respect to latitude from the South Pole to the North Pole, we would once again obtain the generation of total potential energy and the conversion of this form of energy to kinetic energy – that is, a balance between the last two terms in (6.9). The situation is quite different when we consider a specific latitude because the first term is very important.

TRANSPORTS OF SENSIBLE HEAT AND MOMENTUM   57

The general argument in discussing the long-term average of (6.9) has been that since it is known that $H_{z,M}$ is positive at low latitudes and negative at high latitudes, and since the zonal mean temperature is maintained in the long-term average, there must be a systematic heat transport from south to north. This heat transport must be distributed in such a way it has a maximum in middle latitudes and the first term on the right-hand side of (6.9) – the meridional convergence of the heat transport – is negative in low latitudes and positive in high latitudes. The question is then: how does the required heat transport take place? As seen from the expression

$$(Tv)_z = T_z v_z + (T_E v_E)_z, \qquad (6.10)$$

there are two possibilities:

1. The heat transport is carried out by the mean meridional circulation.

2. The heat transport is carried out by the eddies.

The results of observational studies are that the eddy transport dominates at least in the middle and high latitudes, while the transport by the mean meridional circulation plays an important role in low latitudes in connection with the Hadley circulation. We may understand these results by noting that the middle and high latitudes are dominated by relatively weak mean meridional circulations and a strong eddy activity generated more or less constantly by baroclinic instability. On the other hand, the circulation in the low latitudes does not contain eddies to the same extent as in higher latitudes and the Hadley circulation is the stronger part of the mean meridional circulation.

We may also understand the small contribution by the mean meridional circulation in (6.9) because in this equation we have to take the vertical average. Since the mean meridional circulation consists of cells where the flow in the upper part has the opposite sign of the flow in the lower part, it is seen that a compensating effect appears in the vertical average. The situation is different in (6.8) where we consider an individual level. Schematically we may summarize the balance in Fig. 6.1, where the top part shows $H_{z,M}$ as a function of latitude and the lower part $(Tv)_z$ shows a similar arrangement. The figure therefore illustrates the balance

$$\frac{1}{a \cos \varphi} \frac{\partial (Tv)_{z,M} \cos \varphi}{\partial \varphi} \approx \frac{1}{c_p} H_{z,M}. \qquad (6.11)$$

We shall also mention that in a nondivergent model we have

$$\frac{\partial \alpha}{\partial t} + \vec{v}_\psi \cdot \nabla \alpha - \overline{\sigma} \omega = \frac{R}{c_p} \frac{1}{p} H \qquad (6.12)$$

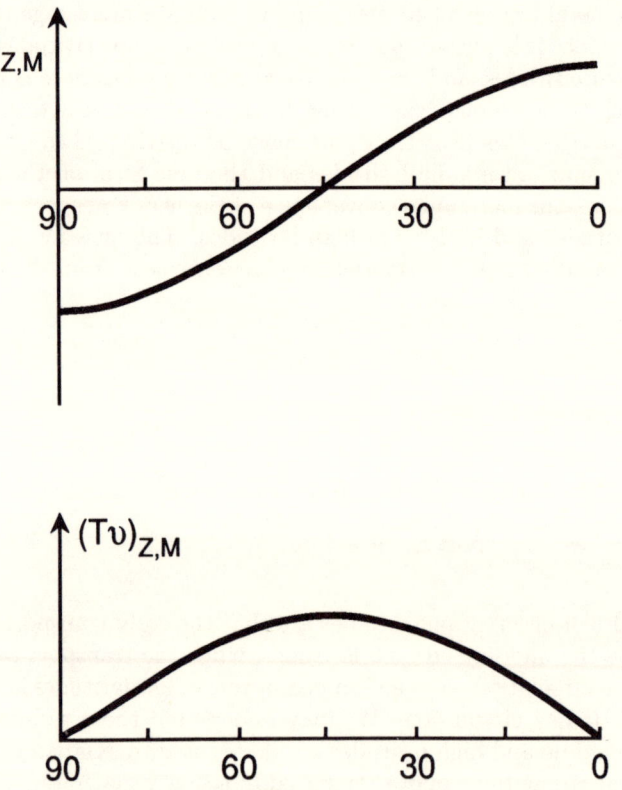

Fig. 6.1: The top part shows a very schematic, meridional distribution of the vertical mean of the heating, while the lower part is an equally schematic curve of the vertical mean of the meridional transport of sensible heat.

or

$$\frac{\partial T}{\partial t} + \nabla \cdot \left(T \vec{v}_\psi \right) - \frac{\overline{\sigma}p}{R}\omega = \frac{1}{c_p} H. \tag{6.13}$$

The zonally averaged equation is

$$\frac{\partial T_z}{\partial t} = -\frac{1}{a\cos\varphi} \frac{\partial \left(T_E v_E\right)_z \cos\varphi}{\partial \varphi} + \frac{\overline{\sigma}p}{R}\omega_z + \frac{1}{c_p} H_z. \tag{6.14}$$

When we are dealing with the advection term in such a model where $v_z = 0$, we get only the first term in the meridional transport by the eddies. The vertical heat transport disappears in this models because $\overline{\sigma}$ is a function of pressure only. It is therefore to be expected that such a model may not perform too well in low latitudes.

The maintenance of the zonal flow may be discussed in a similar way.

TRANSPORTS OF SENSIBLE HEAT AND MOMENTUM

For this purpose we consider the first equation of motion, which is

$$\frac{\partial u}{\partial t} + \frac{u}{a\cos\varphi}\frac{\partial u}{\partial \lambda} + \frac{v}{a}\frac{\partial u}{\partial \varphi} + \omega\frac{\partial u}{\partial p} = -\frac{1}{a\cos\varphi}\frac{\partial \phi}{\partial \lambda} + fv + \frac{\tan\varphi}{a}uv + F_\lambda. \quad (6.15)$$

We shall also need the continuity equation

$$\frac{1}{a\cos\varphi}\left(\frac{\partial u}{\partial \lambda} + \frac{\partial v\cos\varphi}{\partial \varphi}\right) + \frac{\partial \omega}{\partial p} = \frac{1}{a\cos\varphi}\frac{\partial u}{\partial \lambda} + \frac{1}{a}\frac{\partial v}{\partial \varphi} + \frac{\partial \omega}{\partial p} - \frac{\tan\varphi}{a}v = 0. \quad (6.16)$$

Combining (6.15) and (6.16) we obtain an equation in the form

$$\frac{\partial u}{\partial t} + \frac{\partial u^2}{a\cos\varphi\,\partial \lambda} + \frac{\partial(uv)}{a\,\partial \varphi} - 2\frac{\tan\varphi}{a}(uv) + \frac{\partial u\omega}{\partial p} = -\frac{1}{a\cos\varphi}\frac{\partial \phi}{\partial \lambda} + fv + F_\lambda \quad (6.17)$$

or

$$\frac{\partial u}{\partial t} + \frac{\partial u^2}{a\cos\varphi\,\partial \lambda} + \frac{1}{a\cos^2\varphi}\frac{\partial(uv)\cos^2\varphi}{\partial \varphi} + \frac{\partial(u\omega)}{\partial p} = -\frac{1}{a\cos\varphi}\frac{\partial \phi}{\partial \lambda} + fv + F_\lambda. \quad (6.18)$$

The next step is to obtain the zonal average of each term with the following result:

$$\frac{\partial u_z}{\partial t} = -\frac{1}{a\cos^2\varphi}\frac{\partial(uv)_z\cos^2\varphi}{\partial \varphi} - \frac{\partial(u\omega)_z}{\partial p} + fv_z + F_{\lambda,z}. \quad (6.19)$$

The rate of change of u_z is thus determined by

1. the convergence of the meridional transport,
2. the convergence of the vertical transport,
3. the contribution from the mean meridional circulation, and
4. the zonally averaged value of the friction.

Once again we shall first consider the vertical average. Using the same approximate lower boundary condition as before, we see that there will be no contribution from the vertical transport of momentum. The zonal average of the continuity equation is

$$\frac{1}{a\cos\varphi}\frac{\partial v_z\cos\varphi}{\partial \varphi} + \frac{\partial \omega_z}{\partial p} = 0 \quad (6.20)$$

and the vertical average of this equation gives

$$\frac{\partial v_{z,M} \cos\varphi}{\partial \varphi} = 0, \tag{6.21}$$

from which it follows that $v_{z,M} = 0$. Consequently, there is no contribution from the mean meridional circulation. The result is, therefore,

$$\frac{\partial u_{z,M}}{\partial t} = -\frac{1}{a\cos^2\varphi} \frac{\partial (uv)_{z,M} \cos^2\varphi}{\partial \varphi} + F_{\lambda,z,M}. \tag{6.22}$$

As we have seen before we may write F_λ in the form

$$F_\lambda = g\frac{\partial \tau_\lambda}{\partial p}. \tag{6.23}$$

Using this formulation the last term in (6.20) becomes

$$F_{\lambda,z,M} = \frac{g}{p_0}\tau_{\lambda,z,0}, \tag{6.24}$$

where $\tau_{\lambda,z,0}$ is the zonal average of the surface stress, which in turn is approximated as follows:

$$\tau_{\lambda,z,0} = -c_d \rho V_0 u_{z,0}. \tag{6.25}$$

Equation (6.22) finally may be written in the following form:

$$\frac{\partial u_{z,M}}{\partial t} = -\frac{1}{a\cos^2\varphi} \frac{\partial (uv)_{z,M} \cos^2\varphi}{\partial \varphi} - \varepsilon u_{z,0}, \tag{6.26}$$

where

$$\varepsilon = \frac{g}{p_0} c_d \rho V_0 \approx 3 \times 10^{-6} \text{ s}^{-1}. \tag{6.27}$$

Assuming that we consider the average over a long time in which the time derivative of $u_{z,M}$ becomes very small, we have

$$\frac{1}{a\cos^2\varphi} \frac{\partial (uv)_{z,M} \cos^2\varphi}{\partial \varphi} = -\varepsilon u_{z,0}, \tag{6.28}$$

where $u_{z,0}$ is the surface distribution of the zonally averaged wind. $u_{z,0}$ may be characterized by having easterlies ($u_{z,0} < 0$) in low latitudes, westerlies ($u_{z,0} > 0$) in middle latitudes, and weak easterlies in very high latitudes.

If (6.28) applies, we can deduce in general terms that $(uv)_z$ should increase with latitude over regions with easterly flow and decrease with latitude over regions with westerly flow. The terms of $\cos^2 \varphi$ will modify this statement a little.

To illustrate these points we may construct a simple example. Let us suppose that the surface zonal wind is given by

$$u_{0z} = -U_0 \cos(4\varphi), \tag{6.29}$$

which has zeros at $\varphi = 22.5°$ and $\varphi = 67.5°$. Between these points $u_{0z} > 0$, while $u_{0z} < 0$ for $0 < \varphi < 22.5°$ and $67.5° < \varphi < 90°$ N (see Fig. 6.2). We may then use (6.28) to compute

$$M = (uv)_{z,M} \cos^2 \varphi \tag{6.30}$$

by integration using the fact that $M = 0$ for $\varphi = \pi/2$. We find

$$M(\varphi) = -\varepsilon a U_0 \int_\varphi^{\frac{\pi}{2}} \cos(4\varphi) \cos^2 \varphi \, d\varphi \tag{6.31}$$

or

$$M(\varphi) = \frac{1}{24} \varepsilon a U_0 \left(3 \sin 4\varphi + \sin 6\varphi + 3 \sin 2\varphi \right). \tag{6.32}$$

For $\varepsilon = 3 \times 10^{-6}$ s^{-1}, $a = \frac{2}{\pi} \times 10^7$ m, $U_0 = 2$ m s^{-1} we find that

$$\frac{1}{24} \varepsilon a U_0 = 1.6 \text{ m}^2 \text{ s}^{-2} \tag{6.33}$$

The lower diagram in Fig. 6.2 shows M as a function of latitude. We thus have seen that the momentum transport, which is consistent with distribution of the surface zonals, increases from the equator to the zero of u_{z0} and thereafter decreases to a minimum at the other zero. Knowing the surface wind distribution, we can thus compute the vertical average of momentum transport M or vice versa in a steady state of the circulation.

Returning to (6.19), which applies to a single level and latitude, we stress that in such a case one must also pay attention to the convergence of the vertical momentum transport and the effect of the Coriolis term. Both of these terms, however, are considered to give minor contributions especially in middle and higher altitudes because they contain either the small vertical velocity or the small mean meridional velocity. For this reason the convergence of the meridional transport of momentum is considered to be a result of transport by the eddies – that is,

$$-\frac{\partial (uv)_z \cos^2 \varphi}{a \cos^2 \varphi \, \partial \varphi} = -\frac{\partial u_z v_z \cos^2 \varphi}{a \cos^2 \varphi \, \partial \varphi} - \frac{\partial (u_E v_E)_z \cos^2 \varphi}{a \cos^2 \varphi \, \partial \varphi}$$

$$\approx -\frac{\partial (u_E v_E)_z \cos^2 \varphi}{a \cos^2 \varphi \, \partial \varphi}. \tag{6.34}$$

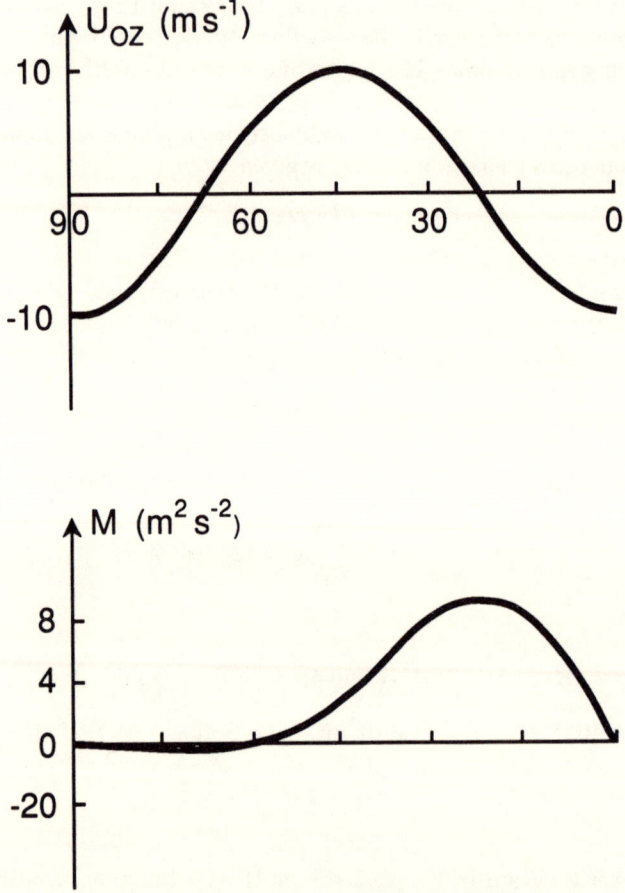

Fig. 6.2: The top curve shows a schematic distribution of the surface zonal wind in m s^{-1}, while the lower curve shows the main features of the vertical mean of the meridional momentum transport in m^2 s^{-2}.

All these arguments lead to the following approximate form of (6.19):

$$\frac{\partial u_z}{\partial t} = -\frac{1}{a \cos^2 \varphi} \frac{\partial (u_E v_E)_z \cos^2 \varphi}{\partial \varphi} + F_{\lambda, z}. \qquad (6.35)$$

We have seen in the discussion in this chapter that for purposes of the balance requirements in the general circulation, a northward transport of heat and momentum is necessary, at least in the low and middle latitudes. We have also shown it to be plausible that the major part of these transports are carried out by the eddies. It is therefore important to recall that a northward transport of sensible heat will be accomplished if the temperature field is lagging behind the geopotential field on an isobaric surface.

TRANSPORTS OF SENSIBLE HEAT AND MOMENTUM

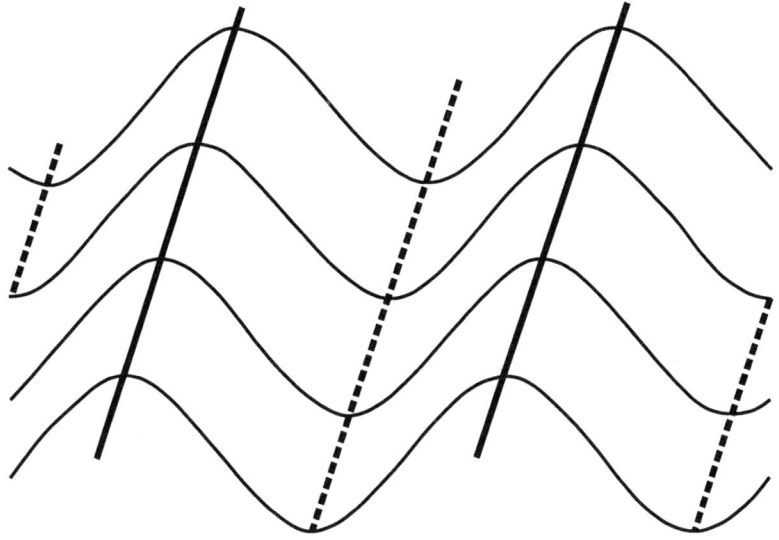

Fig. 6.3: The typical structure of an atmospheric wave that is transporting momentum to the north.

Suppose that the geopotential is given by

$$\phi = \phi_A \sin kx, \tag{6.36}$$

while the temperature field is

$$T = T_A \sin(ks + \gamma). \tag{6.37}$$

We find then that

$$(T_E v_E)_z = \left(T_A \sin(kx + \gamma)\frac{k}{f} \cos kx\right)_z, \tag{6.38}$$

where we have used the geostrophic assumption to calculate v_E. Evaluation of (6.38) gives

$$(T_E v_E)_z = \frac{1}{2}\frac{k}{f}T_A \phi_A \sin \gamma. \tag{6.39}$$

The transport is thus positive if $0 < \gamma < 180°$ and otherwise proportional to the two amplitudes.

Turning to the momentum transport, it is easy to see that simple waves such as the one given by (6.36) have no momentum transport. To have a nonvanishing momentum transport, it is necessary to have sloping trough and ridge lines. One may see qualitatively that if these lines slope from southwest to northeast there will be a positive (northward) momentum transport (see Fig. 6.3). Negative values of the transport are obtained for the opposite slope from southeast to northwest (Wiin-Nielsen, 1973). Simple mathematical expressions may also be used to calculate the geostrophic transport of momentum. Consider the following expression for the geopotential:

$$\phi = a\cos(kx - \alpha y). \tag{6.40}$$

For each constant value of y we have a sinusoidal wave. For $y = 0$ we find the ridge line at $x = 0$. For another value of y the ridge line is found at a location x_r determined by

$$x_r = \frac{\alpha}{k} y. \tag{6.41}$$

The ridge line is therefore a straight line with the slope α. Similar remarks can be made with respect to the trough line, which is parallel to the ridge line but half a wavelength away. For the momentum transport, calculated geostrophically, we find

$$(u_g v_g)_z = \frac{1}{2} \frac{A^2}{f^2} k\alpha. \tag{6.42}$$

It is seen that the transport is positive for $\alpha > 0$ and negative for $\alpha < 0$ in agreement with the discussion given above.

We note finally that in some investigations it has been preferred to make the analysis in terms of the total angular momentum. The total velocity is

$$u + \Omega a \cos\varphi, \tag{6.43}$$

where u is the relative velocity and the second term is the velocity of the earth's rotation. The momentum around the axis of rotation is

$$(u + \Omega a \cos\varphi) a \cos\varphi. \tag{6.44}$$

When we finally calculate by integration the total angular momentum along a latitude circle, we get

$$M_A = \int_0^{2\pi} (u + \Omega a \cos\varphi) a^2 \cos^2\varphi \, d\lambda \tag{6.45}$$

or

$$M_A = 2\pi a^2 \cos^2\varphi u_z + 2\pi\Omega a^3 \cos^3\varphi. \tag{6.46}$$

We note that
$$\frac{\partial M_A}{\partial t} = 2\pi a^2 \cos^2 \varphi \frac{\partial u_z}{\partial t}, \qquad (6.47)$$

which indicates that the two investigations are closely related.

The transports of sensible heat and momentum have been investigated in great detail by many. Early results were obtained by Buch (1954) for the momentum transport and by Peixóto (1960) for the heat transport. These publications were based on data samples of one year. Later investigations covered longer periods, such as Starr and Oort (1973) who used five years of data. Probably the most extensive study was carried out by Oort (1983) who used the 15 year period 1958 - 73.

Calculations based on operationally obtained nondivergent winds have been presented by Wiin-Nielsen, Brown, and Drake (1963,1964) and by Wiin-Nielsen (1967). A summary of early results can be found in a review paper by Lorenz (1967), and Lau (1984a) presented the results of these transports and several other transports based on the analyses made during the period of the First GARP Global Experiment.

To understand the results presented by Oort, Lau and several others, it is important to understand the subdivision made by them. They distinguish between two kinds of averages: one in time and one with respect to longitude. For an arbitrary quantity a the time average is defined by

$$\bar{a} = \frac{1}{T} \int_0^T a \, dt. \qquad (6.48)$$

The deviation from the time average is denoted by a prime – that is,

$$a = \bar{a} + a'. \qquad (6.49)$$

The zonal average is here denoted by the subscript z [see (6.4)] and the deviation by a_E – that is,
$$a = a_z + a_E. \qquad (6.50)$$

Equation (6.7) gives the zonal average of a product of two scalar quantities. When we take the time average and make use of (6.49), we obtain

$$\overline{(ab)}_z = \bar{a}_z \bar{b}_z + \overline{a'_z b'_z} + \left(\bar{a}_E \bar{b}_E\right)_z + \overline{a'_E b'_E}_z. \qquad (6.51)$$

The first two terms in (6.51) relate to zonally averaged quantities. The first may be called the stationary part, and the second the transient part. Together they may be written

$$\overline{(a_z b_z)} = \bar{a}_z \bar{b}_z + \overline{(a'_z b'_z)} \qquad (6.52)$$

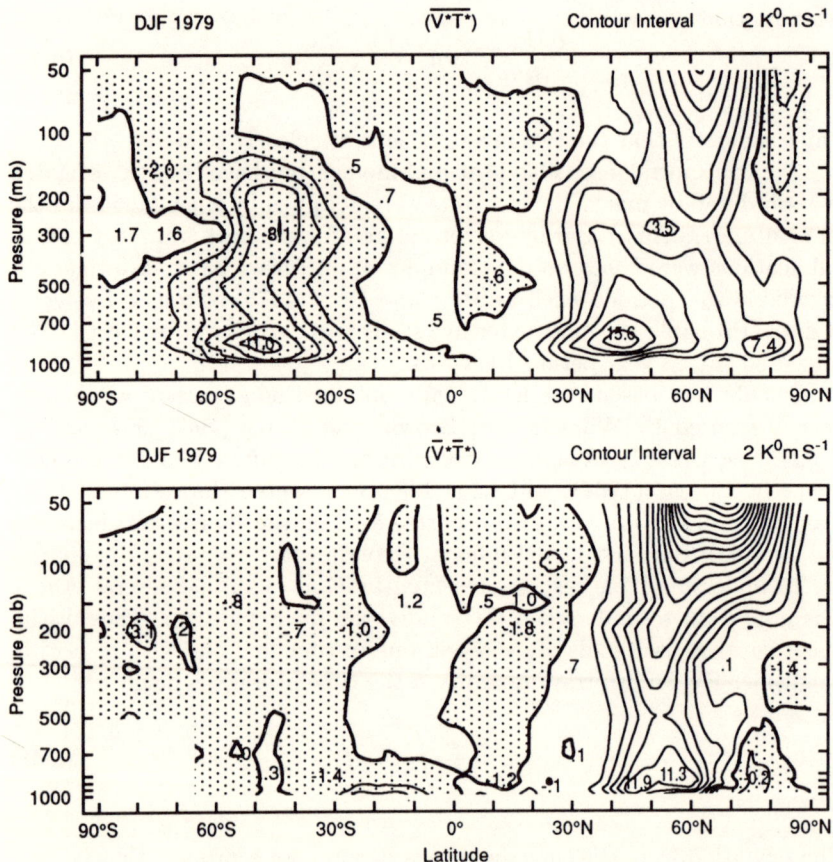

Fig. 6.4: The heat transports by the transient and standing eddies as a mean for December, January, and February.

corresponding to the time average of the product of a_z and b_z. Similarly,

$$\overline{(a_E b_E)}_z = \left(\overline{a}_E \overline{b}_E\right)_z + \overline{(a'_E b'_E)}_z, \tag{6.53}$$

of which the first is the stationary part and the second the transient part. Noting finally that these investigators use the notation

$$a_z = [a]$$

and

$$a_E = a^*,$$

one should remember to use the additions (6.52) and (6.53) to obtain the quantities discussed here.

TRANSPORTS OF SENSIBLE HEAT AND MOMENTUM

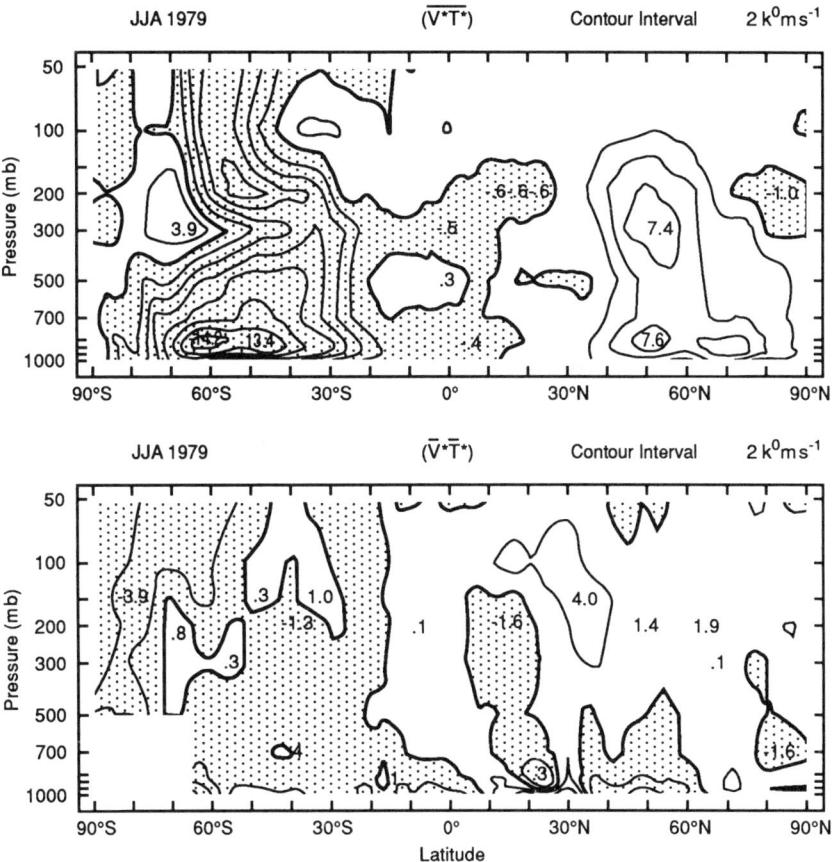

Fig. 6.5: The heat transports by the transient and standing eddies as a mean for June, July, and August.

As one might suspect there is not complete agreement between the results obtained in all of these investigations because of interannual and seasonal variations of the transports, as documented by Oort (1983). Rather large differences are even obtained when transports are calculated from different analyses of the same data, as demonstrated in great detail by Lau (1984b). However, these differences do not change the main conclusion that the eddy transports are by far the major part of the total transport, especially in extratropical regions. As examples, Figs. 6.4 and 6.5 show the heat transports of the transient and stationary (standing) eddies for the winter season (December, January, February) and the summer season (June, July, August). We note the strong polarward transports in both hemispheres with maxima in mid-latitudes and rather low levels around 80 kPa with secondary maxima at about 20 kPa. The standing waves have a much larger contribution in the Northern Hemisphere winter than in that

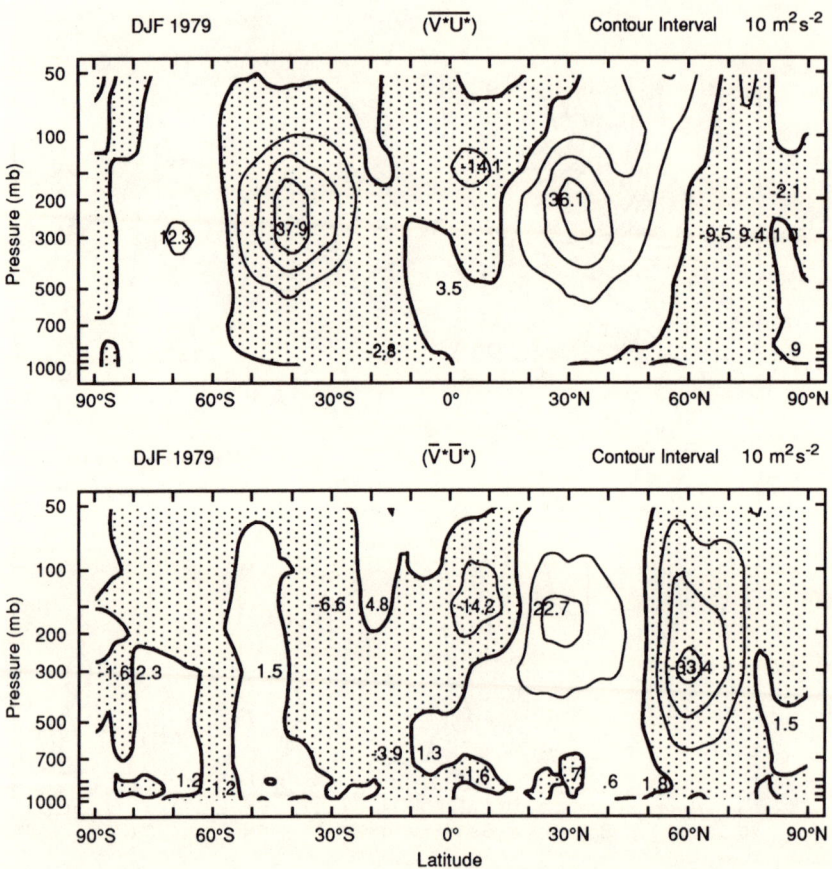

Fig. 6.6: The momentum transports by the transient and standing eddies as a mean for December, January, and February.

of the Southern Hemisphere where the standing wave contribution is almost negligible in all seasons.

Figs. 6.6 and 6.7 show the momentum transports in a similar arrangement. This transport is dominated by maxima in the upper troposphere around 20 kPa. We also see the weak equatorward transports in high latitudes giving the convergence in the region between the maxima and the minima.

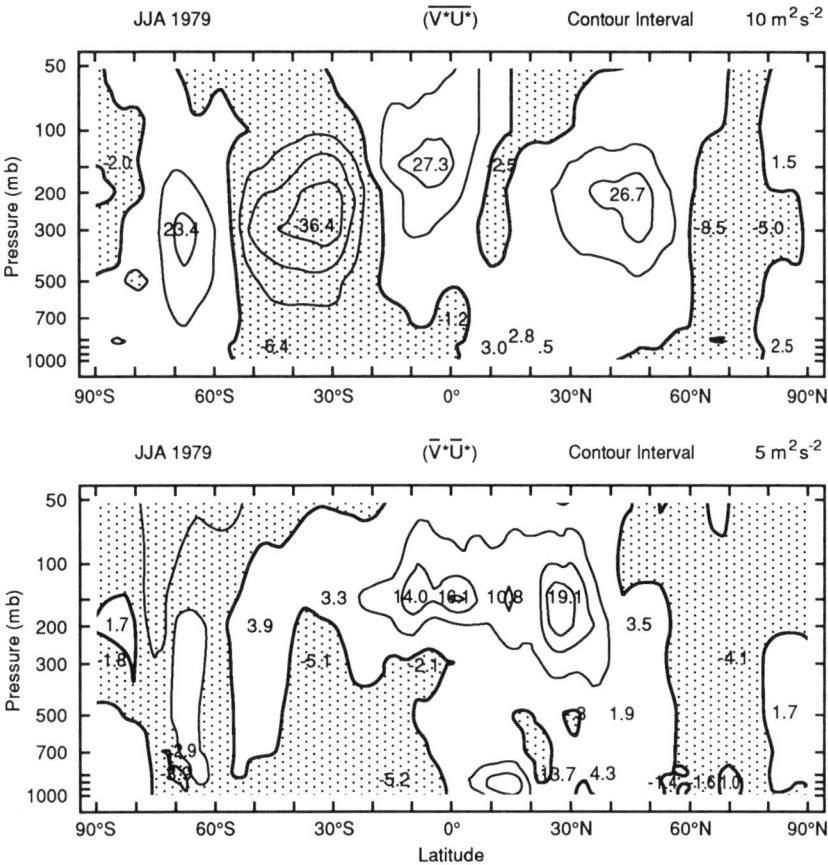

Fig. 6.7: The momentum transports by the transient and standing eddies as a mean for June, July, and August.

7

ZONAL AND EDDY ENERGIES

In this chapter we shall consider the most common way to study the energetics of the atmosphere. In Chapter 6 we introduced the zonal average and the deviation from the zonal mean – the eddy. These quantities are of course artificial and can be obtained only from complete latitude circles. We may define the energies of these zonal averages and eddies in the same way as before. We shall be concerned with available potential and kinetic energy. If such a division into the zonal average and the eddy makes sense, then the energy of the total must be equal to the sum of the energies of the parts. This is the case for the available potential energy and the kinetic energy as long as we use the approximate form of the definition of available potential energy. We have

$$A = \int_M \frac{1}{2\sigma} \alpha'^2 \, dm \qquad (7.1)$$

and

$$K = \int_M \frac{1}{2} \vec{v} \cdot \vec{v} \, dm. \qquad (7.2)$$

Following the remarks made above we define the zonal and eddy energies as follows:

$$A_z = \int_M \frac{1}{2\sigma} \alpha'^2_z \, dm \qquad A_E = \int_M \frac{1}{2\sigma} \alpha'^2_E \, dm, \qquad (7.3)$$

$$K_z = \int_M \frac{1}{2} \vec{v}_z \cdot \vec{v}_z \, dm \qquad K_E = \int_M \frac{1}{2} \vec{v}_E \cdot \vec{v}_E \, dm. \qquad (7.4)$$

It is immediately clear that

$$A = A_z + A_E \qquad K = K_z + K_E \qquad (7.5)$$

when we note for example that $\alpha' = \alpha'_z + \alpha'_E$ and that the contribution from the term $2\alpha'_z \alpha'_E$ integrates to zero because the integration with respect to mass contains an integration with respect to longitude. As we showed earlier,

$$(a_z a_E)_z = a_z (a_E)_z = 0. \qquad (7.6)$$

ZONAL AND EDDY ENERGIES

We note further from (7.5) that

$$\frac{dA}{dt} = \frac{dA_z}{dt} + \frac{dA_E}{dt} \qquad \frac{dK}{dt} = \frac{dK_z}{dt} + \frac{dK_E}{dt}. \tag{7.7}$$

Since we already have expressions for dA/dt and dK/dt [see (4.65) and (3.76), respectively], it will be sufficient to calculate for example dA_z/dt and dK_z/dt.

It would of couse have been preferable to deal with the so-called exact formulation of the available potential energy [see (4.9)], but such a formalism has not been developed and, in addition, the observational studies to be described later are mostly carried out using (7.1) for the available potential energy. Furthermore we have shown that (7.1) is the proper form for a quasi-nondivergent model. It is therefore natural to use this model for the development of the energetics of A_z, A_E, K_z, and K_E. We recall that

$$\frac{dA}{dt} = \frac{1}{g}\int_0^{p_0}\int_S \omega\alpha'\, dS\, dp + \frac{1}{g}\int_0^{p_0}\int_S \frac{R}{c_p\overline{\sigma}p}\alpha'H'\, dS\, dp \tag{7.8}$$

and

$$\frac{dK}{dt} = -\frac{1}{g}\int_0^{p_0}\int_S \omega\alpha'\, dS\, dp + \frac{1}{g}\int_0^{p_0}\int_S \vec{v}\cdot\vec{F}\, dS\, dp. \tag{7.9}$$

The thermodynamic equation to be used is

$$\frac{\partial \alpha'}{\partial t} + \vec{v}_\psi\cdot\nabla\alpha' - \overline{\sigma}\omega = \frac{R}{c_p p}H'. \tag{7.10}$$

Taking the zonal average of (7.10) we obtain

$$\frac{\partial \alpha'_z}{\partial t} + \frac{1}{a\cos\varphi}\frac{\partial(\alpha'v)_z\cos\varphi}{\partial \varphi} - \overline{\sigma}\omega_z = \frac{R}{c_p p}H'_z. \tag{7.11}$$

Equation (7.11) is multiplied by α'_z and divided by $\overline{\sigma}$ to create the integrand in dA_z/dt. After integration we find

$$\frac{dA_z}{dt} = \frac{1}{g}\int_0^{p_0}\int_S \frac{\alpha'_z}{\overline{\sigma}}\frac{\partial \alpha'_z}{\partial t}\, dS\, dp \tag{7.12}$$

$$= -\frac{1}{g}\int_0^{p_0}\int_S \frac{\alpha'_z}{\overline{\sigma}a\cos\varphi}\frac{\partial\left(\alpha'_E v_E\right)_z\cos\varphi}{\partial \varphi}\, dS\, dp$$

$$+\frac{1}{g}\int_0^{p_0}\int_S \alpha'_z\omega_z\, dS\, dp$$

$$+\frac{1}{g}\int_0^{p_0}\int_S \frac{R}{c_p\overline{\sigma}p}\alpha'_z H'_z\, S\, dp.$$

The first integral on the right-hand side of (7.12) originates from the non-linear advection term in the thermodynamic equation. This term gives no contribution to the rate of change of A as given in (7.8). Noting (6.7) we subtract (7.12) from (7.8) and obtain

$$\frac{dA_E}{dt} = \frac{1}{g}\int_0^{p_0}\int_S \frac{\alpha'_z}{\overline{\sigma}a\cos\varphi}\frac{\partial\left(\alpha'_E v_E\right)_z\cos\varphi}{\partial\varphi}\,dS\,dp \qquad (7.13)$$
$$+ \frac{1}{g}\int_0^{p_0}\int_S \alpha'_E \omega_E\,dS\,dp + \frac{1}{g}\int_0^{p_0}\int_S \frac{R}{c_p\overline{\sigma}p}\alpha'_E H'_E\,dS\,dp.$$

The treatment of the kinetic energy follows. The starting point is the vorticity equation for a quasi-nondivergent model:

$$\frac{\partial\zeta}{\partial t} + \vec{v}_\psi\cdot\nabla\zeta + \beta v_\psi = f_0\frac{\partial\omega}{\partial p} + \vec{k}\cdot\nabla\times\vec{F}. \qquad (7.14)$$

We take next the zonal average of (7.14) recalling that

$$v_\psi = \frac{\partial\psi}{a\cos\varphi\,\partial\lambda}. \qquad (7.15)$$

The result is

$$\frac{\partial\zeta_z}{\partial t} = -\frac{1}{a\cos\varphi}\frac{\partial(\zeta_E v_E)_z\cos\varphi}{\partial\varphi} + f_0\frac{\partial\omega_z}{\partial p} - \frac{1}{a\cos\varphi}\frac{\partial F_{\lambda,z}\cos\varphi}{\partial\varphi}. \qquad (7.16)$$

Recalling that

$$\frac{dK_z}{dt} = -\frac{1}{g}\int_0^{p_0}\int_S \psi_z\frac{\partial\zeta_z}{\partial t}\,dS\,dp, \qquad (7.17)$$

it is seen that an equation for the rate of change of K_z is obtained from (7.16) by multiplying by $-\psi_z$ and integrating over the total mass. We obtain

$$\frac{dK_z}{dt} = \frac{1}{g}\int_0^{p_0}\int_S \frac{\psi_z}{a\cos\varphi}\frac{\partial(\zeta_E v_E)_z\cos\varphi}{\partial\varphi}\,dS\,dp \qquad (7.18)$$
$$- \frac{1}{g}\int_0^{p_0}\int_S f_0\psi_z\frac{\partial\omega_z}{\partial p}\,dS\,dp$$
$$+ \frac{1}{g}\int_0^{p_0}\int_S \frac{\psi_z}{a\cos\varphi}\frac{\partial F_{\lambda,z}\cos\varphi}{\partial\varphi}\,dS\,dp.$$

Each of the three integrals in (7.18) is considered separately to obtain a form which is familiar from previous considerations. For the first integral

ZONAL AND EDDY ENERGIES

we get integrating by parts

$$\frac{1}{g}\int_0^{p_0}\int_{-\pi/2}^{\pi/2}\int_0^{2\pi}\frac{\psi_z}{a\cos\varphi}\frac{\partial\left(\zeta_E v_E\right)_z\cos\varphi}{\partial\varphi}a^2\cos\varphi\,d\lambda\,dp \quad (7.19)$$

$$=-\frac{a}{g}\int_0^{p_0}\int_{-\pi/2}^{\pi/2}\int_0^{2\pi}\left(\zeta_E v_e\right)_z\cos\varphi\frac{\partial\psi_z}{\partial\varphi}\,d\lambda\,d\varphi\,dp.$$

We note that
$$u_z=-\frac{1}{a}\frac{\partial\psi_z}{\partial\varphi} \quad (7.20)$$

and that

$$(\zeta_E v_E)_z = \left[\frac{1}{a\cos\varphi}\left(\frac{\partial v_E}{\partial\lambda}-\frac{\partial u_E\cos\varphi}{\partial\varphi}\right)v_E\right]_z \quad (7.21)$$

$$= -\left[\frac{v_E}{a\cos\varphi}\frac{\partial u_E\cos\varphi}{\partial\varphi}\right]_z = -\left[\frac{v_E\cos\varphi}{a\cos^2\varphi}\frac{\partial u_E\cos\varphi}{\partial\varphi}\right]_z$$

$$= -\frac{\partial\left(u_E v_E\right)_z\cos^2\varphi}{a\cos^2\varphi\,\partial\varphi}+\left[\frac{u_E}{a\cos\varphi}\frac{\partial v_E\cos\varphi}{\partial\varphi}\right]_z$$

$$= -\frac{\partial\left(u_E v_E\right)_z\cos^2\varphi}{a\cos^2\varphi\,\partial\varphi}-\left[\frac{u_E}{a\cos\varphi}\frac{\partial u_E}{\partial\lambda}\right]_z$$

$$= -\frac{\partial\left(u_E v_E\right)_z\cos^2\varphi}{a\cos^2\varphi\,\partial\varphi}.$$

The value of the first integral is therefore

$$-\frac{1}{g}\int_0^{p_0}\int_S \frac{u_z}{a\cos^2\varphi}\frac{\partial\left(u_E v_E\right)_z\cos^2\varphi}{\partial\varphi}\,dS\,dp. \quad (7.22)$$

For the second integral we need only follow the steps used previously to show that

$$-\frac{1}{g}\int_0^{p_0}\int_S f_0\frac{\partial\omega_z}{\partial p}=-\frac{1}{g}\int_0^{p_0}\int_S \omega_z\alpha'_z\,dS\,dp. \quad (7.23)$$

Finally, for the third integral we find

$$\frac{1}{g}\int_0^{p_0}\int_{-\pi/2}^{\pi/2}\int_0^{2\pi}\frac{\psi_z}{a\cos\varphi}\frac{\partial F_{\lambda,z}\cos\varphi}{\partial\varphi}a^2\cos\varphi\,d\lambda\,d\varphi\,dp \quad (7.24)$$

$$=-\frac{a}{g}\int_0^{p_0}\int_{-\pi/2}^{\pi/2}\int_0^{2\pi}F_{\lambda,z}\cos\varphi\frac{\partial\psi_z}{\partial\varphi}\,d\lambda\,d\varphi\,dp.$$

But again using (7.20) we may write

$$\frac{1}{g}\int_0^{p_0}\int_S \frac{\psi_z}{a\cos\varphi}\frac{\partial F_{\lambda,z}\cos\varphi}{\partial\varphi}\,dS\,dp = \frac{1}{g}\int_0^{p_0}\int_S u_z F_{\lambda,z}\,dS\,dp. \quad (7.25)$$

The final result is therefore

$$\begin{aligned}\frac{dK_z}{dt} =\ & -\frac{1}{g}\int_0^{p_0}\int_S \frac{u_z}{a\cos^2\varphi}\frac{\partial (u_E v_E)\cos^2\varphi}{\partial\varphi}\,dS\,dp \quad (7.26)\\ & -\frac{1}{g}\int_0^{p_0}\int_S \omega_z \alpha'_z\,dS\,dp \\ & +\frac{1}{g}\int_0^{p_0}\int_S u_z F_{\lambda,z}\,dS\,dp.\end{aligned}$$

Finally, (7.26) is subtracted from (7.9) with the result that

$$\begin{aligned}\frac{dK_E}{dt} =\ & \frac{1}{g}\int_0^{p_0}\int_S \frac{u_z}{a\cos^2\varphi}\frac{\partial (u_E v_E)_z\cos^2\varphi}{\partial\varphi}\,dS\,dp \quad (7.27)\\ & -\frac{1}{g}\int_0^{p_0}\int_S \omega_E \alpha'_E\,dS\,dp \\ & +\frac{1}{g}\int_0^{p_0}\int_S \vec{v}_E\cdot\vec{F}_E\,dS\,dp.\end{aligned}$$

The energetics of the system consisting of A_z, A_E, K_z and K_E is contained in (7.12), (7.13), (7.26), and (7.27). An inspection of these four equations reveals that they may be written in a symbolic form using the notations introduced and used in the previous chapters. These equations are

$$\begin{aligned}\frac{dA_z}{dt} &= G(A_z) - C(A_z, A_E) - C(A_z, K_z), \quad (7.28)\\ \frac{dA_E}{dt} &= G(A_E) + C(A_z, A_E) - C(A_E, K_E),\\ \frac{dK_E}{dt} &= C(A_E, K_E) - C(K_E, K_z) - D(K_E),\\ \frac{dK_z}{dt} &= C(A_z, K_z) + C(K_E, K_z) - D(K_z),\end{aligned}$$

where the terms are defined from the four equations using the following conventions:

$$G(A_z) = \frac{1}{g}\int_0^{p_0}\int_S \frac{R}{c_p \overline{\sigma} p}\alpha'_z H'_z\,dS\,dp$$

$$G(A_E) = \frac{1}{g}\int_0^{p_0}\int_S \frac{R}{c_p \overline{\sigma} p}\alpha'_E H'_E\,dS\,dp$$

ZONAL AND EDDY ENERGIES

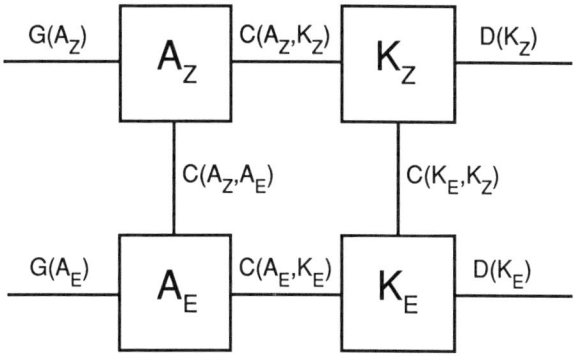

Fig. 7.1: The most common energy diagram as designed originally by Lorenz in 1955. The corresponding equations are (7.28) and (7.29).

$$C(A_z, K_z) = -\frac{1}{g} \int_0^{p_0} \int_S \omega_z \alpha'_z \, dS \, dp$$

$$C(A_E, K_E) = -\frac{1}{g} \int_0^{p_0} \int_S \omega_E \alpha'_E \, dS \, dp$$

$$C(A_z, A_E) = \frac{1}{g} \int_0^{p_0} \int_S \frac{\alpha'_z}{\overline{\sigma} a \cos \varphi} \frac{\partial \left(\alpha'_E v_E \right)_z \cos \varphi}{\partial \varphi} \, dS \, dp$$

$$C(K_E, K_z) = -\frac{1}{g} \int_0^{p_0} \int_S \frac{u_z}{a \cos^2 \varphi} \frac{\partial (u_E v_E)_z \cos^2 \varphi}{\partial \varphi} \, dS \, dp$$

$$D(K_z) = -\frac{1}{g} \int_0^{p_0} \int_S u_z F_{\lambda, z} \, dS \, dp$$

$$D(K_E) = -\frac{1}{g} \int_0^{p_0} \int_S \vec{v}_E \cdot \vec{F}_E \, dS \, dp. \tag{7.29}$$

A diagram showing schematically the content of (7.28) and (7.29) is given in Fig. 7.1. The general problem is to determine the directions and magnitudes of all the generations, conversions and dissipations – that is, all the integrals in (7.29) – either theoretically or from observations. These quantities, or parameters, are difficult to obtain mostly because they are not observed directly, and it is thus necessary to compute them from other observations. There are no general, well-established procedures for these calculations. The most difficult quantities to obtain are H'_z, H'_E, ω_z, ω_E, $F_{\lambda,z}$, and $\vec{F}_E$.

In this chapter we shall limit our discussion to general considerations that will help to establish the direction of some of the processes, while studies based on observations will be treated later. Furthermore, in this first treatment we may limit the discussions to the long-term means, in which case we may assume that each of the four energy components A_z, A_E, K_z, and K_E are unchanged, which also means that the net change for each component is zero.

Starting with the generations we note that they may also be written as follows:

$$G(A_z) = \frac{1}{g} \int_0^{p_0} \int_S \frac{R^2}{c_p \overline{\sigma} p^2} T_z' H_z' \, dS \, dp,$$

$$G(A_E) = \frac{1}{g} \int_0^{p_0} \int_S \frac{R^2}{c_p \overline{\sigma} p^2} T_E' H_E' \, dS \, dp. \quad (7.30)$$

The fraction in each integrand is a positive number at each pressure level. Considering the first integral $G(A_z)$, we may conclude that it is most likely positive because H_z' is positive in low latitudes where the zonally averaged temperature deviation is high, while H_z' is negative in the higher latitudes where T_z' is also negative.

The sign of $G(A_E)$ is very difficult, if not impossible, to determine from such simple arguments because it depends on the heating in the atmospheric waves. We may point to some processes that should give a negative contribution and to others where the contribution is of the opposite sign. The exchange of heat between the underlying surface and the atmosphere is an example of a process where warm air masses are cooled and cold air masses are heated because the transfer of sensible heat is proportional to the temperature difference between the surface and the lower part of the atmosphere. The oceans are heat sources in the fall and winter while the continents are heat sources in the summer and vice versa. On the other hand, the release of heat due to condensation and the formation of rain will mainly be important in moist air masses that contain especially large amounts of moisture if they are warm in the lower layers. Other examples connected with the various radiative processes could also be mentioned. In view of this situation it is difficult to determine in advance if $G(A_E)$ is positive or negative.

Turning next to the conversion $C(A_z, K_z)$ and $C(A_E, K_E)$ we have already shown at the end of Chapter 3 that $C(A_E, K_E)$ is positive in a typical baroclinic wave in which the temperature field is lagging behind the height field. This structure of a baroclinic wave is the same as the one found from baroclinic instability theory for the growing baroclinically unstable wave. Furthermore, since this structure agrees with the common arrangement found from observational studies, it is plausible on these grounds that $C(A_E, K_E)$ on average is positive.

With respect to $C(A_z, K_E)$ we should look at the mean meridional circulation. For the purposes of the present discussion we shall assume that the mean meridional circulation consists of three cells: a Hadley cell in the low latitudes, a Ferrel cell in the middle latitudes, and a weak polar cell in the high latitudes. At tropospheric isobaric surfaces we have in general a temperature field that decreases from equator to pole. On these grounds we can expect that the Hadley cell is a direct cell in the sense that warm air is rising and (relatively) colder air is sinking. The Ferrel cell, on the other hand, is an indirect cell where warmer air is sinking and the colder air is rising. We may thus expect a positive contribution from the Hadley cell and the polar and a negative contribution from the Ferrel cell leaving the sign of $C(A_z, K_z)$ uncertain. It may, however, be expected that this energy conversion is small because the vertical velocities in the mean meridional circulation, converted to the unit of LT^{-1}, are of the order of magnitude mm s^{-1} while the total vertical velocity is of the order cm s^{-1}. It is therefore likely that $|C(A_z, K_z)| < |C(A_E, K_E)|$.

The next problems are connected with the energy conversions between the zonal averages and the eddies. We consider first

$$C(A_z, A_E) = \frac{1}{g} \int_0^{p_0} \int_S \frac{R^2}{\overline{\sigma} p^2} \frac{T_z'}{a \cos \varphi} \frac{\partial \left(T_E' v_E\right)_z \cos \varphi}{\partial \varphi} \, dS \, dp. \qquad (7.31)$$

We may argue that this energy conversion ought to be positive. The eddy transport of sensible heat is northward in baroclinic waves. These waves occur mainly in a broad band of middle latitudes. We may thus expect that $(T_E' v_E)_z \cos \varphi$ is mainly positive in the troposphere with a maximum somewhere in the middle latitudes. The derivative is positive in the low latitudes and negative in the high latitudes. T_z' follows the same rule, thus making $C(A_z, A_E)$ positive.

We next consider the exchange of kinetic energy expressed in the integral

$$C(K_E, K_z) = -\frac{1}{g} \int_0^{p_0} \int_S \frac{u_z}{a \cos^2 \varphi} \frac{\partial (u_E v_E)_z \cos^2 \varphi}{\partial \varphi} \, dS \, dp. \qquad (7.32)$$

From the point of view of balance in the energy reservoir K_z, we may note that the integral in (7.32) should be positive because the other processes that may influence K_z are $C(A_z, K_z)$ and $D(K_z)$. We have reasoned that $C(A_z, K_z)$ in general should be quite small although the sign was uncertain, while $D(K_z)$ ought to be a sink for K_z. If this is so, then the only source left for K_z is the energy conversion from K_E as calculated from (7.32).

Such a balance argument, however, is not completely satisfying because it rests on earlier assumptions such as the smallness of $C(A_z, K_z)$, and it applies only for stationary states. The existing simple theories of baroclinic instability do not give the same guidance in this case as they did for the case of $C(A_z, A_E)$. The reason is that simple baroclinic theories assume that the waves are of a sinusoidal nature, and that the wave and the basic current u_z do not vary with latitude. As we have seen earlier, there is no geostrophic momentum transport by such waves. More complicated theories of baroclinic instability such as the one proposed by Charney (1959) produce waves with a nonzero momentum transport, and it is found in

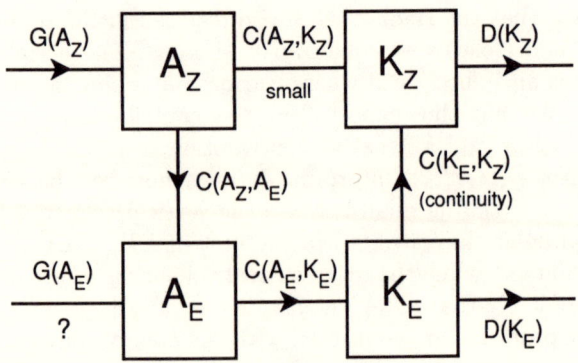

Fig. 7.2: The directions of generations, conversions, and dissipations as deduced from simple arguments concerning the heat and momentum transports.

these studies that $C(K_E, K_z)$ is positive. But these studies based on perturbation theory are applied to finite amplitude waves, and this cannot be done without further hypotheses, which may or may not be justified. We are therefore forced to conclude that we cannot on general grounds deduce the sign of the conversion $C(K_E, K_z)$.

Finally, although we feel that $D(K_z)$ and $D(K_E)$ as formulated ought to be positive, we must admit that such a statement cannot be firmly supported because it would require a formulation of the frictional force which in turn can be given only from a well-founded turbulence theory. This turbulence theory does not exist at the present time.

Fig. 7.2 shows the deduced directions of the processes discussed above as well as those where the directions are uncertain. In any case, studies based on observations and models are needed to determine the values and the missing directions.

It is appropriate in this chapter to summarize calculations of energy conversions based on the model behind the derivations used here. Fig. 7.3 is based on calculations carried out by Wiin-Nielsen (1964) and applies to winter conditions. We recall that the diagnostic computations are carried out using standard operational analyses of isobaric height fields up to 100 mb. The height fields were converted to streamfunctions using the linear part of the balance equation – that is,

$$\nabla \cdot [f \nabla \psi] = \nabla^2 \phi. \tag{7.33}$$

At the time operational analyses were available in an octagonal region centered on the North Pole, but not extending to the equator. For practical reasons the calculations were performed in a region north of 20° N. The details are described in the references. The calculations of the vertical ve-

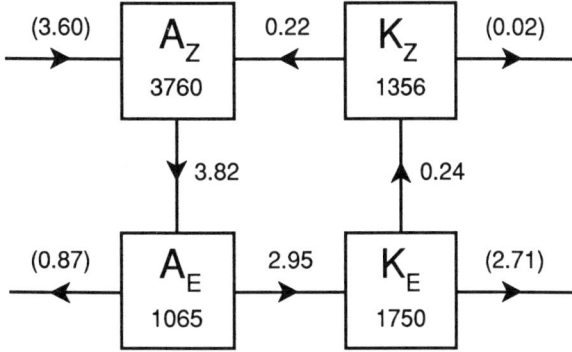

Fig. 7.3: The quasi-geostrophic energy diagram for the winter season (Wiin-Nielsen, 1964).

locity needed for the computation of $C(A, K)$ were obtained as solutions of the quasi-nondivergent ω-equation. This equation is derived by elimination of the time derivatives from the vorticity equation and the thermodynamic equation – that is,

$$\sigma \nabla^2 \omega + f_0^2 \frac{\partial^2 \omega}{\partial p^2} = f_0 \left[\frac{\partial}{\partial p} \left[\vec{v} \cdot \nabla (f + \zeta) \right] - \nabla^2 \left(\vec{v} \cdot \nabla \frac{\partial \psi}{\partial p} \right) \right], \quad (7.34)$$

where we have included only the adiabatic, nonviscous part. Fig. 7.3 shows the amounts of energy in the reservoirs A_z, A_E, K_z, and K_E, measured in kJ m^{-2}. The energy exchanges, given in W m^{-2}, between the four boxes are obtained from data. The generations $G(A_z)$ and $G(A_E)$ and the dissipations $D(K_z)$ and $D(K_E)$ are obtained by assuming steady-state conditions for each box. Fig. 7.4 is arranged in a similar way, but applies to summer conditions in the Northern Hemisphere.

Independent calculations of $G(A_z)$ and $G(A_E)$ have been carried out by Wiin-Nielsen and Brown (1960), Brown (1964), and Lawniczak (1970). $G(A_E)$ is negative in these studies. For January 1959 used in the first study, it was found that $G(A_z) = 5$ W m^{-2} and $G(A_E) = -3.5$ W m^{-2}, while the corresponding numbers for a study using January 1969 data were: $G(A_z) = 5.5$ W m^{-2} and $G(A_E) = -2.9$ W m^{-2}. However, later calculations using global data and much improved parametrizations do not agree with this result. The reason is probably that the early calculations do not catch the contribution from the release of latent heat and the contributions from the tropics.

A definite weakness of the above calculation is the conversion $C(A_z, K_z)$ which depends on the meridional temperature and the mean meridional circulation. There is no doubt that the tropical Hadley circulation will give a positive contribution to this energy conversion, and it is equally obvious

FUNDAMENTALS OF ATMOSPHERIC ENERGETICS

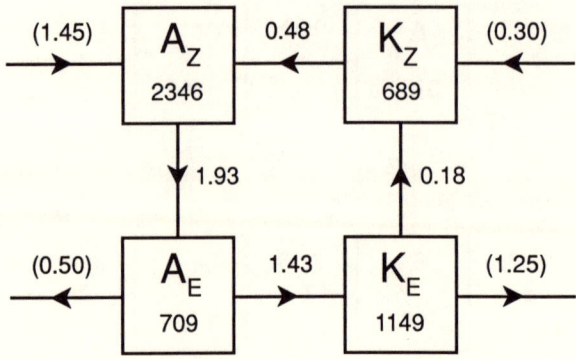

Fig. 7.4: The quasi-geostrophic energy diagram for the summer season (Wiin-Nielsen, 1964).

that this contribution is not included properly (if at all) in calculations stopping at 20° N. We may get an idea of the sensitivity of the conversion $C(A_z, K_z)$ by considering an idealized example. Per unit area we have

$$C(A_z, K_z) = -\frac{1}{gS} \int_0^{p_0} \int_S \alpha'_z \omega_z \, dS \, dp. \qquad (7.35)$$

Using 50 cb as a representative level we have

$$C(A_z, K_z) = -\frac{p_0}{gS} \cdot 2\pi \int_{\varphi_1}^{\varphi_2} \alpha'_z \omega_z a^2 \cos\varphi \, d\varphi. \qquad (7.36)$$

We consider a region from the equator to the latitude φ_*, and we have

$$S = 2\pi a^2 \sin\varphi_*. \qquad (7.37)$$

In addition, we write

$$\alpha'_z = \frac{2R}{p_0} T'_z \quad T'_z = A\left[\cos(2\varphi) - \frac{1}{3}\right]. \qquad (7.38)$$

We have thus assumed that $T'_z(\varphi)$ can be represented as a simple trigonometric function. As a first approximation this is justified in the winter season as can be seen by the cross sections given by Palmén and Newton (1967). With a temperature of 258 K at the equator and 50 kPa and 232 K at the North Pole, we find that $A = 13.5$ gives the correct temperature difference.

For the vertical velocity we write

$$\omega_z = -g\rho w_z = -\frac{gp_0}{2R\overline{T}_{50}}w_z \qquad w_z = B\left(\cos 6\varphi - \frac{6}{35}\right). \qquad (7.39)$$

The numerical constants in the expressions for T'_z and w_z are needed to have zero area averages for the two quantities. Using these quantities we calculate the integral (7.36) for $\varphi_1 = 0$ and $\varphi_2 = \varphi_*$. We find that the contribution from the Hadley cell is

$$C(A_z, K_z) = \qquad (7.40)$$
$$\frac{p_0}{\overline{T}_{50}}\frac{AB}{\sin\varphi_*}\left(\frac{\sin(9\varphi_*)}{36} + \frac{\sin(7\varphi_*)}{84} + \frac{\sin(5\varphi_*)}{60} + \frac{23\sin(3\varphi_*)}{420} - \frac{\sin\varphi_*}{35}\right).$$

Setting $\varphi_* = 20°$ N we find a contribution of 1.17 W m^{-2} from (7.41). The corrected value of $C(A_z, K_z)$ is then

$$\begin{aligned}C(A_z, K_z) = \quad & (1-\sin\varphi_*)\,C_C(A_z, K_z) \qquad (7.41)\\ & +\sin\varphi_* C_E(A_z, K_z)\\ & = 0.25\ Wm^{-2},\end{aligned}$$

where C_C is the computed value from data in Fig. 7.3 while C_E is the estimated contribution from the Hadley cell between the equator and 20° N. In the calculation in (7.41) the values are: $p_0 = 100$ kPa, $\overline{T}_{50} = 250$ K, $A = 13.5$ K and $B = 1.2 \times 10^{-3}$ m s^{-1} = 0.5 mm s^{-1}. It is thus seen that the contribution from the Hadley cell is sufficient to reverse the direction of $C(A_z, K_z)$. On the other hand, $C(A_z, K_z)$ is small, and is probably not significantly different from zero considering the uncertainty of the data and the length of the period in question. If we were to set $C(A_z, K_z) = 0$, we should find $G(A_z) = 3.82$ W m^{-2} and $D(K_z) = 0.24$ W m^{-2} in Fig. 7.3 and $G(A_z) = 1.93$ W m^{-2} and $D(K_z) = 0.18$ W m^{-2} in Fig. 7.4.

While Figs. 7.3 and 7.4 have the advantage that the various calculations are consistent with each other because the same procedures have been used in all computations, there have of course been other diagnostic studies. Oort (1964) summarized all values available at that time. His diagram, taken from Lorenz (1967), is reproduced as Fig. 7.5. While differences exist between the various estimates as one would expect, we note that orders of magnitude and directions agree between the diagrams if Figs. 7.3 and 7.4 are corrected as proposed in this section.

Thus Figs. 7.3, 7.4, and 7.5 are believed to give a realistic, but probably not final, picture of the average flow of energy through the various energy reservoirs in the atmosphere. However, one cannot conclude that the atmosphere at any instant or in any given period operates according to these flow diagrams. An example is provided by the winter 1962–63 which

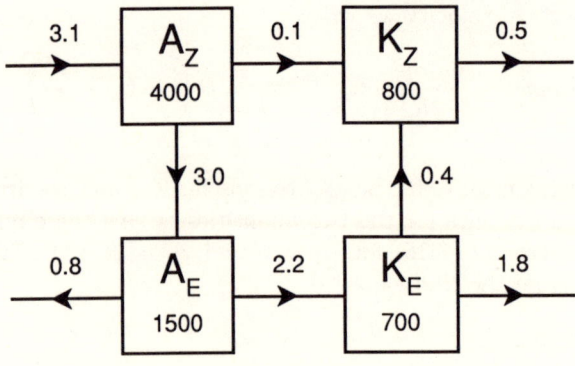

Fig. 7.5: Oort's composite energy diagram (1964).

was abnormal in many ways. We shall consider January 1963. This month was characterized by long lasting blocking situations in both the Pacific and Atlantic regions.

Fig. 7.6 shows the average flow at 70 kPa for January 1963, while Fig. 7.7 is the corresponding map for February 1963. A general description of the extreme nature of these circulations is given by O'Connor (1963). January 1963 can, according to these descriptions, be characterized as "the coldest month in 70 years on a countrywide basis" – that is, for the United States. The circulation is one of extreme amplification with an enormous ridge in the eastern Pacific followed eastward by a deep trough over eastern North America. In the eastern Atlantic a blocking pattern prevailed throughout the month between Iceland and Britain with a deep trough south of the anticyclone near the Azores; see O'Connor (1963) for further details. The blocking patterns continued in February 1963 although less extreme than in January as can be seen from Fig. 7.7. We next shall describe the energetics for January 1963.

Fig. 7.8 displays the energy diagram for this month. One conversion stands out as unusual: the conversion $C(K_z, K_E)$ is positive contrary to the normal conditions. For this month a balanced energy diagram cannot be expected. A time series for January 1963 indicates that A_z decreases through the month; A_E increases in the first half of the month and has a net decrease in the second half of the month. All the forms of kinetic energy vary significantly during the period, but there is a marked tendency for an increase in the eddy energy to be accompanied by a decrease in the corresponding zonal energy. This is an indication that the eddies receive their energy from the zonal current in the greater part of the period, a process that is akin to barotropic processes. With respect to the maintenance of the zonal kinetic energy, K_z, during the period, it would appear from Fig. 7.8 that K_z should decrease. This is not the case as can be seen from the data. It can be shown that K_z is maintained by a flux across the

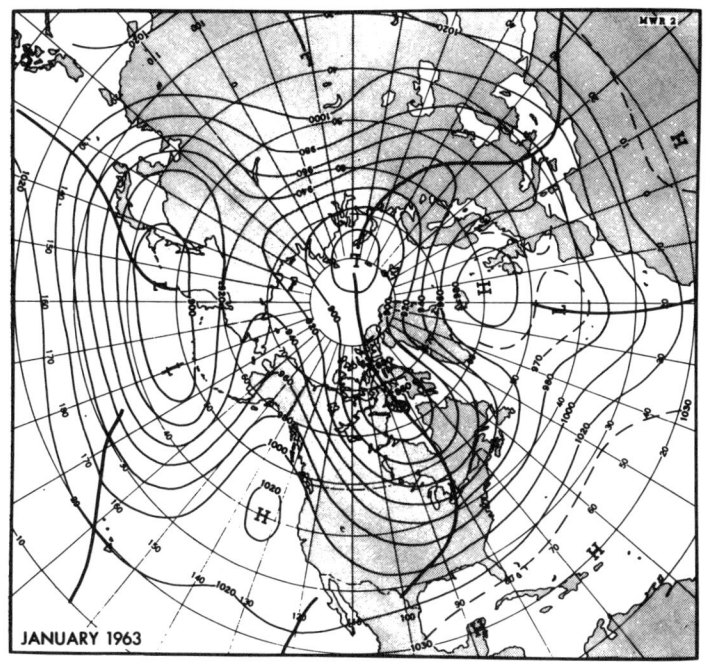

Fig. 7.6: 70 kPa mean map for January 1963.

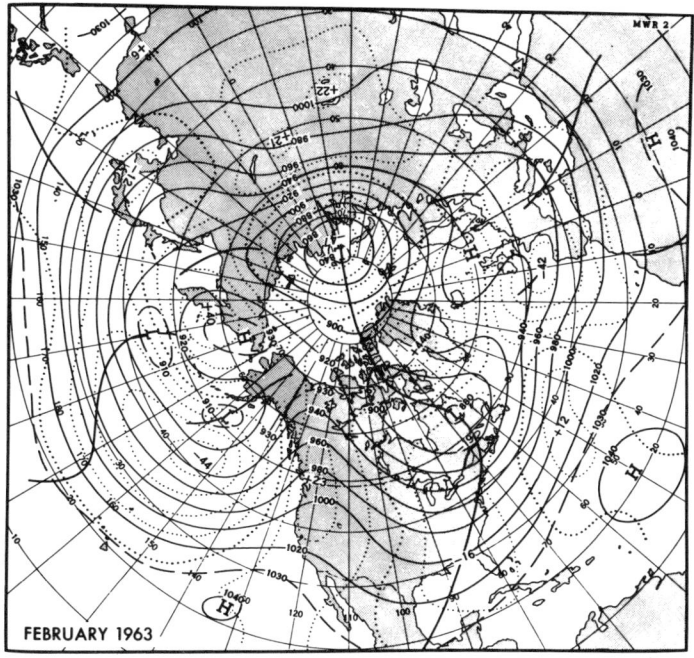

Fig. 7.7: 70 kPa mean map for February 1963.

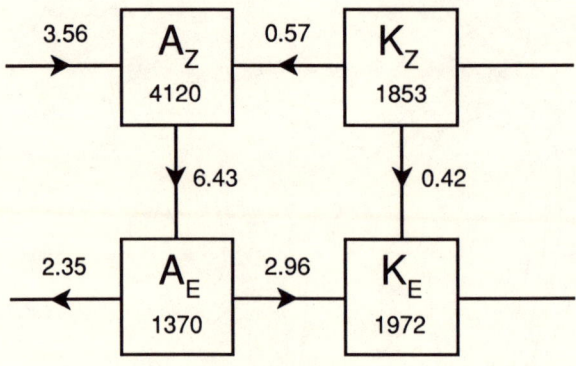

Fig. 7.8: Energy diagram for January 1963.

southern boundary at 20° N. There is thus good reason to believe that the atmosphere during periods of global-type blocking may operate in an energetical mode different from the normal long-term average energy flow through the reservoirs.

8

DIVERGENT AND NONDIVERGENT FLOW

It is evident from the previous chapters that the divergence and the vertical velocity play a most important role in atmospheric energetics. At the same time, they are quantities that can be obtained only indirectly from observations. The gradual development of atmospheric models for both numerical weather prediction and general circulation studies from a simple nondivergent equivalent barotropic model to advanced prediction and climate models has made it possible to study many aspects of divergence and vertical velocity. The development of models based on the primitive equations has also meant that the total horizontal velocity and not the nondivergent part alone has been used for advection purposes.

In view of the importance of divergence and vertical velocity in atmospheric energetics, it has been found interesting to divide the kinetic energy into two parts: nondivergent velocity and divergent velocity. The partitioning of the horizontal velocity into these two parts is made possible by Helmholtz' theorem, which in short says that it is possible to divide a two-dimensional vector into two parts: one part containing all the vorticity and the other all the divergence.

To ease the following theoretical development we shall introduce a scheme of notations helpful in distinguishing the various parameters from each other. The thermodynamic variables (density, specific volume, temperature, potential temperature) and the geopotential, which is related to the specific volume through the hydrostatic equation, will be denoted by the subscript 1. The nondivergent wind and the quantities directly related to it, such as the streamfunction and the vorticity will have the subscript 2. Finally, the divergent wind, the velocity potential, the divergence, and the vertical velocity will be given the subscript 3.

Using Helmholtz' theorem we write

$$\vec{v} = \vec{v}_2 + \vec{v}_3, \qquad (8.1)$$

where $\vec{v}_2 = \vec{k} \times \nabla \psi_2$ and $\vec{v}_3 = \nabla \chi_3$. Here, ψ_2 is the streamfunction, and χ_3 is the velocity potential.

$$K = K_2 + K_3, \qquad (8.2)$$

where

$$K = \frac{1}{g} \int_0^{p_0} \int_S \frac{\vec{v} \cdot \vec{v}}{2} \, dS \, dp, \qquad (8.3)$$

and correspondingly for K_2 and K_3. Inserting (8.1) in (8.3) we find

$$K = \frac{1}{g} \int_0^{p_0} \int_S \left(\frac{1}{2} \vec{v}_2 \cdot \vec{v}_2 + \frac{1}{2} \vec{v}_3 \cdot \vec{v}_3 + \vec{v}_2 \cdot \vec{v}_3 \right) dS\, dp. \tag{8.4}$$

Equation (8.2) will therefore hold if we can show that the last integral in (8.4) vanishes. We find

$$\int_S \vec{v}_2 \cdot \vec{v}_3\, dS = \int_S \left(-\frac{\partial \psi_2}{a \partial \varphi} \frac{\partial \chi_3}{a \cos \varphi\, \partial \lambda} \right. \tag{8.5}$$
$$\left. + \frac{\partial \psi_2}{a \cos \varphi\, \partial \lambda} \cdot \frac{\partial \chi_3}{a\, \partial \varphi} \right) a^2 \cos \varphi\, d\lambda\, d\varphi.$$

Further evaluation of the integrand in (8.6) leads to

$$\int_S \vec{v}_2 \cdot \vec{v}_3\, dS = \int_S \left[\frac{\partial}{\partial \lambda} \left(\psi_2 \frac{\partial \chi_3}{\partial \varphi} \right) - \frac{\partial}{\partial \varphi} \left(\psi_2 \frac{\partial \chi_3}{\partial \lambda} \right) \right] d\lambda\, d\varphi, \tag{8.6}$$

and it is seen that the integral in (8.6) vanishes. It follows from the definitions that

$$\frac{dK_2}{dt} = \frac{1}{g} \int_0^{p_0} \int_S \left(u_2 \frac{\partial u_2}{\partial t} + v_2 \frac{\partial v_2}{\partial t} \right) dS\, dp \tag{8.7}$$
$$\frac{dK_3}{dt} = \frac{1}{g} \int_0^{p_0} \int_S \left(u_3 \frac{\partial u_3}{\partial t} + v_3 \frac{\partial v_3}{\partial t} \right) dS\, dp.$$

We recall from the previous chapters that

$$\frac{dK}{dt} = -\frac{1}{g} \int_0^{p_0} \int_S \vec{v} \cdot \nabla \phi_1\, dS\, dp + \frac{1}{g} \int_0^{p_0} \int_S \vec{v} \cdot \vec{F}\, dS\, dp \tag{8.8}$$

of which the first integral is the conversion from available potential energy or from total potential energy (see Chapters 3 and 4). With our notations we find easily that

$$C(A, K) = -\frac{1}{g} \int_0^{p_0} \int_S (\vec{v}_2 + \vec{v}_3) \cdot \nabla \phi_1\, dS\, dp \tag{8.9}$$
$$= -\frac{1}{g} \int_0^{p_0} \int_S \vec{v}_3 \cdot \nabla \phi_1\, dS\, dp$$

because $\vec{v}_2$ is nondivergent.

After these preparations we are ready to start on the main problem. The equation of motion for the horizontal velocity is

$$\frac{\partial \vec{v}}{\partial t} + \vec{v} \cdot \nabla \vec{v} + \omega \frac{\partial \vec{v}}{\partial p} + \nabla \phi + f\, \vec{k} \times \vec{v} = \vec{F}. \tag{8.10}$$

DIVERGENT AND NONDIVERGENT FLOW

It is easy to see by differentiation that

$$\vec{v} \cdot \nabla \vec{v} = \nabla k + \zeta \vec{k} \times \vec{v}, \tag{8.11}$$

where k is the kinetic energy per unit mass. We then may write (8.10) in the form

$$\frac{\partial \vec{v}}{\partial t} + \nabla k + (\zeta + f) \vec{k} \times \vec{v} + \omega \frac{\partial \vec{v}}{\partial p} + \nabla \phi = \vec{F}. \tag{8.12}$$

The division of $\vec{v}$ into $\vec{v}_2$ and $\vec{v}_3$ is introduced in (8.12). Using the convention on subscripts, we obtain

$$\frac{\partial \vec{v}_2}{\partial t} + \frac{\partial \vec{v}_3}{\partial t} + \nabla k + (\zeta_2 + f) \vec{k} \times (\vec{v}_2 + \vec{v}_3) + \omega_3 \frac{\partial (\vec{v}_2 + \vec{v}_3)}{\partial p} + \nabla \phi_1 = \vec{F}. \tag{8.13}$$

To obtain the change in the kinetic energy of the nondivergent motion we shall multiply (8.13) by $\vec{v}_2$ and integrate over the mass of the atmosphere. It may be convenient to calculate the integrations term by term. The first term gives dK_2/dt. The second term integrates to zero because

$$\int_S \vec{v}_2 \cdot \frac{\partial \vec{v}_3}{\partial t} dS = \int_S \vec{k} \times \nabla \psi_2 \cdot \nabla \left(\frac{\partial \chi_3}{\partial t} \right) dS, \tag{8.14}$$

but the integral (8.14) is analogous to (8.6) when $\nabla \chi_3$ is replaced by $\nabla (\partial \chi_3 / \partial t)$ and will therefore vanish.

The integral of the third term will vanish because $\vec{v}_2$ is nondivergent – that is,

$$\int_S \vec{v}_2 \cdot \nabla k \, dS = \int_S \nabla \cdot \left(k \vec{v}_2 \right) dS = 0. \tag{8.15}$$

The fourth term will give the contribution

$$\frac{1}{g} \int_0^{p_0} \int_S (f + \zeta_2) \vec{v}_2 \cdot \vec{k} \times \vec{v}_3 \, dS \, dp \tag{8.16}$$

because $\vec{v}_2 \cdot \left(\vec{k} \times \vec{v}_2 \right) = 0$.

The contribution from the fifth term may be written as two integrals as follows:

$$\frac{1}{g} \int_0^{p_0} \int_S \omega_3 \vec{v}_2 \cdot \frac{\partial \vec{v}_2 + \vec{v}_3}{\partial p} dS \, dp \tag{8.17}$$

$$= \frac{1}{g} \int_0^{p_0} \int_S \omega_3 \frac{\partial k_2}{\partial p} dS \, dp + \frac{1}{g} \int_0^{p_0} \int_S \omega_3 \vec{v}_2 \cdot \frac{\partial \vec{v}_3}{\partial p} dS \, dp$$

$$= \frac{1}{g} \int_0^{p_0} \int_S k_2 \nabla \cdot \vec{v}_3 \, dS \, dp + \frac{1}{g} \int_0^{p_0} \int_S \omega_3 \vec{v}_2 \cdot \frac{\partial \vec{v}_3}{\partial p} dS \, dp.$$

Noting finally that

$$\int_S \vec{v}_2 \cdot \nabla \phi_1 \, dS = \int_S \nabla \cdot \left(\phi_1 \vec{v}_2 \right) dS = 0, \qquad (8.18)$$

we may write the final result as follows:

$$\begin{aligned}\frac{dK_2}{dt} =\ & -\frac{1}{g} \int_0^{p_0} \int_S \left[(f + \zeta_2) \vec{v}_2 \cdot \left(\vec{k} \times \vec{v}_3 \right) \right. \\ & \left. + k_2 \nabla \cdot \vec{v}_3 + \omega_3 \vec{v}_2 \cdot \frac{\partial \vec{v}_3}{\partial p} \right] dS \, dp \\ & + \frac{1}{g} \int_0^{p_0} \int_S \vec{v}_2 \cdot \vec{F} \, dS \, dp. \end{aligned} \qquad (8.19)$$

The first integral in (8.19) originates from the advection terms and the Coriolis term in (8.10), while these terms give no contribution to (8.8). When we obtain the equation for dK_3/dt by subtracting (8.19) from (8.8), the first integral in (8.19) will appear with the opposite sign in the equation for dK_3/dt. It therefore is clear that the integral represents the conversion from K_3 to K_2 – that is,

$$C(K_3, K_2) = -\frac{1}{g} \int_0^{p_0} \int_S \left[(f + \zeta_2) \vec{v}_2 \cdot \vec{k} \times \vec{v}_3 + k_2 \nabla \cdot \vec{v}_3 + \omega_3 \vec{v}_2 \cdot \frac{\partial \vec{v}_3}{\partial p} \right]$$

$$dS \, dp, \qquad (8.20)$$

and (8.19) can then be written

$$\frac{dK_2}{dt} = C(K_3, K_2) - D(K_2), \qquad (8.21)$$

where

$$D(K_2) = -\frac{1}{g} \int_0^{p_0} \int_S \vec{v}_2 \cdot \vec{F} \, dS \, dp. \qquad (8.22)$$

Carrying out the required subtraction we obtain

$$\frac{dK_3}{dt} = C(A, K_3) - C(K_3, K_2) - D(K_3), \qquad (8.23)$$

where $C(A, K_3)$ is given by (8.10) and

$$D(K_3) = -\frac{1}{g} \int_0^{p_0} \int_S \vec{v}_3 \cdot \vec{F} \, dS \, dp. \qquad (8.24)$$

We thus have obtained the following results:

DIVERGENT AND NONDIVERGENT FLOW 89

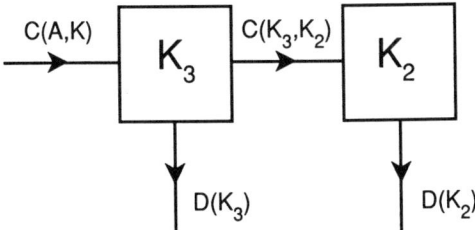

Fig. 8.1: A diagram for the kinetic energy, where K_3 is the kinetic energy of the divergent motion, while K_2 is the kinetic energy of the nondivergent flow.

1. The conversion from available potential energy to kinetic energy goes entirely into the kinetic energy of the divergent part of the flow.

2. The energy transformation from K_3 to K_2 depends on the interaction between $\vec{v}_3$ and $\vec{v}_2$ in a complicated way as given by (8.20).

A diagram showing the energy conversions is given in Fig. 8.1. The directions of the transformations are included based on continuity principles for a long-term average starting from the assumptions that $C(A, K_3) > 0$, $D(K_3) > 0$, and $D(K_2) > 0$.

The integral in (8.20) is as difficult to evaluate as the conversion integral $C(A, K_3)$ because it is a necessity to obtain good estimates of the vertical velocity ω_3 with respect to p. The divergent part of the velocity $\vec{v}_3$ must thus be obtained first by finding the velocity potential through solving the equation

$$\nabla^2 \chi_3 = \nabla \cdot \vec{v}_3 = -\frac{\partial \omega_3}{\partial p}. \tag{8.25}$$

Using the velocity potential we obtain $\vec{v}_3$ from the relation $\vec{v}_3 = \nabla \chi_3$. Such a procedure can be carried out using data from a prediction or general circulation model, but it can hardly be carried out in the above form from observed data.

The expression for $C(K_3, K_2)$ given in (8.20) is, however, very convenient when investigating various simplifications in atmospheric modeling introduced in the past. As it stands, (8.20) applies to a general primitive equation model in which only the hydrostatic assumption has been used. The first term in the integrand in (8.20) comes from the horizontal advection terms while the last two are connected with the vertical advection of momentum. Based on the results of scale analysis (Phillips, 1963), we therefore would expect that the first term in the integrand gives the major contribution. Haltiner (1971) considers two models the first of which is the general hydrostatic model while the other has $\vec{v}_2$ instead of $\vec{v} = \vec{v}_2 + \vec{v}_3$ in the term for the vertical advection of momentum. Considering the derivation of (8.20) it is seen that the conversion in the latter case is

$$C(K_3, K_2) = -\frac{1}{g} \int_0^{p_0} \int_S \left[(f + \zeta_2) \, \vec{v}_2 \cdot \left(\vec{k} \times \vec{v}_3 \right) + k_2 \nabla \cdot \vec{v}_3 \right] dS \, dp. \tag{8.26}$$

If one were to neglect the vertical advection of momentum completely, the last term in the integrand of (8.26) would also disappear. The models just mentioned are still based on the primitive equations or, equivalently, on the vorticity and divergence equations. The main point is that K_3 may still change in time for these models, and the energy diagram in Fig. 8.1 applies in principle although the expression for $C(K_3, K_2)$ is simplified.

On the other hand, if we simplify even further and go to the so-called balanced models, we experience a radical change because in this case the divergence equation is replaced by a balance equation requiring at least the neglect of the time derivative in the divergence equation. In that case, K_3 cannot change, and the conversion $C(A, K_3)$ must be equal to the conversion $C(K_3, K_2)$ because friction is neglected in the balance equation.

We may obtain the balance equation by taking the divergence of (8.27), neglecting the time derivatives, and maintaining only those terms that belong to classes 1 and 2. The result is:

$$\nabla^2 k_2 - \nabla \cdot [(\zeta_2 + f) \nabla \psi_2] + \nabla^2 \phi_1 = 0. \tag{8.27}$$

It may easily be shown by differentiation that (8.27) reduces to the more well-known form

$$\nabla \cdot [f \nabla \psi_2] + 2 \left[\frac{\partial^2 \psi_2}{\partial x^2} \frac{\partial^2 \psi_2}{\partial y^2} - \left(\frac{\partial^2 \psi_2}{\partial x \, \partial y} \right)^2 \right] = \nabla^2 \phi_1. \tag{8.28}$$

Since we know for these models that $C(A, K_3) = C(K_3, K_2)$, we should be able to obtain this result from the balance equation. Noting that $C(A, K_3)$ is related to the pressure force, we write (8.27) in the form

$$\nabla^2 \phi_1 = \nabla \cdot [(\zeta_2 + f) \nabla \psi_2] - \nabla^2 k_2. \tag{8.29}$$

To obtain the energy balance we multiply both sides of (8.29) by χ_3. From the left-hand side we obtain

$$\frac{1}{g} \int_0^{p_0} \int_S \chi_3 \nabla^2 \phi_1 \, dS \, dp = \frac{1}{g} \int_0^{p_0} \int_S [\nabla \cdot (\chi_3 \nabla \phi_1) - \nabla \chi_3 \cdot \nabla \phi_1] \, dS \, dp$$
$$= -\frac{1}{g} \int_0^{p_0} \int_S \vec{v}_3 \cdot \nabla \phi_1 \, dS \, dp, \tag{8.30}$$

but the last expression in (8.30) is $C(A, K_3)$.

From the right-hand side in (8.29) we obtain

$$\frac{1}{g} \int_0^{p_0} \int_S \chi_3 \left[\nabla \cdot \left[(\zeta_2 + f) \nabla \psi_2 \right] - \nabla^2 k_2 \right] dS\, dp \qquad (8.31)$$
$$= -\frac{1}{g} \int_0^{p_0} \int_S \left[(f + \zeta_2)\, \vec{v}_2 \cdot \vec{k} \times \vec{v}_3 + k_2 \nabla \cdot \vec{v}_3 \right] dS\, dp,$$

which is a form of $C(K_3, K_2)$ corresponding to the proposal by Haltiner (1971) mentioned above. Now, the general vorticity equation is obtained in the normal fashion from (8.12) by taking the curl of the equation. The result, written without subscripts is

$$\frac{\partial \zeta}{\partial t} + \nabla \cdot \left[(f + \zeta)\, \vec{v} \right] + \vec{k} \cdot \nabla \times \left(\omega \frac{\partial \vec{v}}{\partial p} \right) = \vec{k} \cdot \nabla \times \vec{F}. \qquad (8.32)$$

We observed earlier that the form (8.32) is obtained if $\vec{v}$ in the vertical advection term is replaced by $\vec{v}_2$. The consistent vorticity equation corresponding to the complete balance equation (8.29) is thus

$$\frac{\partial \zeta_2}{\partial t} + \nabla \cdot \left[(f + \zeta_2)\, \vec{v}_2 \right] + \nabla \cdot \left[(f + \zeta_2)\, \vec{v}_3 \right] + \vec{k} \cdot \nabla \times \left(\omega_3 \frac{\partial \vec{v}_2}{\partial p} \right) = \vec{k} \cdot \nabla \times \vec{F}. \qquad (8.33)$$

Suppose next that we apply the following simplified balance equation:

$$\nabla^2 \phi_1 = \nabla \cdot [f \nabla \psi_2]. \qquad (8.34)$$

The left-hand side of (8.34) still gives (8.30) for $C(A, K_3)$, while the right hand side gives

$$\frac{1}{g} \int_0^{p_0} \int_S \chi_3 \nabla \cdot [f \nabla \psi_2]\, dS\, dp = -\frac{1}{g} \int_0^{p_0} \int_S f \nabla \psi_2 \cdot \nabla \xi_3\, dS\, dp \qquad (8.35)$$
$$= \frac{1}{g} \int_0^{p_0} \int_S f\, \vec{k} \times \vec{v}_2 \cdot \vec{v}_3\, dS\, dp$$
$$= \frac{1}{g} \int_0^{p_0} \int_S f\, \vec{v}_2 \cdot \left(\vec{k} \times \vec{v}_3 \right) dS\, dp.$$

The question once again is which simplifications must we make in the vorticity equation to get (8.36). It is easily verified that it has to be

$$\frac{\partial \zeta_2}{\partial t} + \nabla \cdot \left[(f + \zeta_2)\, \vec{v}_2 \right] + \nabla \cdot \left(f\, \vec{v}_3 \right) = \vec{k} \cdot \nabla \times \vec{F}. \qquad (8.36)$$

To verify that (8.26) and (8.36) are consistent with each other, we must multiply (8.36) by $-\psi_2$ and integrate, in which case we get

$$\frac{dK_2}{dt} = -\frac{1}{g}\int_0^{p_0}\int_S f\,\vec{v}_3\cdot\nabla\psi_2\,dS\,dp - D(K_2), \qquad (8.37)$$

which agrees with (8.26).

The most simple model is the one in which the balance equation is reduced to

$$\nabla^2\phi_1 = f_0\nabla^2\psi_2. \qquad (8.38)$$

In this case we find from the right-hand side that

$$\frac{1}{g}\int_0^{p_0}\int_S f_0\chi_3\nabla^2\psi_2\,dS\,dp = -\frac{1}{g}\int_0^{p_0}\int_S f_0\nabla\psi_2\cdot\nabla\chi_3\,dS\,dp \qquad (8.39)$$

and the consistent vorticity equation is

$$\frac{\partial\zeta_2}{\partial t} + \nabla\cdot\left[(f+\zeta_2)\,\vec{v}_2\right] + f_0\nabla\cdot\vec{v}_3 = \vec{k}\cdot\nabla\times\vec{F}. \qquad (8.40)$$

The three balanced models considered here, having the balance equations (8.27), (8.34), and (8.38) with the corresponding vorticity equations (8.33), (8.36), and (8.40), have the property that K_3 is a constant. The energy reservoir K_3 therefore has a catalytic effect because it participates in the conversions $C(A,K_3)$ and $C(K_3,K_2)$, but K_3 itself remains unchanged. We also know from scale analysis that the amount of kinetic energy in K_3 is small at all times. It is therefore plausible that $C(A,K_3)$ and $C(K_3,K_2)$ in the real atmosphere remain approximately equal because if there was a significant difference for any length of time, K_3 would grow or decay.

The discussion in this chapter follows to a large extent the paper by Chen and Wiin-Nielsen (1976). We shall also use the results from this paper to describe the behavior of $C(K_3,K_2)$. The observational study is based on Eq. (8.20). For reference purposes we define the following integrals:

$$A1: \quad -\frac{1}{g}\int_0^{p_0}\int_S f\,\vec{v}_2\cdot\vec{k}\times\vec{v}_3\,dS\,dp.$$

$$A2: \quad -\frac{1}{g}\int_0^{p_0}\int_S \zeta_2\,\vec{v}_2\cdot\vec{k}\times\vec{v}_3\,dS\,dp.$$

$$B: \quad -\frac{1}{g}\int_0^{p_0}\int_S k_2\nabla\cdot\vec{v}_3\,dS\,dp.$$

$$C: \quad -\frac{1}{g}\int_0^{p_0}\int_S \omega_3\,\vec{v}_2\cdot\frac{\partial\vec{v}_3}{\partial p}\,dS\,dp.$$

As seen from (8.20),

$$C(K_3,K_2) = (A1) + (A2) + (B) + (C).$$

Unfortunately, only a single study based on a short two-week period (1–15 August 1970) is available. In the study all available data from the Northern Hemisphere were used. The observed winds were used to calculate the vorticity, ζ_2, and the equation $\nabla^2 \psi = \zeta_2$ was solved by the spectral method to obtain the streamfunction ψ_2. Finally, $\vec{v}_2 = \vec{k} \times \nabla \psi_2$ was used to calculate $\vec{v}_2$, and $\vec{v}_3$ was obtained by the relation $\vec{v}_3 = \vec{v} - \vec{v}_2$. The amounts of kinetic energy were

$$K_2 = 639 \text{ kJ m}^{-2},$$
$$K_3 = 71 \text{ kJ m}^{-2},$$
$$K = 739 \text{ kJ m}^{-2}.$$

The fact that $K \neq K_2 + K_3$ is due to the fact that for a limited region the integral involving the scalar product of $\vec{v}_2$ and $\vec{v}_3$ does not vanish exactly. As we see, the difference is quite small. The major result is that $K_3 \ll K_2$. In round numbers we have $K_2/(K_2+K_3) \approx 90\%$ and $K_3/(K_2+K_3) \approx 10\%$. The amount of kinetic energy is rather representative for this time of the year. For August 1963 a mean value of 911 kJ m^{-2} was found by Wiin-Nielsen (1967). Oort and Peixóto (1974) found from a study using five years of data that $K = 737$ kJ m^{-2} for the month of August. For the various energy conversions the following values were obtained:

$$(A1) = 0.588 \text{ W m}^{-2}$$
$$(A2) = -0.030 \text{ W m}^{-2}$$
$$(B) = -0.032 \text{ W m}^{-2}$$
$$C(K_3, K_2) = 0.526 \text{ W m}^{-2}$$

These values are probably not representative due to the short period. As expected $(A1)$ gives by far the largest contribution. We may compare that contribution with the energy conversion $C(A, K_3)$ calculated for the same period. It turns out to be 0.620 W m^{-2}.

Earlier in the discussion we made the point that K_3 should be quasi-catalytic. We may investigate this point by estimating the residence times and find that $K_3/[C(K_3, K_2)] \approx 1.4$ days while $K_2/[C(K_3, K_2)] \approx 12.3$ days, verifying that the first is an order of magnitude smaller than the second.

Another study has been made, more representative in time and space but using a global general circulation model rather than observed data. The model is global, based on the primitive, hydrostatic equations with height as the vertical coordinate and in the form in which it existed in 1975. For such a model it is not strictly true that the framework used in the theoretical derivations in this chapter applies. The reason is that one obtains certain additional terms due to the fact that the density varies in a constant height surface. It is this density variation that is avoided by using the isobaric surfaces as reference surfaces. However, these extra terms are probably negligible.

The calculations of $C(K_3, K_2)$ and related quantities were obtained from a simulated 30-day period consisting of days 71–100 of a 100-day simulation. The horizontal spacing was 2.5° in longitude and latitude. The data were u, v, w, p, ρ, and the components of the frictional force per unit volume in the longitudinal and latitudinal directions. An iterative scheme suggested by Endlich (1967) was used to obtain the nondivergent and divergent parts of the horizontal wind fields – that is, $\vec{v}_2$ and $\vec{v}_3$. We may assume that the simulation applies to January conditions.

Table 8.1:

K	K_2	K_3	K_2/K	K_3/K
1470	1452	18	98.8%	1.2%

Energy amounts in kJ m^{-2}.

The total kinetic energy (see Table 8.1) compares favorably with the value 1849 kJ /rmm^{-2} obtained by Wiin-Nielsen (1967) which was based on data from one year north of 20° N and the value of 1239 kJ m^{-2} from the study for a five year period by Oort and Peixóto (1974). The ratios K_2/K and K_3/K indicate quite clearly that $K_3 \ll K_2$, even more than in the data study. The major contributions to K_2 come naturally from the two major jet stream systems in the two hemispheres while K_3 is influenced mostly by the divergent region above the equator and by contributions from the region in and above the planetary boundary layer.

Table 8.2:

(A1)	(A2)	(B)	(C)	$C(K_3, K_2)$	$C(A, K_3)$	$D(K_2)$	$D(K_3)$
1.72	-0.29	0.27	0.04	1.74	2.85	2.52	0.22

Unit: W m^{-2}.

The energy conversions are given in Table 8.2. It is obvious that the model is not in a balanced state during the period used for the calculations of the mean values in Table 8.2. It is clear, however, that the main result is the same for the model study and the data study because the major contribution to $C(K_3, K_2)$ comes from the integral (A1) with a very minor contribution from $(A2) + (B) + (C)$. The result is also in agreement with (8.26) in the sense that both the integrals (A2) and (B) are needed for consistency, while the vertical advection term $\omega(\partial \vec{v})/(\partial p)$ apparently could be approximated by $\omega(\partial \vec{v}_2)/(\partial p)$ without any major damage.

We finally shall make a comment on the implied residence times. For the energy of the nondivergent flow we find $K_2/[C(K_3, K_2)] \approx 9.7$ days while $K_3/[C(K_3, K_2)] \approx 2.9$ hours. Obviously, K_3 is a quasi-catalytic quantity in the model and in the atmosphere.

9

WAVENUMBER REPRESENTATIONS

In studies of atmospheric energetics we have so far considered a global representation: (1) a division between the energetics of the zonally averaged fields and those of the deviation from the zonal average, the so-called eddies, (2) a division between the energetics of the vertical mean flow (the barotropic component) and the vertical shear flow (the baroclinic component), and (3) the energetics of the divergent flow and the nondivergent flow. The purpose of these more or less artificial partitions in each case is to throw some light on the energy flow through the atmosphere and to investigate various processes from an energetical point of view. One may naturally visualize that several of these partitions are used successively. One may for example consider first a division into the barotropic and baroclinic components, where each of these subsequently is divided into the zonal average and the eddies. Such studies have indeed been done by Wiin-Nielsen (1962) from observations and by Smagorinsky (1963) from model integrations. For such studies it is of course necessary to expand the theoretical framework.

In addition to the interests just mentioned there have also been considerable efforts to determine the atmospheric scales that give a major or at least a significant contribution to a given energy amount or to a generation, conversion, or dissipation. For this purpose it is necessary to employ a procedure that can be used to isolate the contribution from a given scale. A representation of the atmospheric fields in terms of a series expansion in orthogonal functions is a procedure that can be used for this purpose. It was first introduceed in the field of atmospheric energetics by Saltzman (1957) who selected a Fourier analysis along latitude circles as his representation.

We shall briefly review such a representation. Let us consider a given latitude φ. Considering first the real domain we want to write an atmospheric parameter $a(\lambda, \varphi, p_1, t_1) = a(\lambda, \varphi)$ at a given instant t_1 and at the isobaric surface p_1 as a series of the form

$$a = B_0(p) + \sum_{m=1}^{\infty} [B_m(\varphi) \cos(m\lambda) + C_m(\varphi) \sin(m\lambda)]. \tag{9.1}$$

λ is longitude varying from 0 to 2π. We note for the trigonometric functions in (9.1) that they are periodic with the period of 2π. Over this period each of them integrates to zero. It is also easy to verify that the solutions to the following integrals are true:

$$\frac{1}{2\pi} \int_0^{2\pi} \cos(m\lambda)\cos(n\lambda)\, d\lambda = \begin{cases} 0, & m \neq n \\ \frac{1}{2}, & m = n \end{cases} \qquad (9.2)$$

$$\frac{1}{2\pi} \int_0^{p_0} \sin(m\lambda)\sin(n\lambda)\, d\lambda = \begin{cases} 0, & m \neq n \\ \frac{1}{2}, & m = n \end{cases} \qquad (9.3)$$

$$\frac{1}{2\pi} \int_0^{2\pi} \sin(m\lambda)\cos(n\lambda)\, d\lambda = 0. \qquad (9.4)$$

When (9.1) is integrated from 0 to 2π we get

$$B_0(\varphi) = a_z, \qquad (9.5)$$

where a_z is the zonal average.

To get an expression for the computation of $B_{m_*}(\varphi)$ we multiply (9.1) by $\cos(m_*\lambda)$ and integrate from 0 to 2π. Using the relations (9.2) – (9.4) we find that

$$B_{m_*}(\varphi) = \frac{1}{\pi} \int_0^{2\pi} a \cos(m_*\lambda)\, d\lambda. \qquad (9.6)$$

Similarly $C_{m_*}(\varphi)$ is obtained by this procedure using $\sin(m_*\lambda)$ as a multiplier, with the result that

$$C_{m_*}(\varphi) = \frac{1}{\pi} \int_0^{2\pi} a \sin(m_*\lambda)\, d\lambda. \qquad (9.7)$$

The formulas (9.5) – (9.7) can be used to calculate $B_0(\varphi)$, $B_m(\varphi)$, and $C_m(\varphi)$ for all integers m if $a(\lambda,\varphi)$ is available along the latitude circle. The integrals are normally replaced by finite sums using a suitable integration formula. Equation (9.1) is naturally also truncated at some point $m = M$. The magnitude of M depends on the nature of the variable $a = a(\lambda,\varphi)$. If a is the temperature along a latitude circle of a long-term climatic mean, it contains only the larger scales, and M will be comparatively small, say $M = 8$ to 10. On the other hand, if we are dealing with instantaneous synoptic maps, M should be somewhat larger – its magnitude being determined by the spacing of observation points. No general rule can be given, but numbers used in various investigations are of the order of $M = 15$ to 20. A wave of wavenumber 15 in middle latitudes will correspond to a wavelength of about 2000 km, while $M = 20$ corresponds to 1500 km. Such wavelengths can be resolved over continental regions with good accuracy, but this is not the case over oceanic regions such as the Pacific Ocean or in many parts of the Southern Hemisphere. In any case, as we shall see later that the Fourier coefficients of most atmospheric parameters as represented in synoptic, objective analyses tend to become quite small for large values of M.

WAVENUMBER REPRESENTATIONS

It is sometimes convenient to compress the series representation into Fourier series by using the complex domain. In this case we write

$$a = \sum_{m=-\infty}^{\infty} A_m e^{im\lambda}, \qquad (9.8)$$

where A_m is a complex number, and where m is a positive or negative integer (but $m \neq 0$). We note that

$$\frac{1}{2\pi} \int_0^{2\pi} e^{im\lambda} \, d\lambda = \begin{cases} 0, & m \neq 0 \\ 1, & m = 0 \end{cases}. \qquad (9.9)$$

It is thus seen that

$$A_0 = a_z. \qquad (9.10)$$

The following orthogonality relation holds:

$$\frac{1}{2\pi} \int_0^{2\pi} e^{im\lambda} e^{in\lambda} \, d\lambda = \frac{1}{2\pi} \int_0^{2\pi} e^{i(m+n)\lambda} \, d\lambda = \begin{cases} 0, & m+n \neq 0 \\ 1, & m+n = 0 \end{cases}. \qquad (9.11)$$

We find therefore the value of A_{m_*} by multiplying (9.8) by $e^{in\lambda}$ and integrating from 0 to 2π, giving

$$A_{m_*} = \frac{1}{2\pi} \int_0^{2\pi} a e^{im\lambda} \, d\lambda. \qquad (9.12)$$

The parameter a in (9.8) is a real variable. We must thus assure that the right-hand side in (9.8) is real as well. This is accomplished by the convention that A_m is the complex conjugate of A_{-m}. To get agreement between (9.1) and (9.8) it is necessary to define

$$\begin{aligned} A_m &= \frac{1}{2}(B_m - iC_m), & m &> 0, \\ A_{-m} &= \frac{1}{2}(B_m + iC_m), & m &< 0. \end{aligned} \qquad (9.13)$$

At this point it is important to recall the discussion in Chapter 7, especially the generations, conversions, and dissipations defined in (7.29) as well as the amounts of energy in A and K. We note that each of these quantities are integrals over the mass of the atmosphere. One of the parts in the triple integral is an integration with respect to longitude. The integrand contains in our formulation a function of pressure, in some instances a function of latitude and pressure such as u_z or T_z, and not more than two factors that are functions of longitude as well. The integrals are then of the form

$$C = \frac{1}{g} \int_0^{p_o} \int_{-\pi/2}^{\pi/2} \int_0^{2\pi} F(\varphi,p) r(\lambda,\varphi,p) s(\lambda,\varphi,p) \, dS \, dp. \qquad (9.14)$$

In evaluating such an integral in the wavenumber domain we write the integral as follows:

$$C = \frac{2\pi}{g} \int_0^{p_o} \int_{-\pi/2}^{\pi/2} F(\varphi,p) \left\{ \frac{1}{2\pi} \int_0^{2\pi} rs \, d\lambda \right\} a^2 \cos\varphi \, d\varphi \, dp. \qquad (9.15)$$

The inner parenthesis presents therefore the problem of expressing the zonal average of a product in the wavenumber domain – that is,

$$(rs)_z = r_z s_z + (r_{ES}s_E)_z. \qquad (9.16)$$

Let r and s be described by

$$r = \sum_{m=-\infty}^{\infty} R_m e^{im\lambda} \quad m \neq 0; \qquad s = \sum_{n=-\infty}^{\infty} S_n e^{in\lambda}; \; n \neq 0. \qquad (9.17)$$

We have then

$$(r_{ES}s_E)_z = \frac{1}{2\pi} \int_0^{2\pi} \left\{ \sum_{m=-\infty}^{\infty} R_m e^{im\lambda} \right\} \left\{ \sum_{n=-\infty}^{\infty} S_n e^{in\lambda} \right\} d\lambda, m \neq 0, n \neq 0. \qquad (9.18)$$

Multiplying these series and integrating we notice that each term in the first series gives a nonzero contribution with only one term in the second series – that is, index m combined with $(-m)$. The result is therefore

$$(r_{ES}s_E)_z = \sum_{m=-\infty}^{\infty} R_m S_{-m}, \quad m \neq 0, \qquad (9.19)$$

or in real numbers

$$(r_{ES}s_E)_z = \sum_{m=1}^{\infty} (R_m S_{-m} + R_{-m} S_m) = \frac{1}{2} \sum_{m=1}^{\infty} (a_m b_m + c_m d_m), \qquad (9.20)$$

when $R_m = \frac{1}{2}(r_m - iq_m)$ and $S_m = \frac{1}{2}(s_m - it_m)$. The expression

$$(rs)_z = r_z s_z + \frac{1}{2} \sum_{m=1}^{\infty} (r_m s_m + q_m t_m) \qquad (9.21)$$

is known as Parseval's generalized identity.

We should recall that the Fourier coefficients are functions of φ and p. Substituting (9.21) into (9.15), we may write:

$$C = \frac{2\pi}{g} \int_0^{p_0} \int_{-\pi/2}^{\pi/2} F(\varphi,p) \left[r_z s_z + \frac{1}{2} \sum_{m=1}^{\infty} (r_m s_m + q_m t_m) \right] a^2 \cos\varphi \, d\varphi \, dp, \tag{9.22}$$

which in turn can be written in the form

$$C = C_0 + \sum_{n=1}^{\infty} C_n \tag{9.23}$$

where

$$C_0 = \frac{2\pi}{g} \int_0^{p_0} \int_{-\pi/2}^{\pi/2} F(\varphi,p) r_z s_z a^2 \cos\varphi \, d\varphi \, dp, \tag{9.24}$$

$$C_n = \frac{\pi}{g} \int_0^{p_0} \int_{-\pi/2}^{\pi/2} F(\varphi,p) (r_n s_n + q_n t_n) a^2 \cos\varphi \, d\varphi \, dp. \tag{9.25}$$

It is thus understandable from (9.25) that all the generations and conversions in the energy diagram containing A_z, A_E, K_E and K_z can be evaluated as a function of wavenumber m.

The series (9.23) represents a one-dimensional spectrum or histogram of the process C. This representation is a spectral decomposition with respect to wavenumbers in the longitudinal direction, and has been widely used in studies of atmospheric energetics. It is popular because we normally think of atmospheric waves as moving in the west–east direction with the wavelength measured in the same direction. In addition, global maps have become available only recently. Before this time many studies of atmospheric energetics were made with hemispheric or even smaller regions. If the maps were reaching from the North Pole to latitude φ_*, we should replace the limits of the second integration by φ_* and $\pi/2$. It is also for this reason that C, C_0, and C_n normally were divided by the total area, which in our example would be $S = 2\pi^2(1 - \sin\varphi_*)$. The formulas developed in Chapter 7 and other chapters are of course not always correct in these limited area cases because in many instances they were obtained making repeated use of the fact that the global integrals of a divergence or a Jacobian are zero. Whenever areas of less than global dimensions are used, it is necessary to carry all the boundary integrals.

We may give an example. The integral for $C(K_E, K_z)$ in (7.29) is actually obtained without any integrations by part. Expanded it becomes

$$C(K_E, K_z) = -\frac{1}{g} \int_0^{p_0} \int_{\varphi_*}^{\pi/2} \int_0^{2\pi} \frac{u_z}{a \cos^2\varphi} \frac{\partial (u_E v_E)_z \cos^2\varphi}{\partial \varphi} a^2 \cos\varphi \, d\lambda \, d\varphi \, dp$$

$$= -\frac{a}{g}\int_0^{p_0}\int_{\varphi_*}^{\pi/2}\int_0^{2\pi} \frac{u_z}{\cos\varphi}\frac{\partial(u_E v_E)_z \cos^2\varphi}{\partial\varphi} d\lambda\, d\varphi\, dp. \tag{9.26}$$

We can integrate by parts with respect to latitude and obtain

$$C(K_E, K_z) = \frac{a}{g}\int_0^{p_0}\int_0^{2\pi} u_z(\varphi_*)(u_E v_E)_z \cos\varphi_*\, d\lambda\, dp \tag{9.27}$$
$$+ \frac{a}{g}\int_0^{p_0}\int_{\varphi_*}^{\pi/2}\int_0^{2\pi} (u_E v_E)_z \cos^2\varphi \frac{\partial}{\partial\varphi}\left(\frac{u_z}{\cos\varphi}\right) d\lambda\, d\varphi\, dp.$$

In this case one may commit an error if the first integral in (9.27), which is a boundary integral at $\varphi = \varphi_*$, is omitted. This boundary integral is

$$I = \frac{2\pi a}{g}\int_0^{p_0} u_z(\varphi_*)[u_E(\varphi_*) v_E(\varphi_*)]_z \cos\varphi_*\, dp, \tag{9.28}$$

and it could be particularly large if φ_* were about 30° N, where $u_z(\varphi_*)$ and the momentum transport are both large.

We may give an example. The data presented by Obasi (1965) give the value of $C(K_E, K_z)$ in (9.26) and the last integral in (9.27). These two integrals would be equal if the boundary term were zero. Denoting $I_1 = C(K_E, K_z)$ and I_2 for the last integral in (9.27), we find the following values for these integrals if we assume that the boundary was at φ_* as indicated in Table 9.1.

Table 9.1:

	Summer 1958			Winter 1958		
φ_*	I_1	I_2	Bndry.	I_1	I_2	Bndry.
30° S	1.0048	0.4746	0.5302	1.1446	1.7105	-0.5659
25° S	0.8021	0.5302	0.2672	0.8912	0.3000	0.5912
20° S	0.6341	0.5488	0.0853	0.6584	0.4025	0.2559
15° S	0.5223	0.5243	-0.0020	0.5194	0.4494	0.0700
10° S	0.4542	0.4822	-0.0280	0.4497	0.4357	0.0140
5° S	0.4130	0.4389	-0.0259	0.4133	0.3918	0.0215
0° S	0.3802	0.4009	-0.0207	0.3886	0.3687	0.0199

Unit: W m^{-2}

It is seen that a boundary at 30° S or 25° S gives very large differences in the two estimates. The differences become small for $\varphi \leq 15°$ S.

Although Fourier analysis along latitude circles is as far as one can go with less than global regions, the procedure is nevertheless a mixture of spectral representations and finite differences. It is more natural to use orthogonal functions well suited to the spherical shape of the earth. The spherical harmonic functions come to mind in this regard because they form an orthogonal set over the complete sphere. In addition, they are widely used in the so-called spectral models applied in numerical weather prediction. These functions have been applied in the field of atmospheric energetics, but far less than the Fourier analysis along latitude circles.

The series expansion of a function $a(\lambda, \varphi)$ in spherical harmonic functions is

$$a(\lambda, \varphi) = \sum_{m=-\infty}^{\infty} \sum_{n \geq |m|}^{\infty} A_n^m P_n^m(\varphi) e^{im\lambda}, \qquad (9.29)$$

in which $P_n^m(\varphi)$ are the associated Legendre functions, while A_n^m are constants. In most applications it is convenient to introduce the variable $\mu = \sin \varphi$. The associated Legendre functions satisfy the equation

$$\frac{d}{d\mu}\left((1-\mu^2)\frac{dZ}{d\mu}\right) + \left(n(n+1) - \frac{m^2}{1-\mu^2}\right) Z = 0, \qquad (9.30)$$

where m and n are integers. In the special case of $m = 0$, the functions are called Legendre functions, and they are polynomials. The associated Legendre functions are orthogonal over the sphere, and they satisfy the following integral relations:

$$\int_{-1}^{+1} P_n^m(\mu) \cdot P_{n_*}^m(\mu) \, d\mu = \begin{cases} 0, & n \neq n_* \\ \frac{2}{2n+1} \cdot \frac{(n+m)!}{(n-m)!}, & n = n_*. \end{cases} \qquad (9.31)$$

Based upon (9.30), it is possible to introduce normalized associated Legendre functions in the sense that the integral of the square of a function is constant. We shall use a normalized function defined by

$$\overline{P}_n^m(\mu) = \left((2n+1)\frac{(n-m)!}{(n+m)!}\right)^{\frac{1}{2}} P_n^m(\mu). \qquad (9.32)$$

In that case we get

$$\int_{-1}^{+1} \left[\overline{P}_n^m(\mu)\right]^2 d\mu = 2. \qquad (9.33)$$

It is convenient to introduce a notation for the spherical harmonic functions. We write

$$Y_\gamma = Y_n^m = P_n^m e^{im\lambda} \qquad (9.34)$$

in which γ is a shorthand notation for the complex number $\gamma = n + im$. It is clear already from (9.29) that we need a definition for the associated

Legendre function when m is negative. We define

$$P_n^{-m}(\mu) = P_n^m(\mu). \tag{9.35}$$

We have then

$$Y_n^{-m} = P_n^m(\mu)e^{-im\lambda} = Y_{\bar{\gamma}}, \tag{9.36}$$

where $\bar{\gamma}$ is the complex conjugate $\bar{\gamma} = n - im$. The selection of the constant 2 in (9.33) is justified because the following integral over the whole sphere is then unity:

$$\frac{1}{4\pi}\int_{-1}^{+1}\int_0^{2\pi} Y_\gamma Y_{\bar{\gamma}}\, d\lambda\, d\mu = \frac{1}{4\pi}\int_{-1}^{+1}\int_0^{2\pi}\left[\overline{P}_n^m(\mu)\right]^2 d\lambda\, d\mu$$

$$= \frac{1}{2}\int_{-1}^{+1}\left(\overline{P}_n^m\right)^2 d\mu = 1. \tag{9.37}$$

In the following discussion we shall assume that the associated Legendre functions are normalized, and we shall drop the overbar. With these preparations we see from (9.29) that the coefficient $A_\gamma = A_n^m$ can be computed from the integral

$$A_\gamma = \frac{1}{4\pi}\int_{-1}^{+1}\int_0^{2\pi} a(\lambda,\mu)Y_{\bar{\gamma}}\, d\lambda\, d\mu. \tag{9.38}$$

A_γ is of course a complex number. If we use the same convention as in the case of the Fourier analysis, that is

$$A_\gamma = A_n^m = \frac{1}{2}(b_n^m - ic_n^m), \qquad m > 0, \tag{9.39}$$

$$A_{\bar{\gamma}} = A_n^{-m} = \frac{1}{2}(b_n^m + ic_n^m), \qquad m > 0,$$

we find from (9.39) that

$$b_n^m = \frac{1}{2\pi}\int_{-1}^{+1}\int_0^{2\pi} a(\lambda,\mu) P_n^m(\mu)\cos(m\lambda)\, d\lambda\, d\mu,$$

$$c_n^m = \frac{1}{2\pi}\int_{-1}^{+1}\int_0^{2\pi} a(\lambda,\mu) P_n^m(\mu)\sin(n\lambda)\, d\lambda\, d\mu. \tag{9.40}$$

After this brief summary of those properties of the spherical harmonic functions that we shall need immediately, we may continue with a couple of examples. The properties just described have been used to consider the spectral representation of atmospheric energy, mostly the kinetic energy.

WAVENUMBER REPRESENTATIONS

We shall start by considering the approximate form of available potential energy. We recall that

$$A = \frac{1}{gS} \int_0^{p_0} \int_S \frac{1}{2\overline{\sigma}} \alpha'^2 \, dS \, dp. \tag{9.41}$$

The spectral representation applies as shown above to functions of λ and μ, but there is no orthogonal representation of the vertical direction in the present formulation. It is thus necessary in practice to treat the integration with respect to pressure using finite sums. Since α, the specific volume, is inconvenient in practice, we replace it with temperature using the gas equation. Therefore

$$A = \frac{1}{gS} \int_0^{p_0} \int_S \frac{R^2}{2\overline{\sigma}p^2} T'^2 \, dS \, dp. \tag{9.42}$$

Let us define

$$I(p) = \frac{1}{S} \int_S T'^2 \, dS. \tag{9.43}$$

Furthermore, let T' be expanded in spherical harmonic functions:

$$T' = \sum_\gamma T_\gamma Y_\gamma. \tag{9.44}$$

We then have

$$I = \frac{1}{4\pi} \int_{-1}^{+1} \int_0^{2\pi} \left(\sum_\gamma T_\gamma Y_\gamma \right) \left(\sum_\beta T_\beta Y_\beta \right) d\lambda \, d\mu, \tag{9.45}$$

which using (9.37) becomes,

$$I = \sum_\gamma T_\gamma T_{\overline{\gamma}}. \tag{9.46}$$

We finally may write

$$A = \sum_\gamma \left(\frac{1}{g} \int_0^{p_0} \frac{R^2}{2\overline{\sigma}p^2} T_\gamma T_{\overline{\gamma}} \, dp \right), \tag{9.47}$$

where the integral is calculated as a finite sum using the values of T_γ at the various pressure levels.

We may visualize the sum in (9.47) as contributions from all the components, $\gamma = n + im$ and $\overline{\gamma} = n - im$, where we recall that $n \geq |m|$. It is customary to plot a point for each component in a coordinate system with m as the abscissa and n as the ordinate. The sums in (9.47) and the other

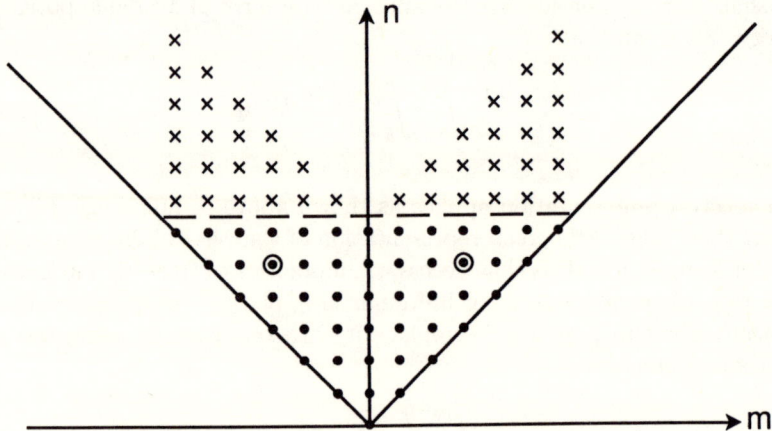

Fig. 9.1: Two different truncation schemes in a diagram with the longitudinal wavenumber as the abscissa and the meridional index as the ordinate. The dots indicate the components included in a triangular truncation, and the two small circles remind us that a total component consists of (m, n) and $(-m, n)$. The crosses are the additional components included in a rhomboidal truncation.

formulas are of course finite, and there have been at least two different ways to make the truncation. They are illustrated in Fig. 9.1.

Each component and its complex conjugate (see the two circles) are denoted by a dot or a cross. One truncation widely used is to set $n = n_{\max} = N$, in which case only the dots are included. This is called the triangular truncation. Another truncation is to include all points where $n - m \leq N$ in which all the components are those indicated by dots and crosses. This is called the rhomboidal truncation. The components with a cross, which are those added to the triangular truncation to obtain the rhomboidal, are all characterized by a large value of n. They therefore have a small meridional scale. If N is sufficiently large, they will not contain much energy and will not add very much to the accuracy.

Each term in (9.36) may be plotted in a diagram such as Fig. 9.1 using only the sector where $m > 0$. From an analysis using isolines, one may visualize the two-dimensional spectral distribution and locate those scales (if any) that give the major contributions. One may also construct two one-dimensional spectra by either summing all the contributions for constant n, and thereby getting a histogram as a function of n, or taking vertical sums in Fig. 9.1 for constant m and plotting the spectral distribution as a function of this variable. The former procedure is generally preferred because the Laplacian of $b(\lambda, \varphi)$ in (9.29) is

$$\nabla^2 b = \sum_{\gamma} B_\gamma \nabla^2 Y_\gamma. \qquad (9.48)$$

According to (9.30) the result is

$$\nabla^2 Y_\gamma = -\frac{n(n+1)}{a^2} Y_\gamma = -\frac{c_\gamma}{a^2} Y_\gamma. \qquad (9.49)$$

The quantity $n(n+1)/a^2$ is thus analogous to the square of the one-dimensional wavenumber. It is thus in this sense natural to call $[n(n+1)]^{\frac{1}{2}}/a$ the two-dimensional wavenumber.

As our next illustration we shall consider the kinetic energy of the non-divergent flow, and we shall limit ourselves to the contribution from one level with the understanding that such contributions are added in accordance with the integration with respect to pressure. Therefore

$$K_2 = \frac{1}{S}\int_S \frac{1}{2}\nabla\psi \cdot \nabla\psi \, dS. \qquad (9.50)$$

Integration by parts gives

$$K_2 = \frac{1}{8\pi}\int_{-1}^{+1}\int_0^{2\pi} \left(-\psi\nabla^2\psi\right) \, d\lambda \, d\mu. \qquad (9.51)$$

Letting

$$\psi = \sum_\gamma \psi_\gamma Y_\gamma, \qquad (9.52)$$

we have

$$\nabla^2\psi = \sum_\gamma -\frac{c_\gamma}{a^2}\psi_\gamma Y_\gamma. \qquad (9.53)$$

Inserting in (9.51) we obtain

$$K_2 = \frac{1}{2}\frac{1}{4\pi}\int_{-1}^{+1}\int_0^{2\pi} \left(\sum_\gamma \psi_\gamma Y_\gamma\right)\left(\sum_\beta \frac{c_\beta}{a^2}\psi_\beta Y_\beta\right) \, d\lambda \, d\mu, \qquad (9.54)$$

$$K_2 = \sum_\gamma \left(\frac{1}{2}\frac{c_\gamma}{a^2}\psi_\gamma\psi_{\overline{\gamma}}\right). \qquad (9.55)$$

An integral quantity, called enstrophy, was introduced in dynamic meteorology some years ago in connection with studies of two-dimensional, large-scale turbulence. Enstrophy is defined as the area integral of one half of the square of the vorticity. For a given isobaric level the vorticity is defined as

$$\zeta = \nabla^2\psi = \sum_\gamma \left(-\frac{c_\gamma}{a^2}\psi_\gamma Y_\gamma\right), \qquad (9.56)$$

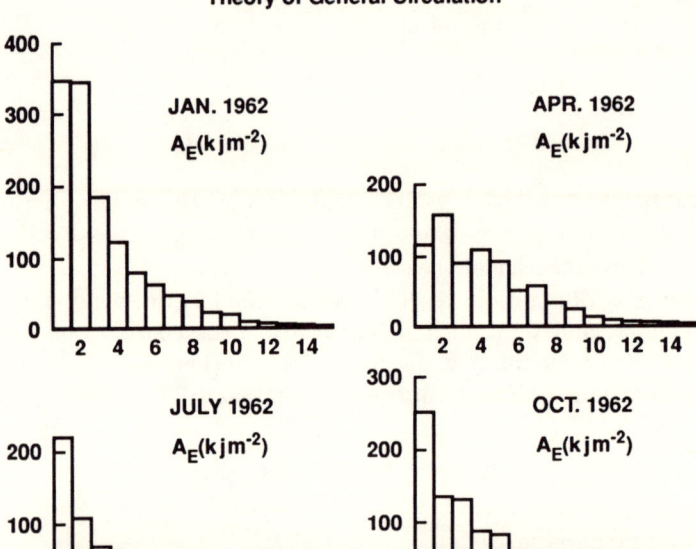

Fig. 9.2: The spectral distribution of the eddy available potential energy for January, April, July, and October 1962.

and therefore, following the same procedure as in (9.54), we find

$$E_n = \sum_\gamma \left(\frac{1}{2} \frac{c_\gamma^2}{a^2} \psi_\gamma \psi_{\overline{\gamma}} \right), \qquad (9.57)$$

$$E_n(\gamma) = c_\gamma K_n(\gamma). \qquad (9.58)$$

The techniques described in this chapter have been used extensively in both diagnostic studies and model simulations. We conclude the chapter by showing some examples. Fig. 9.2 shows an example of the spectral distribution of the available potential energy for January, April, July, and October 1962. The energy is computed for the total wave disregarding a separation into standing and transient components. We observe that the largest contributions originate from the longest waves. The same observation can be made from Fig. 9.3, which displays the spectra of the kinetic energy for the same four months. While the spectra in Figs. 9.2 and 9.3 may be rather typical, it should be pointed out that on occasion there are large deviations. As an example we take January 1963, where the spectra were as shown in Fig. 9.4. In this month wavenumbers 2 and 3 made very large contributions. January 1963 was a very unusual month with blocking

Theory of General Circulation

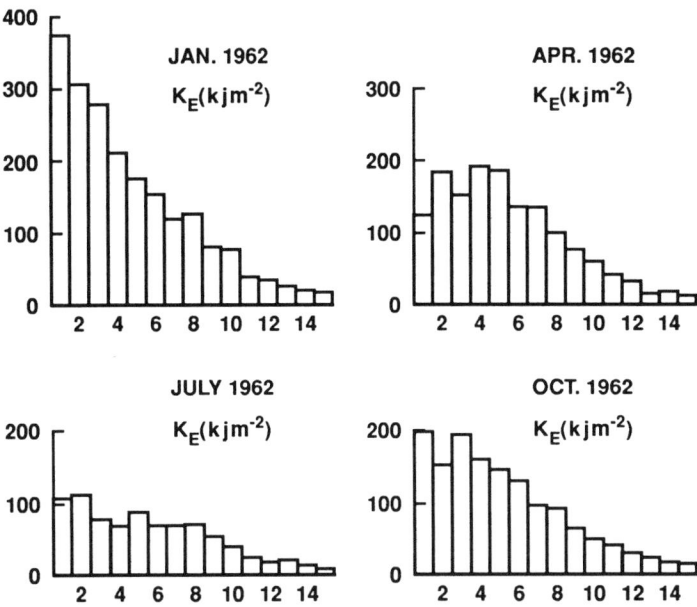

Fig. 9.3: The spectral distribution of the eddy kinetic energy for January, April, July, and October 1962.

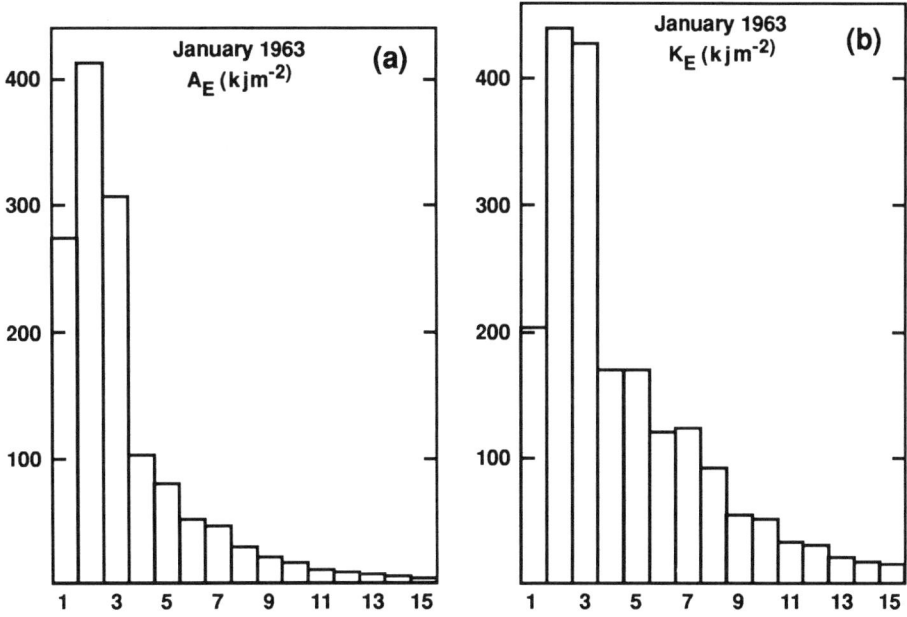

Fig. 9.4: The spectral distribution of the eddy available potential and the eddy kinetic energies for January 1963.

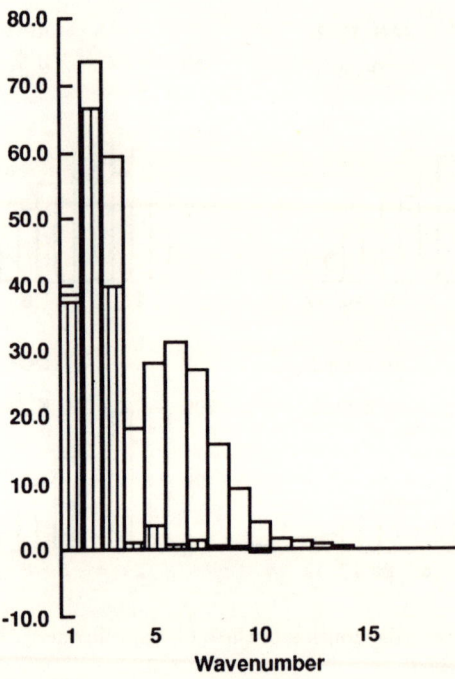

Fig. 9.5: The energy conversion from the zonal available to the eddy available potential energy as a function of the longitudinal wavenumber. Data from December (1976) and January and February (1977). The contribution from the standing eddies is shaded. Unit: 10^{-2} W m^{-2}.

patterns in both the Pacific Ocean and the Atlantic Ocean.

The contribution from an individual wavenumber to the spectrum can be computed in any energy integral where the integrand is a product of two factors. It is thus seen that all quantities in quasi-geostrophic energetics can in principle be represented as spectra. As examples we show some selected quantities for the winter season 1976–77 for the Northern Hemisphere. Fig. 9.5 shows the the conversion $C(A_z, A_E)$ as a function of wavenumber. In this investigation the author (Chen, 1982) has separated the total wave into the standing wave and the transient wave. The contribution from the standing waves is shaded in this and the following figures. Fig. 9.5 shows that the standing waves contribute their largest amounts to the conversion from the longest waves (that is, wavenumbers 1, 2, and 3) with very small contributions in the other waves. The standing waves are apparently forced by heat sources and sinks and by the influence of the mountains on the atmospheric flow. Fig. 9.5 also shows that the transient waves, taken by themselves, give a maximum contribution around wavenumber 6, which in mid-latitudes corresponds to a wavelength of about 6000 km. The contribution by the transient waves can then be ascribed to the baroclinic waves created by the instability mechanism.

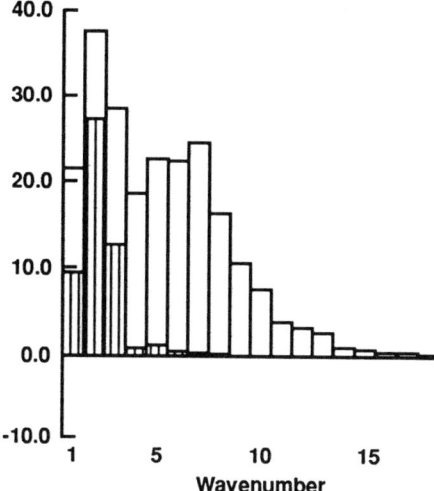

Fig. 9.6: The energy conversion from eddy available potential energy to eddy kinetic energy as a function of longitudinal wavenumber. Data and units as in Fig. 9.5.

Fig. 9.6 shows the energy conversion from available potential energy to kinetic energy as a function of wavenumber from the same study. This figure is consistent with the previous figure with respect to the contributions from the standing and transient waves, and may be taken as a verification of baroclinic instability theory. Fig. 9.7 shows finally the spectrum of the energy conversion from the eddies to the zonal flow. The contributions from the standing waves are again to be found at the largest wavenumbers. All the shorter waves contribute to the conversion from the eddies to the zonal flow.

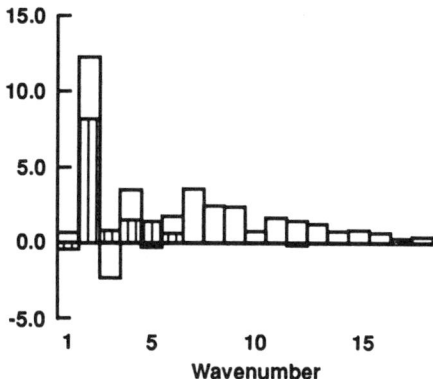

Fig. 9.7: The energy conversion from the eddy to the zonal kinetic energy as a function of wavenumber. Data and units as in Fig. 9.5.

10

INTERACTION AMONG WAVES

In addition to the energy exchange between the waves and the zonal average, it is also possible to consider exchange of energy among waves. We recall from Chapter 9 that it was possible to calculate the energy exchange between the zonal average and the eddies as a function of wavenumber, or in other words, we can compute the conversion between an individual wavenumber $m \neq 0$ and wavenumber 0, which is the zonal average.

In this chapter we shall be interested in the interaction between waves, which is a process originating in the nonlinear advection terms. One of the early contributions to this subject is given by Fjørtoft (1953), who considered the changes in the spectral distribution of kinetic energy in nondivergent flow. The constraints on such a flow are that the total kinetic energy and the enstrophy are conserved. As shown by Platzman (1960) these constraints are true also for systems with a finite number of components. A low-order system of special interest is the three component system. We denote the components by subscripts 1, 2, and 3, where 1 is the large, 2 the middle, and 3 the small scale. If K_1, K_2, and K_3 are the kinetic energies and E_1, E_2, and E_3 the corresponding enstrophies, we have

$$K_1 + K_2 + K_3 = K_0, \qquad (10.1)$$
$$E_1 + E_2 + E_3 = E_0.$$

We know from (9.55) and (9.57) that

$$E_i = \frac{c_i}{a^2} K_i. \qquad (10.2)$$

The system (10.1) then becomes

$$K_1 + K_2 + K_3 = K_0, \qquad (10.3)$$
$$c_1 K_1 + c_2 K_2 + c_3 K_3 = a^2 E_0.$$

Equation (10.1) has been used to create a mechanical equivalent, first proposed by Charney in a lecture. The equivalence to K_1, K_2, and K_3 will be three mass points. The first equation in (10.3) is then an expression for conservation of mass. Setting $a = 1$, the second equation then can be

INTERACTION AMONG WAVES

considered as a balance (see Fig. 10.1) where E_0 has the distance 1 and K_1, K_2, and K_3 the distances c_1, c_2, and c_3, respectively. Suppose then that we change K_2 by ΔK_2. There must then be changes in K_1 and K_3 such that the equations in (10.3) continue to be fulfilled. The changes can of course be calculated from (10.3). They are

$$\Delta K_1 = -\Delta K_2 \cdot \frac{c_3 - c_2}{c_3 - c_1}, \qquad (10.4)$$

$$\Delta K_3 = -\Delta K_2 \cdot \frac{c_2 - c_1}{c_3 - c_1}.$$

It is thus seen that ΔK_1 and ΔK_3 have the same sign. It has furthermore been reasoned that the change in K_1 must be numerically larger than the change in K_3. In other words, a given change on the middle component should result in changes of the opposite sign on the other components in such a way that the larger of the two changes will take place on the larger scale. Unfortunately, this agreement does not hold completely. It is of course true that ΔK_1 and ΔK_3 have the same sign, which will be positive if ΔK_2 is negative, but it is not always true that $|\Delta K_1| > |\Delta K_3|$. Consider the following example. Let the three components be (2,2), (3,8), and (5,9). This triplet satisfies all the selection rules of Platzman (1960) because

$$m_3 = m_1 + m_2,$$
$$n_2 - n_1 < n_3 < n_2 + n_1,$$
$$n_1 + n_2 + n_3 \text{ is odd.}$$

However, for the ratio $\Delta K_1 / \Delta K_3$ we get

$$\frac{\Delta K_1}{\Delta K_3} = \frac{c_3 - c_2}{c_2 - c_1} = \frac{18}{66},$$

and thus

$$\Delta K_1 < \Delta K_3.$$

We next shall consider the interaction among waves in general. It is possible to formulate all the problems concerning the exchange of available potential energy and kinetic energy among waves by using Fourier analysis along latitude circles. Such a formulation has indeed been carried through by Yang (1967), and we shall later consider his results. The derivation of the formulas to calculate these energy exchanges is, however, very laborious and complicated. We shall be satisfied to consider the problems using spherical harmonic functions, and we start by considering the contribution from the advection of vorticity in nondivergent flow. We recall that the kinetic energy of such a flow is

$$K_2 = \frac{1}{4\pi a^2} \int_S \frac{1}{2} \left(-\psi \nabla^2 \psi \right) \, dS. \qquad (10.5)$$

The series expansion is

$$\psi = \sum_\gamma \psi_\gamma Y_\gamma, \tag{10.6}$$

where the meaning of the symbols is given in Chapter 9.

$$K_2 = \sum_\gamma \frac{1}{2}\frac{c_\gamma}{a^2}\psi_\gamma \psi_{\overline{\gamma}}, \tag{10.7}$$

from which it follows that

$$\frac{dK_2}{dt} = \sum_\gamma \frac{1}{2}\frac{c_\gamma}{a^2}\left(\psi_{\overline{\gamma}}\frac{d\psi_\gamma}{dt} + \psi_\gamma \frac{d\psi_{\overline{\gamma}}}{dt}\right). \tag{10.8}$$

From (10.8) it is then seen that the problem is to obtain an equation for $d\psi_\gamma/dt$. To illustrate the procedure we consider the equation

$$\frac{\partial \nabla^2 \psi}{\partial t} = -\vec{v}\cdot\nabla(\nabla^2\psi) \tag{10.9}$$

$$= \frac{1}{a^2}\left(\frac{\partial \nabla^2\psi}{\partial \lambda}\frac{1}{\cos\varphi}\frac{\partial \psi}{\partial \varphi} - \frac{\partial \psi}{\partial \lambda}\frac{1}{\cos\varphi}\frac{\partial \nabla^2 \psi}{\partial \varphi}\right).$$

Using $\mu = \sin\varphi$ as before, we find that

$$\frac{\partial \nabla^2\psi}{\partial t} = \frac{1}{a^2}\left(\frac{\partial \nabla^2\psi}{\partial \lambda}\frac{\partial \psi}{\partial \mu} - \frac{\partial \psi}{\partial \lambda}\frac{\partial \nabla^2\psi}{\partial \mu}\right). \tag{10.10}$$

We recall that

$$\nabla^2 \psi = \sum_\gamma -\frac{c_\gamma}{a^2}\psi_\gamma Y_\gamma. \tag{10.11}$$

Equation (10.10) is then transformed by using (10.6) and (10.11). For this purpose we introduce the counters β and α to distinguish between the various summations. We get

$$\sum_\gamma -\frac{c_\gamma}{a^2}\frac{d\psi_\gamma}{dt}Y_\gamma \tag{10.12}$$

$$= \frac{1}{a^2}\left[\left(\sum_\beta -\frac{c_\beta}{a^2}im_\beta \psi_\beta Y_\beta\right)\left(\sum_\alpha \psi_\alpha \frac{dY_\alpha}{d\mu}\right)\right.$$

$$\left. - \left(\sum_\beta -\frac{c_\beta}{a^2}\psi_\beta \frac{dY_\beta}{d\mu}\right)\left(\sum_\alpha im_\alpha \psi_\alpha Y_\alpha\right)\right].$$

INTERACTION AMONG WAVES

To obtain nonredundant sums we can add the contribution from (α, β) since (10.13) gives the contribution from (β, α). We then get

$$\sum_\gamma -\frac{c_\gamma}{a^2}\frac{d\psi_\gamma}{dt}\psi_\gamma Y_\gamma \tag{10.13}$$

$$= \frac{1}{a^2}\left[\sum_\beta \sum_\alpha \frac{c_\beta}{a^2} i m_\alpha \psi_\beta \psi_\alpha Y_\alpha \frac{dY_\beta}{d\mu} - \frac{c_\beta}{a^2} i m_\beta \psi_\beta \psi_\alpha Y_\beta \frac{dY_\alpha}{d\mu}\right]$$

$$+ \frac{1}{a^2}\left[\sum_\beta \sum_\alpha \frac{c_\alpha}{a^2} i m_\beta \psi_\beta \psi_\alpha Y_\beta \frac{dY_\alpha}{d\mu} - \frac{c_\alpha}{a^2} i m_\alpha \psi_\beta \psi_\alpha Y_\alpha \frac{dY_\beta}{d\mu}\right]$$

$$= \frac{1}{a^2}\left[\sum_\beta \sum_\alpha i\frac{c_\beta - c_\alpha}{a^2}\psi_\beta \psi_\alpha \left(m_\alpha Y_\alpha \frac{dY_\beta}{d\mu} - m_\beta Y_\beta \frac{dY_\alpha}{d\mu}\right)\right].$$

Equation (10.13) is multiplied by $Y_{\bar{\gamma}}$ and integrated over the complete sphere. Recalling that

$$\frac{1}{4\pi}\int_{-1}^{+1}\int_0^{2\pi} Y_\gamma Y_{\bar{\gamma}}\,d\lambda\,d\mu = 1, \tag{10.14}$$

we find

$$\frac{d\psi_\gamma}{dt} = \frac{i}{a^2}\sum_\beta \sum_\alpha \frac{c_\beta - c_\alpha}{c_\gamma}\psi_\beta \psi_\alpha K(\gamma, \beta, \alpha), \tag{10.15}$$

where

$$K(\gamma, \beta, \alpha) \tag{10.16}$$

$$= \frac{1}{4\pi}\int_{-1}^{+1}\int_0^{2\pi} Y_{\bar{\gamma}}\left(m_\beta Y_\beta \frac{dY_\alpha}{d\mu} - m_\alpha Y_\alpha \frac{dY_\beta}{d\mu}\right) d\lambda\,d\mu$$

$$= \frac{1}{4\pi}\int_{-1}^{+1}\int_0^{2\pi} P_\gamma\left(m_\beta P_\beta \frac{dP_\alpha}{d\mu} - m_\alpha P_\alpha \frac{dP_\beta}{d\mu}\right) e^{i(m_\alpha + m_\beta - m_\gamma)\lambda}\,d\lambda\,d\mu.$$

We have then that $K(\gamma, \beta, \alpha) = 0$ unless $m_\gamma = m_\alpha + m_\beta$, in which case

$$K(\gamma, \beta, \alpha) = \frac{1}{2}\int_{-1}^{+1} P_\gamma\left(m_\beta P_\beta \frac{dP_\alpha}{d\mu} - m_\alpha P_\alpha \frac{dP_\beta}{d\mu}\right) d\mu. \tag{10.17}$$

We seek now an equation for $d\psi_{\bar{\gamma}}/dt$. This equation is obtained by taking the complex conjugate of both sides. According to the established convention we have in general

$$\bar{\psi}_\gamma = \psi_{\bar{\gamma}}. \tag{10.18}$$

The result is

$$\frac{d\psi_{\bar{\gamma}}}{dt} = -\frac{i}{a^2}\sum_{\beta}\sum_{\alpha}\frac{c_\beta - c_\alpha}{c_\gamma}\psi_{\bar{\beta}}\psi_{\bar{\alpha}}K(\gamma,\beta,\alpha). \qquad (10.19)$$

Referring to (10.8) we find

$$\frac{dK_\gamma}{dt} = \sum_{\beta}\sum_{\alpha}\frac{1}{2}i\frac{c_\beta - c_\alpha}{a^4}K(\gamma,\beta,\alpha)\left(\psi_\beta\psi_\alpha\psi_{\bar{\gamma}} - \psi_{\bar{\beta}}\psi_{\bar{\alpha}}\psi_\gamma\right). \qquad (10.20)$$

The expression in (10.20) can be considered as the sum of contributions from all pairs (β,α) that give a nonzero value of $K(\gamma;\beta,\alpha)$. We may denote it as the energy conversion between the component γ and all other pairs (β,α) that have an active interaction with γ. The notation $c(\gamma;\beta,\alpha)$ is used and defined as follows:

$$c_K(\gamma;\beta,\alpha) = \sum_{\beta}\sum_{\alpha}\frac{1}{2}i\frac{c_\beta - c_\alpha}{a^4}K(\gamma,\beta,\alpha)\left(\psi_\beta\psi_\alpha\psi_{\bar{\gamma}} - \psi_{\bar{\beta}}\psi_{\bar{\alpha}}\psi_\gamma\right).$$

(10.21)

$c_K(\gamma;\beta,\alpha)$ should naturally be a real number. It is easy to show that this condition is fulfilled due to the fact that the complex conjugate of the general term in (10.21) is equal to the term itself because

$$\overline{i\left[\psi_\beta\psi_\alpha\psi_{\bar{\gamma}} - \psi_{\bar{\beta}}\psi_{\bar{\alpha}}\psi_\gamma\right]} = -i\left[\psi_{\bar{\beta}}\psi_{\bar{\alpha}}\psi_\gamma - \psi_\beta\psi_\alpha\psi_{\bar{\gamma}}\right]$$
$$= i\left[\psi_\beta\psi_\alpha\psi_{\bar{\gamma}} - \psi_{\bar{\beta}}\psi_{\bar{\alpha}}\psi_\gamma\right]. \qquad (10.22)$$

Equation (10.21) can be used to calculate the kinetic energy exchange among the components included in the expansion (10.6). The calculation is cumbersome because it is necessary in this formulation to have calculated all the interaction coefficients $K(\gamma;\beta,\alpha)$. It is probably for this reason that formulas such as (10.21) have not been used in extensive observational studies of atmospheric energetics and that these calculations have instead been restricted to the formulation in Fourier analysis along latitude circles that may be applied to less than global regions.

At this point it should be stressed that the development just given can be used to calculate the wave–wave interaction in quasi-nondivergent models because the formalism applies to each level. We may write

$$K_2 = \frac{1}{g}\int_0^{p_0}\left[\frac{1}{S}\int_S \frac{1}{2}(-\psi)\nabla^2\psi\,ds\right]dp \qquad (10.23)$$

and

$$\frac{dK_2}{dt} = \frac{1}{g}\int_0^{p_0}\left[\sum_{\gamma}\frac{1}{2}\frac{c_\gamma}{a^2}\left(\psi_{\bar{\gamma}}\frac{d\psi_\gamma}{dt} + \psi_\gamma\frac{d\psi_{\bar{\gamma}}}{dt}\right)\right]dp. \qquad (10.24)$$

INTERACTION AMONG WAVES

The formalism developed in the section from (10.10) to (10.22) is applicable to the advection of relative vorticity in the barotropic equation or a quasi-nondivergent model. We have then only to remember that the expression for the nonlinear energy exchange in (10.21) should be evaluated at each pressure level – that is, $c_K(\gamma; \beta, \alpha)$ is a function of pressure, say $c_{K_p}(\gamma; \beta, \alpha)$. The final result is then

$$c_K(\gamma; \beta, \alpha) = \frac{1}{g} \int_0^{p_0} c_{K_p}(\gamma; \beta, \alpha) \, dp. \tag{10.25}$$

The nonlinear advection term in the thermodynamic equation will also give rise to an exhange of available potential energy among the components in the spectrum. Using once again the approximate form of the available potential energy,

$$\begin{aligned}
A &= \frac{1}{gS} \int_0^{p_0} \int_S \frac{1}{2\overline{\sigma}} {\alpha'}^2 \, dS \, dp \\
&= \frac{1}{g} \int_0^{p_0} \frac{R^2}{2\overline{\sigma}p^2} \left\{ \frac{1}{4\pi a^2} \int_S T'^2 \, ds \right\} dp,
\end{aligned} \tag{10.26}$$

we notice that the amount of available potential energy is from (9.47):

$$A = \frac{1}{g} \int_0^{p_0} \frac{R^2}{2\overline{\sigma}p^2} \left(\sum_\gamma T_\gamma T_{\overline{\gamma}} \right) dp \tag{10.27}$$

or

$$\frac{dA}{dt} = \frac{1}{g} \int_0^{p_0} \frac{R^2}{2\overline{\sigma}p^2} \left(\sum_\gamma \left[T_{\overline{\gamma}} \frac{dT_\gamma}{dt} + T_\gamma \frac{dT_{\overline{\gamma}}}{dt} \right] \right) dp. \tag{10.28}$$

As in the previous case it is now a question of finding the equation for dT_γ/dt that must come from the thermodynamic equation, which is

$$\frac{\partial T'}{\partial t} = - \vec{v}_\psi \cdot \nabla T' + \frac{\overline{\sigma}}{R} \omega + \frac{1}{c_p} H'. \tag{10.29}$$

In our case, where we are interested in the nonlinear exchange of available potential energy among the waves, we shall be interested only in the advection term and its contribution. We know that the remaining two terms will result in the conversion integral from available potential energy and the generation integral for available potential energy due to differential heating. To isolate the effect in which we are interested, we write

$$\left(\frac{\partial T'}{\partial t} \right)_{\text{adv.}} = \frac{1}{a^2} \left[\frac{\partial T'}{\partial \lambda} \frac{\partial \psi}{\partial \mu} - \frac{\partial T'}{\partial \mu} \frac{\partial \psi}{\partial \lambda} \right]. \tag{10.30}$$

As before we use the series expansions

$$T' = \sum_\gamma T_\gamma Y_\gamma,$$

$$\psi = \sum_\gamma \psi_\gamma Y_\gamma, \qquad (10.31)$$

and obtain by substitution of (10.31) into (10.30)

$$\sum_\gamma \left(\frac{dT_\gamma}{dt}\right)_{\text{adv.}} Y_\gamma \qquad (10.32)$$

$$= \frac{1}{a^2}\left[\left\{\sum_\beta im_\beta T_\beta Y_\beta\right\}\left\{\sum_\alpha \psi_\alpha \frac{\partial Y_\alpha}{\partial \mu}\right\}\right]$$

$$- \frac{1}{a^2}\left[\left\{\sum_\beta T_\beta \frac{\partial Y_\beta}{\partial \mu}\right\}\left\{\sum_\alpha im_\alpha \psi_\alpha Y_\alpha\right\}\right]$$

$$+ \frac{1}{a^2}\left[\left\{\sum_\alpha im_\alpha T_\alpha Y_\alpha\right\}\left\{\sum_\beta \psi_\beta \frac{\partial Y_\beta}{\partial \mu}\right\}\right]$$

$$- \frac{1}{a^2}\left[\left\{\sum_\alpha T_\alpha \frac{\partial Y_\alpha}{\partial \mu}\right\}\left\{\sum_\beta im_\beta \psi_\beta Y_\beta\right\}\right]$$

where we have already written (10.32) in the nonredundant form. Equation (10.32) is next multiplied by $Y_{\overline\gamma}$ and integrated over the whole sphere. Making use of the orthogonality relation for the functions Y_γ, we get

$$\left(\frac{dT_\gamma}{dt}\right)_{\text{adv.}} = \frac{i}{a^2}\sum_\beta\sum_\alpha (T_\beta \psi_\alpha - T_\alpha \psi_\beta) \cdot K(\gamma; \beta, \alpha) \qquad (10.33)$$

in which $K(\gamma; \beta, \alpha)$ is defined by (10.17). Taking the complex conjugate of (10.33), we get

$$\left(\frac{dT_{\overline\gamma}}{dt}\right)_{\text{adv.}} = \frac{i}{a^2}\sum_\beta\sum_\alpha \left(T_{\overline\beta}\psi_{\overline\alpha} - T_{\overline\alpha}\psi_{\overline\beta}\right) K(\gamma; \beta, \alpha). \qquad (10.34)$$

After these derivations we have the necessary formulas to calculate the brackets in (10.28), which become

$$T_{\overline\gamma}\left(\frac{dT_\gamma}{dt}\right)_{\text{adv.}} + T_\gamma\left(\frac{dT_{\overline\gamma}}{dt}\right)_{\text{adv.}} \qquad (10.35)$$

$$= \frac{i}{a^2}\sum_\beta\sum_\alpha \left[(T_\beta\psi_\alpha - T_\alpha\psi_\beta)\psi_{\overline\gamma} - \left(T_{\overline\beta}\psi_{\overline\alpha} - T_{\overline\alpha}\psi_{\overline\beta}\right)\psi_\gamma\right] K(\gamma; \beta, \alpha).$$

INTERACTION AMONG WAVES 117

The expression on the right-hand side of (10.35) is proportional to the conversion of available potential energy in component γ due to its interaction with all pairs of components (β, α), which have a nonzero interaction coefficient $K(\gamma; \beta, \alpha)$. In analogy with the notation used for the kinetic energy, we may write $C_{Ap}(\gamma; \beta, \alpha)$ for the right-hand side of (10.35). The final result is then obtained from (10.28):

$$C_A(\gamma; \beta, \alpha) = \frac{1}{g} \int_0^{p_0} \frac{R^2}{2\overline{\sigma} p^2} C_{Ap}(\gamma; \beta, \alpha) \, dp. \qquad (10.36)$$

The results from a couple of data studies conclude this chapter. The first study was carried out by Yang (1967) who used a quasi-nondivergent formulation and data from one year, February 1963 – January 1964. However, an earlier study by Saltzman and Teweles (1964) had calculated the kinetic energy exchanges for a nine-year period, but the study was based on only 50 kPa data. Yang's study used daily calculations summerized as monthly averages. We however shall be content with the annual averages.

As mentioned before, Yang used a Fourier decomposition along latitude circles, but finite differences in the meridional direction. In the following figures the total wave spectrum has been compressed to three scales: long (1–5), medium (6–10) and short (11–15) wavenumbers. Such a division cannot be fully justified because all latitudinal contributions are taken together. However, realizing that the polar regions cannot contribute a major part due to their small areas and, furthermore, remembering that latitudes south of 20° N are not included, we may look on the wavenumbers 1–5 as a reasonable representation of the long period waves. Using the same argument, wavenumbers 6–10 should represent the transient, baroclinic waves while wavenumbers 11–15 should be representative of the short stable waves.

Fig. 10.1 shows the results for the annual average of available potential energy. The energy of the zonally averaged field, A_z, is shown to the left while the wave groups are represented by three boxes on the right. The three arrows going from A_z to each of the other boxes indicate the conversion from A_z to the wave groups. They are all in W m^{-2}. The arrows on the right-hand side of the figure indicate the exchange of available potential energy from the other groups into the group in question. We see that the very long waves export available potential energy received by the other two groups. For the available potential energy we may thus say that the energy is transferred from the larger to the smaller scales.

For the kinetic energy the corresponding diagram is shown in Fig. 10.2. As expected, all three wave groups feed energy into the zonally averaged state from the waves. The interaction among the waves is an export from the medium-long waves to the very long and short waves. All numbers are again given in W m^{-2}. Fig. 10.2 is in agreement with the results obtained by Saltzman and Teweles (loc. cit.) with respect to the directions of the energy exchanges.

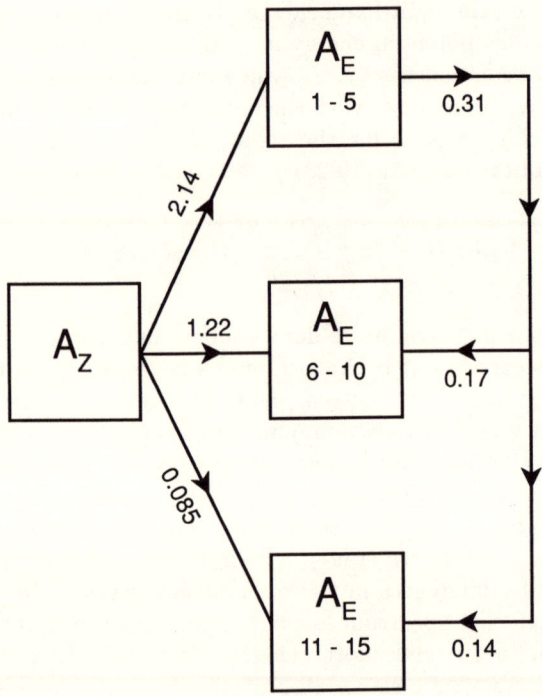

Fig. 10.1: The exchanges of available potential energy on an annual basis between the zonal amount A_z and the three wave groups: $m = 1$ to 5, $m = 6$ to 10 and $m = 11$ to 15, as well as the exchanges among the three wave groups. Note that the cascade goes from small to large wavenumbers.

A summary of the whole study is shown in Fig. 10.3 in which additional information has been included. The annually averaged amounts of energy in the zonally averaged state and the three wave groups are shown by the histograms. Another new piece of information is the arrows indicating the energy conversion from A_E to K_E, where we notice that the major conversion is between the medium long waves. Regarding the input of the kinetic energy, we see that the larger amount comes on the scale of the baroclinic, transient waves, which are amplified by the instability of the zonal currents. The kinetic energy received on this scale is then cascaded toward both larger and smaller scales.

Another study using data from the complete Northern Hemisphere for 10 levels (100, 85, 70, 50, 40, 30, 25, 20, 15, and 10 kPa) during a two month period (December – February 1971–72) has been carried out by Chen and Wiin-Nielsen (1978). The decomposition in orthogonal functions was in this case done using spherical harmonic functions. The proper measure of scale is n, where n is the meridional index of the associated Legendre func-

INTERACTION AMONG WAVES

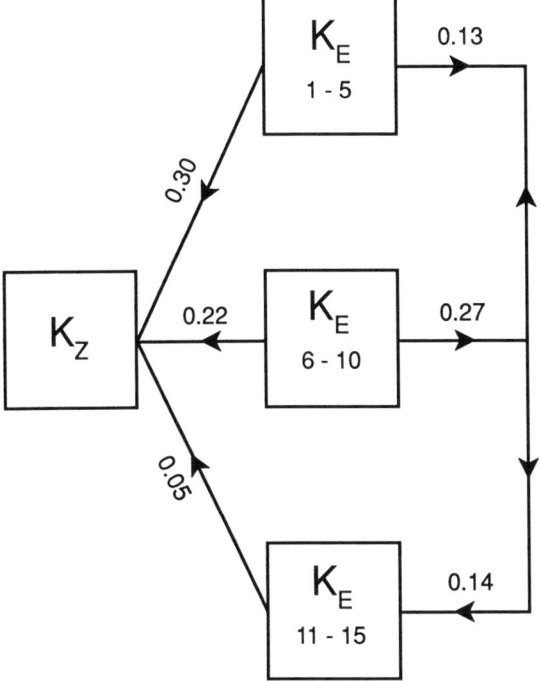

Fig. 10.2: The exchanges of kinetic energy on an annual basis arranged as in Fig. 10.2. Note that kinetic energy is transferred from the medium scale to the large and small scale.

tion. The resolution is from $n = 1$ to $n = 31$, and the three wave groups were in this study defined by $1 \leq n \leq 7$, $8 \leq n \leq 14$, and $15 \leq n \leq 31$.

The results for available potential energy and kinetic energy are shown in Fig. 10.4. There is agreement between the directions of the energy exchanges in the two studies, as can be seen comparing Fig. 10.5 with Figs. 10.2 and 10.3. In Fig. 10.5 the amounts of energy exchange are larger because the second study is for two winter months only. We note also that the differences in the wavenumbers in three wave groups account for the difference in the division of the energy exchanges.

We shall finally look at the nonlinear exchange of enstrophy, which was included in the study by Chen and Wiin-Nielsen. The results using a unit of 10^{-18} s^{-3} are shown in Fig. 10.5. It is seen that the exchange of enstrophy has the same direction as the exchange of kinetic energy. The partitioning is, however, very different because for enstrophy the major transfer is from the middle to the small scale while the transfer for kinetic energy largely goes to the largest scales.

It turns out that the spectra of available potential and kinetic energies

120 FUNDAMENTALS OF ATMOSPHERIC ENERGETICS

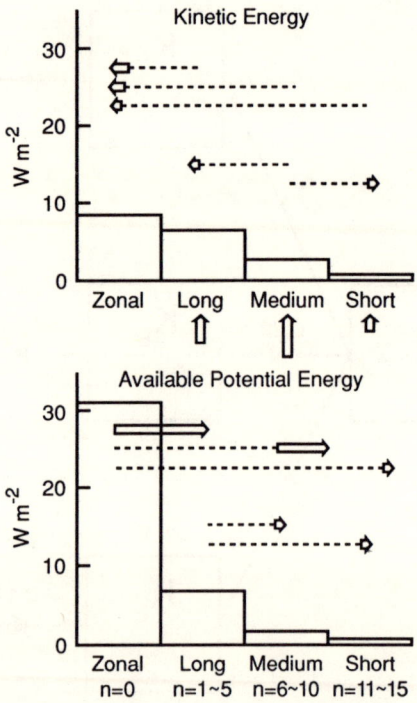

Fig. 10.3: The lower part is a schematic spectrum for the available potential energy, while the upper part is the spectrum for the kinetic energy. The vertical arrows indicate the conversion from eddy available potential energy to eddy kinetic energy, while the horizontal arrows indicate the transfer of energy within each spectrum.

are such that in the region of the high wavenumbers the spectra can be approximated by a power law, where the exponent is close to (-3). The spectrum of the enstrophy can also be approximated by a power law, but here the exponent is (-1). If spherical harmonic functions had been used, in which case the scale would be given by the meridional index $n(n+1)$, it would be obvious that a difference by 2 units would apply; but the studies which we refer to use m as the measure of scale. However, the empirical determination of the slope of the enstrophy spectrum characterized by the exponent (-1) shows that the difference of 2 in the power law holds even when finite differences are used in the meridional direction.

At about the same time as these empirical determinations of the power laws were made, Kraichnan (1967) and Leith (1968) were interested in the properties of so-called two-dimensional turbulence. They demonstrated that two different regimes were possible. One regime would have a constant flux of kinetic energy and a vanishing flux of enstrophy, while the

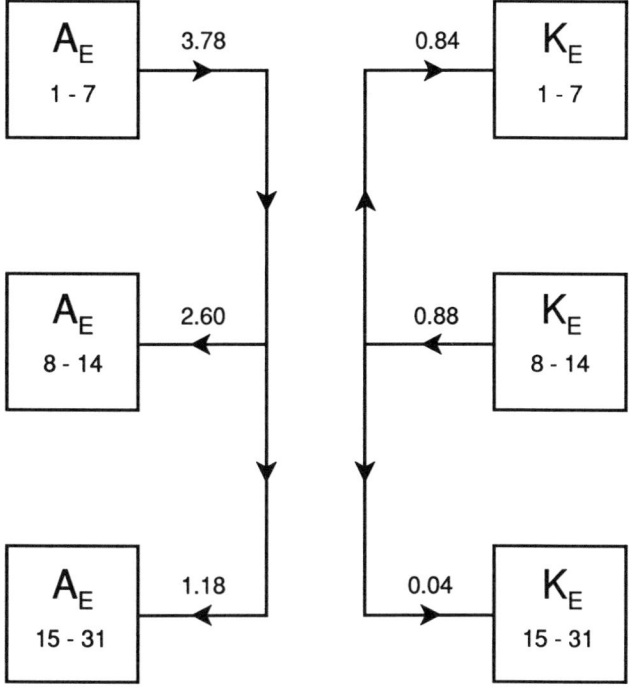

Fig. 10.4: A summary of a study of nonlinear exchanges of available potential and kinetic energy for the Northern Hemisphere during two winter months. The three groups are arranged according to the meridional index. Numbers are given in W m^{-2}.

other would be characterized by a constant flux of enstrophy, but a vanishing flux of kinetic energy. The latter regime would be characterized by a (-3) exponent in the power law for the kinetic energy, but a (-1) exponent for the enstrophy. When compared with the observed spectra, the high wavenumber end of the spectrum has the correct slopes for an inertial subrange of the second kind. The flux functions for both kinetic energy and enstrophy have been calculated to investigate if the first flux function is essentially zero and the other essentially constant. However, it was not possible to verify the existence of such an inertial subrange; see Steinberg, Wiin-Nielsen, and Yang (1971) for details. The lack of verification does not prove that the inertial subrange does not exist because the studies included only wavenumbers up to 15.

Barros and Wiin-Nielsen (1974) have made a numerical experiment with a quasi-geostrophic, two-level model with forcing and dissipation completing a long-term integration. The results were used primarily to investigate the spectra and the nonlinear cascade processes. The model results show also that the eddy kinetic energy is generated by conversion from eddy

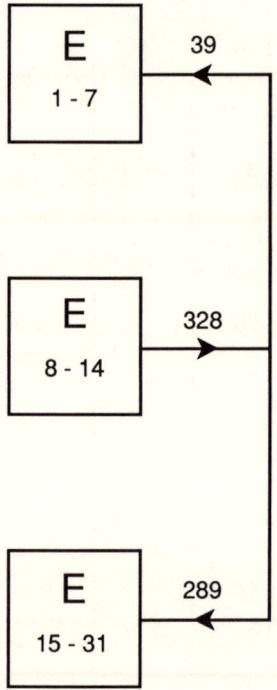

Fig. 10.5: The nonlinear exchange of enstrophy among the same wave groups as in Fig. 10.4. Note that enstrophy is transferred from the middle scale to the large and small scale with the large amount going to the small scale. Unit: 10^{-18} s^{-3}.

available potential energy mainly on wavenumbers 4 and 5. The flux function for enstrophy is almost constant for the interval $5 < m < 12$. The interval in wavenumbers from 6 to 11 may therefore be an inertial subrange; it does not extend to higher wavenumbers. This is explained in the experiment by the fact that the model equations include a lateral diffusion term that dissipates the kinetic energy effectively for wavenumbers higher than 12. This dissipation takes place in such a way that in this region of the spectrum there is a balance between the convergence of the flux and the dissipation. It may be of interest in the future to repeat an experiment of a similar nature, but to avoid the use of lateral diffusion.

11

ENERGETICS AND PREDICTABILITY

The study of cascade processes gives the major result that the kinetic energy enters the atmospheric spectrum on the scale of baroclinic instability. It is then cascaded to longer as well as shorter waves. In the region of the larger wavenumbers we have a cascade of both available potential and kinetic energy and enstrophy to higher and higher wavenumbers or smaller and smaller scales. At sufficiently high wavenumbers the energy is converted to heat.

The shape of the atmospheric spectra for the higher wavenumbers is also important for the nature of the cascade processes. Empirically we find that the available potential and kinetic energy vary as the wavenumber to the (-3) power for the waves on a scale that is smaller than the scale of baroclinic instability but larger than the scale on which the (-5/3) power law applies. The (-3) power law was first found by Wiin-Nielsen (1967) in an analysis of spectra calculated from observations, and the distribution has reappeared in numerous studies of both data and model results. Superficially the (-3) power law is in agreement with the predictions from two-dimensional turbulence theory as formulated by Kraichnan (1967) and Leith (1968), where it corresponds to a region of constant enstrophy cascade and vanishing cascade of energy, but it has not so far been possible to show clearly from simulations or data that such an inertial subrange exists, where the (-3) law is found. This agreement may thus be accidental. The first studies of the atmosphere spectra used the longitudinal wavenumber as a measure of scale, but later studies using spherical harmonic functions have used the so-called two-dimensional index as a scale measure (Chen and Wiin-Nielsen, 1978).

Any numerical prediction model has a smallest scale that it can resolve with satisfactory accuracy. In the following discussion it will be an advantage to think about a model with spectral representation. If the model does not have any frictional processes, which gradually remove the kinetic energy, we will, due to the cascade processes, see an accumulation of energy and enstrophy at the small scales. This accumulation will change the cascade processes, and there may then be a feedback from the accumulated energy to the smaller wavenumbers. If the cut-off scale is close to the significant scales, which should be predicted accurately, it may destroy the forecast and thus limit the predictability. This qualitative reasoning is made plausible from the fact that a numerical prediction with higher resolution in general gives a better forecast, even if the initial analysis remains the same, and the description of the physical processes in the model is unchanged. Generally speaking, a higher resolution gives better forecasts, everything else remaining unchanged.

The fact that atmospheric energy spectra fall off so rapidly, as shown by the (-3) law, indicates, on the other hand, that there should be a limit to how high the resolution can be and still give advantageous results. This limit is in a general sense determined by the scale through which only a very small energy cascade takes place. Knowing the spectra of the atmospheric energies, one may ask if it is possible to estimate such a scale. The remaining part of this chapter is devoted to this question. We shall divide the discussion into two parts. The first part will deal with the most simple model – that is, the barotropic model – while the second part will treat the quasi-geostrophic models without sources and sinks. The omission of sinks and sources allows the use of the conservation theorems, which apply in these adiabatic, frictionless models.

11.1 Nondivergent, Horizontal Flow

The barotropic model conserves total kinetic energy and total enstrophy. With reference to Chapters 9 and 10 we start by developing the streamfunction in a series of associated Legendre functions. We can then express the total kinetic energy and the total enstrophy as infinite series. If n is the meridional index, we may write the kinetic energy in the form

$$K = \sum_{n=1}^{\infty} K(n). \qquad (11.1)$$

It follows then [see Eq. 9.58] that the infinite series for the enstrophy is

$$a^2 E = \sum_{n=1}^{\infty} c(n) K(n), \qquad (11.2)$$

where a is the radius of the earth and $c(n) = n(n+1)$.

For the considerations to follow it will be necessary to define both the kinetic energy and the enstrophy for certain wave groups. For a wave group starting at n' and ending at n'' we define the kinetic energy in that group as

$$K(n', n'') = \sum_{n=n'}^{n''} K(n), \qquad (11.3)$$

while the following equation, which on the right-hand side has the enstrophy for the same wave group, defines the averaged scale for the wave group:

$$cK(n', n'') = \sum_{n=n'}^{n''} c(n)K(n). \tag{11.4}$$

In the following discussion we shall divide the whole spectrum in two parts. The first part will consist of the waves from $n' = 1$ to $n'' = n$, where n so far is an arbitrary number. The second part will contain all other waves in the spectrum. Since it is our intention to study the effect of the cascade processes, we will in general place n in a region where the enstrophy goes from the larger to the smaller scales with certainty – that is, at large wavenumbers. The general plan is then to use the two conservation theorems which hold for the two-dimensional, nondivergent flow to obtain information on the intensity of the cascade through a given wavenumber or, to be exact, through the wavenumber which divides the spectrum in the two parts.

The conservation theorems can be written in the following manner:

$$K_* + K_{**} = K,$$
$$c_* K_* + c_{**} K_{**} = cK. \tag{11.5}$$

The first of these equations expresses the conservation of total kinetic energy, while the second does the same for the enstrophy. K_* and K_{**} are the kinetic energies for the low and high wavenumber parts of the spectrum, respectively, while c_* and c_{**} are the corresponding averaged scales. K is the total kinetic energy in the spectrum, and c is the averaged scale for the total kinetic energy. We notice that the conservation theorems can be stated also as a conservation of scale for the total spectrum. Due to the cascade processes, we cannot expect that the scales for the two parts of the spectrum will do the same. On the contrary, since we select the dividing wavenumber in the region of cascade processes toward smaller scales, we should expect that both of the partial scales will change.

The solution of (11.5) is

$$\frac{K_*}{K} = \frac{c_{**} - c}{c_{**} - c_*} \qquad \frac{K_{**}}{K} = \frac{c - c_*}{c_{**} - c_*}. \tag{11.6}$$

The second ratio measures the relative amount of kinetic energy which is contained in the high wavenumber part of the spectrum. We shall now proceed to show that there is an upper limit to the amount of energy which can reside in this part of the spectrum. The following two inequalities are seen to be correct because the first reduces to the statement that $c_1 < c_*$, while the second is true because $c_{**} > c(n+1)$:

$$\frac{K_{**}}{K} \leq \frac{c - c_1}{c_{**} - c_1} \leq \frac{c - c_1}{c(n+1) - c_1} = r. \tag{11.7}$$

To work with (11.7) it is necessary to use an observed spectrum of the kinetic energy. In the following we use an averaged spectrum for January, obtained as an average for the three years 1983–85. Furthermore we use only the spectrum for the transient waves. The spectrum covers the range from $n = 1$ to $n = 30$. The averaged scale for the whole spectrum is $c = 177.43$, corresponding approximately to $n = 13$. An upper limit to the cascade across $n = 30$, estimated from (11.7), is 0.18. This number indicates that not more than 18% of the original spectrum can be cascaded to wavenumbers higher than 30. If the streamfunction field corresponding to the spectrum is used in a barotropic forecast with a resolution $n = 30$, as much as 18% of the energy could be accumulating at the highest wavenumber in the prediction model.

We may also use (11.7) in a different way. Let r denote the value of the last ratio in (11.7). We can then write

$$c(n+1) = \frac{1}{r}\left[c - (1-r)c_1\right]. \tag{11.8}$$

To take a specific example we may assume that the observations in our judgment do not justify a resolution higher than, say $n = 30$. Due to the cascade processes, the prediction model should have a higher resolution. Equation (11.8) can then be used to calculate $c(n+1)$ in such a way that not more than the ratio r will be cascaded to wavenumbers higher than $c(n+1)$. Suppose that we select $r = 0.01$. We find $c(n+1) = 17,545$, corresponding to a value of $n = 132$. It is encouraging to see that this value of n is of the same order of magnitude as the resolution presently being used in the most advanced and best prediction models. The same argument as above has been applied earlier in a qualitative sense by Wiin-Nielsen (1985) to explain why an increase in resolution without any improved quantity of observations or any change in the various parameterization schemes can give a better forecast.

11.2 The Quasi-geostrophic Case

For the general case of the primitive equations we do not have two conservation theorems which are essential for the arguments in this chapter. However, for the intermediate case of quasi-geostrophic flow with its two theorems, we may hope to employ analogous considerations to obtain some more general results. For the quasi-geostrophic case we have conservation of the potential vorticity expressed as

$$\frac{dQ}{dt} = 0, \tag{11.9}$$

where Q is defined by

$$Q = \eta + \frac{\partial}{\partial p}\left(\frac{f_0^2}{\sigma}\frac{\partial \psi}{\partial p}\right), \qquad \eta = f + \zeta. \tag{11.10}$$

As indicated by (11.9) it has been assumed that the flow is adiabatic and frictionless. In that case we have conservation of the total energy, which is the sum of the available potential and kinetic energy, provided we use the boundary condition at the surface of the earth that the vertical derivative of the streamfunction vanishes. Under the same boundary condition it can be shown (Wiin-Nielsen, 1990) that the enstrophy is conserved in an integral sense. We therefore have two conservation theorems in analogy with the barotropic case.

In the following discussion we shall express each of the two theorems in wavenumber space. For each isobaric surface we use an expansion of the streamfunction in spherical harmonic functions. In addition it will be necessary for our purposes to introduce the vertical dimension in the problem by defining a set of vertical structure functions, which should be orthogonal and normalized over the vertical interval. As we have recalled several times, the static stability in the quasi-geostrophic model is a function of pressure only. There is, however, some degree of freedom, because the theory does not specify the exact nature of the function. Several possibilities are open, but we shall prefer to select the most simple vertical structure function, which appears when it is assumed that the static stability is inversely proportional to the square of the pressure. In that case we must also abstain from letting the top surface of the model be at $p = 0$. The top surface will be at $p = p_T$. In that case the structure functions are solutions to the equation

$$\frac{d}{dp_*}\left(p_*^2 \frac{dE_q}{dp_*}\right) + \Lambda_q^2 E_q = 0 \tag{11.11}$$

with

$$\Lambda_q^2 = \lambda(q)^2 \frac{\sigma_0 p_0^2}{f_0^2}. \tag{11.12}$$

In the above equations we have used a normalized pressure $p_* = p/p_0$. The required solutions to (11.11) can be shown to be (Wiin-Nielsen, 1989)

$$E_q(p^*) = \left(\frac{2}{1 + 4\mu(q)^2 \xi_T}\right)^{1/2} e^{\xi/2}\left[\sin(\mu(q)\xi) - 2\mu(q)\cos(\mu(q)\xi)\right], \tag{11.13}$$

where

$$\xi = -\ln p^*, \qquad \xi_T = -\ln p_T, \qquad \mu(q) = \frac{q\pi}{\xi_T}, \tag{11.14}$$

and where the first factor in (11.13) has been determined in such a way that the structure functions are normalized. We note also that

$$\mu(q)^2 = \Lambda(q)^2 - \frac{1}{4}, \tag{11.15}$$

which can be converted to

$$\lambda^2(q) = \frac{f_0^2}{\sigma_0 p_0^2}\left(\frac{1}{4} + \frac{q^2\pi^2}{\xi_T^2}\right). \tag{11.16}$$

In complete analogy to the nondivergent, horizontal case we get in this case

$$K + A = \frac{p_0}{2g}\sum_q\sum_n\sum_m \left[s(n)^2 + \lambda(q)^2\right]\Psi(m,n,q)^2, \tag{11.17}$$

$$E = \frac{p_0}{2g}\sum_q\sum_n\sum_m \left[s(n)^2 + \lambda(q)^2\right]^2 \Psi(m,n,q)^2. \tag{11.18}$$

Setting

$$T = K + A = \frac{p_0}{2g}\sum_q\sum_n T(n,q), \tag{11.19}$$

we find

$$E = \frac{p_0}{2g}\sum_q^n \alpha(n,q)^2 T(n,q) \tag{11.20}$$

with

$$\alpha(n,q)^2 = s(n)^2 + \lambda(q)^2, \qquad s(n)^2 = \frac{n(n+1)}{a^2}. \tag{11.21}$$

For a given truncation in n and q the numbers $\alpha(n,q)^2$ may be arranged as an increasing series. In the following discussion we shall assume that this has been done, but the numbers are not produced here. We shall use r as the counter in this series. We may then define the mean scale by the equation

$$m^2 \cdot \frac{p_0}{2g}\sum_{r=1}^\infty T(r) = \frac{p_0}{2g}\sum_{r=1}^\infty \alpha(r)^2 T(r). \tag{11.22}$$

We realize of course that in practice all spectra have to contain a finite number of terms. When we nevertheless use an infinite upper limit, it is with the understanding that $T(r)$ should be essentially zero for sufficiently large values of r. We can now, just as in the barotropic case, divide the whole spectrum into two parts. Let R be the end of the first group such

ENERGETICS AND PREDICTABILITY

that the two groups are defined by $r = 1$ to $r = R$ and $r > R$. The mean scales are defined by

$$m_*^2 \sum_{r=1}^{\infty} T(r) = \sum_{r=1}^{R} \alpha(r)^2 T(r),$$

$$m_{**}^2 \sum_{r=R+1}^{\infty} T(r) = \sum_{r=R+1}^{\infty} \alpha(r)^2 T(r). \tag{11.23}$$

We may now proceed in a way completely analogous to the barotropic case. The two conservation theorems are

$$\begin{aligned} S_1^R + S_{R+1}^{\infty} &= S_1^{\infty}, \\ m_*^2 S_1^R + m_{**}^2 S_{R+1}^{\infty} &= m^2 S_1^{\infty}. \end{aligned} \tag{11.24}$$

It is then seen that all the arguments in the barotropic case will hold again, and also that the final result will be formally the same. We find therefore that

$$\frac{S_{R+1}^{\infty}}{S_1^{\infty}} < \frac{m^2 - \alpha(1)^2}{\alpha(R+1)^2 - \alpha(1)^2} = b. \tag{11.25}$$

Using (11.25) it is possible to make some statements about cascade processes in the quasi-geostrophic case following the same reasoning as in the previous case. For example, let us assume that it has been decided that $n = 30$ and $q = 3$ is the highest resolution justified by the data. We can then use (11.25) to calculate $\alpha(R+1)^2$ for $b = 0.05$. We find $\alpha(R+1)^2 = 8.34 \times 10^{-10}$ m^{-2}. Having this value we go to the table mentioned before, where we have arranged the three-dimensional wavenumbers according to size. Naturally we can always find a value in the table closest to the computed value. In that sense the problem has a unique solution. On the other hand, we have an inaccurate estimate, and we therefore have a number of values in the table which could just as well be used. For example, the following values of (n, q) are all very close to the computed value: (112,15), (98,16), (80,17), and (56,18). Even more extreme cases could be given with extremely large values of n but very small values of q, or vice versa. The practical resolution in those places where it is needed – that is, close to the solution – is to select the appropriate value of q to secure a sufficient vertical resolution at the tropopause and then to determine the value of n. Based on these considerations we would probably prefer the first of the values given above.

A data study has been carried out in order to make further statements. These data (Wiin-Nielsen, 1990) had a horizontal resolution of $n = 63$ as the largest value of n, and the vertical resolution permitted use of the first five vertical structure functions. The mean scale for the whole spectrum was 2.715×10^{-11} m^{-2} corresponding very nearly to $n = 15$ and $q = 3$. Using the spectrum a number of estimates were made for the expanded scale in a prediction model. Using $b = 0.02$ this time we found the following series of "close" solutions:

n	q
89	21
109	20
124	19
138	18

With respect to the horizontal resolution we find values comparable to those presently used in the global prediction models with the highest resolution. On the other hand, the estimate of the vertical resolution in terms of structure functions is considerably higher than those used at the moment. The reason for this is most probably that the structure functions used in this study have relatively few zeros in the lower part of the troposphere. To get the large variability in the lower part of the atmosphere, especially the planetary boundary layer, it is necessary to use structure functions with a large value of q.

12

ENERGETICS OF AN OPEN DOMAIN

So far we have discussed only the atmospheric energetics of a closed domain. The spectral energetics study, however, revealed that synoptic-scale disturbances are the most effective atmospheric motions, releasing atmospheric available potential energy (Chen, 1982). Moreover, Palmén and Newton (1969) pointed out that *"only about four to five developing cyclones of typical size and intensity are required to account for the entire kinetic energy generated in the extratropical cap of the hemisphere."* Therefore, in view of the importance of extratropical cyclone energetics, it is desirable to formulate the energetics scheme in an open domain.

In dealing with the atmospheric energetics of an open system, we face two unavoidable difficulties: (1) selection of the system's lateral boundary and (2) more than one possible expression of certain energy conversion. The first difficulty is related to the evolution of the energy transport process. However, it may be cumbersome to develop a mathematical rule for the definition of the lateral boundary of an open system. Practically speaking, a relatively objective selection may be made based upon the synoptic structure of the system in question. On the other hand, the second difficulty should be resolved physically. The necessary condition when discussing the existence of a conversion term between two energy forms is the appearance of an energy conversion term bearing the opposite sign in the energy equations of the two forms. In fact, an energy conversion process in an open system can be expressed using several mathematical expressions. However, a proper conversion term between two energy forms cannot be formed; only a direct physical relationship can be established between them.

12.1   Eulerian Energy Budget Analysis

a. Kinetic Energy Equation and Budget Analysis

The horizontal equation of motion is

$$\frac{\partial \vec{V}}{\partial t} + \vec{V}\cdot\nabla\vec{V} + \omega\frac{\partial \vec{V}}{\partial p} - \hat{k}\times f\,\vec{V} = -\nabla\phi + \vec{F}. \qquad (12.1)$$

Multiplying this equation with $\vec{V}$, one obtains

$$\frac{\partial k}{\partial t} + \vec{V} \cdot \nabla k + \omega \frac{\partial k}{\partial p} = -\vec{V} \cdot \nabla \phi + \vec{V} \cdot \vec{F}. \quad (12.2)$$

This Eulerian kinetic energy equation can be combined with the continuity equation giving the flux form

$$\frac{\partial k}{\partial t} + \nabla \cdot (\vec{V} k) + \frac{\partial}{\partial p}(\omega k) = -\vec{V} \cdot \nabla \phi + \vec{V} \cdot \vec{F} \quad (12.3)$$

where $k = 1/2 \, \vec{V} \cdot \vec{V}$. The vertical integration of (12.3) averaged over an open domain, s, is

$$\frac{\partial K}{\partial t} = \underbrace{\frac{1}{gs} \int_0^{p_0} \int_s -\vec{V} \cdot \nabla \phi \, dp \, ds}_{G(K)} - \underbrace{\frac{1}{gs} \int_0^{p_0} \int_s \nabla \cdot (\vec{V} k) \, dp \, ds}_{B(K)}$$

$$+ \underbrace{\frac{1}{gs} \int_0^{p_0} \int_s \vec{V} \cdot \vec{F} \, dp \, ds}_{D(K)}, \quad (12.4)$$

where $K = [1/(gs)] \int_0^{p_0} \int_s k \, dp \, ds$. The boundary conditions $\omega = 0$ at $p = 0$ and $p = p_0$ are applied to obtain (12.4). $G(K)$ is the generation of kinetic energy due to the cross-contour flow, $B(K)$ the divergence of kinetic energy flux, and $D(K)$ the dissipation of kinetic energy.

Table 12.1: The energy units are 10^5 J m^{-2}, and those of energy conversions are W m^{-2}.

	K	$\partial k/\partial t$	G(K)	B(K)	D(K)
Extratropical cyclones	15	0.5	8.5	-1.7	-6.3
Hemispheric general circulation	12	0	4	0	-4

Table 12.1 shows the climatology of the extratropical-cyclone kinetic energy budget, computed by Kung and Baker (1975) for 780 cyclones occurring in North America over a five year period, and for the general circulation in the Northern Hemisphere, as described by Smith (1980). The energy budget values cover the atmosphere from the surface to 100 mb. Note that the negative values of $B(K)$ and $D(K)$ indicate energy sinks. $G(K)$ may be changed so the time variation of kinetic energy can be related to that of available potential energy in the following manner:

$$G(K) = \underbrace{\frac{1}{gs} \int_0^{p_0} \int_s -\nabla \cdot (\vec{V} \phi) \, dp \, ds}_{B(\phi)} + \underbrace{\frac{1}{gs} \int_0^{p_0} \int_s -\alpha \omega \, dp \, ds}_{C(A, K)} \quad (12.5)$$

or
$$G(K) - C(A, K) = B(\phi). \tag{12.6}$$

b. Available Potential Energy Equation and Budget Analysis

Palmén and Newton (1969) found that developing extratropical cyclones are vital systems for the generation of kinetic energy, maintaining hemispheric general circulation. Therefore, after developing an open domain kinetic energy budget equation, we naturally explored the "contribution" of developing cyclones to the globally or hemispherically available potential energy budget, following the Lorenz energy cycle. Note that the average pressure surface determining the minimum potential energy in (4.2) is a sphere – namely, a closed domain. In essence, the available potential energy defined by (4.9) is used to illustrate the *global* general circulation. Therefore, to evaluate the contribution of an open domain to the globally available potential energy, we face the dilemma of defining the average pressure surface.

Adopting an expression of potential energy given in (4.9), we formulate the contribution of an open system to globally available potential energy as

$$A_{so} = \frac{1}{gs} \int_{s_0} \int_0^{p_0} \frac{p^\kappa - \bar{p}^\kappa}{p^\kappa} T \, dp \, ds. \tag{12.7}$$

Also, let us introduce p_{s_0}:

$$\bar{p}_{s_0} = \frac{1}{s_0} \int_{s_0} p(\lambda, \varphi, \Theta) \, ds. \tag{12.8}$$

Using a "local" average pressure on the potential temperature surface (Θ) passing a given point within an open domain, (12.7) can be written as

$$A_{so} = \underbrace{\frac{c_p}{gs_0} \int_{s_0} \int_0^{p_0} \frac{p^\kappa - \bar{p}_{s_0}^\kappa}{p^\kappa} T \, dp \, ds}_{A_{sol}} + \underbrace{\frac{c_p}{gs_0} \int_{s_0} \int_0^{p_0} \frac{\bar{p}_{s_0}^\kappa - \bar{p}_r^\kappa}{p^\kappa} T \, dp \, ds}_{A_{sog}}. \tag{12.9}$$

A_{sol} is the available potential energy of an open domain, defined with the local reference state without any exchange of mass with the surrounding environment, and A_{sog} is the contribution of an open system to the globally available potential energy attributed to the departure of the local reference state from that of the global atmosphere.

To relate open-domain available potential energy to open-domain kinetic energy, we shall focus on the A_{sol} budget. Suppose the area of an open domain s_0 remains constant, then the rate of change of A_{sol} within the concerned open domain should be

$$\frac{\partial A_{sol}}{\partial t} = \frac{c_p}{gs_0} \int_{s_0} \int_0^{p_0} \left(\frac{\partial T}{\partial t} - \frac{\bar{p}_{s_0}}{p_{00}^\kappa} \frac{\partial \Theta}{\partial t} - \frac{\Theta}{p_{00}^\kappa} \frac{\partial \bar{p}_{s_0}}{\partial t} \right) dp \, ds; \quad p_{00} = 100 \text{ cb}.$$

$$\tag{12.10}$$

We have applied the Poisson equation and assumed that $\partial p_0/\partial t = 0$ in (12.10). Using the thermodynamic equation, we may express all local rates of change of T, Θ, and $\bar{p}_{s_0}$ in (12.10) as

$$\frac{\partial T}{\partial t} = \frac{1}{c}\left(\alpha\omega + \dot{Q} - \nabla\bullet(\vec{V}T) - \frac{\partial}{\partial p}(\omega T)\right),$$

$$\frac{\partial \Theta}{\partial t} = \frac{1}{c_p}\left(\frac{p_{00}^\kappa}{p^\kappa}\dot{Q} - \frac{p_{00}^\kappa}{p^\kappa}\nabla\bullet(\vec{V}T) - \frac{\partial}{\partial p}(\omega\Theta)\right),$$

$$\frac{\partial \bar{p}_{s_0}^\kappa}{\partial t} = \frac{d\bar{p}_{s_0}^\kappa}{dt} - \vec{V}\bullet\nabla\bar{p}_{s_0}^\kappa - \omega\frac{\partial \bar{p}_{s_0}^\kappa}{\partial p}, \qquad (12.11)$$

where α is specific volume, and $\dot{Q}$ is the heating rate per unit mass. Substituting (12.11) and (12.10), the latter equation transforms into

$$\frac{\partial A_{s_{ol}}}{\partial t} = \frac{1}{gs_0}\int_{s_0}\int_0^{p_0}\alpha\omega\,dp\,ds + \frac{1}{gs_0}\int_{s_0}\int_0^{p_0}\left(\frac{p^\kappa - \bar{p}_{s_0}^\kappa}{p^\kappa}\right)\dot{Q}\,dp\,ds$$

$$+ \frac{1}{gs_0}\int_{s_0}\int_0^{p_0}\frac{p^\kappa - \bar{p}_{s_0}^\kappa}{p^\kappa}\nabla\bullet(\vec{V}T)\,dp\,ds \qquad (12.12)$$
$$\qquad\qquad (1)$$

$$+ \frac{c_p}{gs_0}\int_{s_0}\int_0^{p_0}\left(\frac{\bar{p}_{s_0}^\kappa}{p_{00}^\kappa}\frac{\partial}{\partial p}(\omega\Theta) - \frac{\partial}{\partial p}(\omega T)\right)dp\,ds \qquad (12.13)$$
$$\qquad\qquad (2)$$

$$- \frac{c_p}{g}\int_{s_0}\int_0^{p_0}\frac{T}{p^\kappa}\left(\frac{d\bar{p}_{s_0}^\kappa}{dt} - \vec{V}\bullet\nabla\bar{p}_{s_0}^\kappa - \omega\frac{\partial \bar{p}_{s_0}^\kappa}{\partial p}\right)dp\,ds.$$
$$\qquad\qquad (3)\qquad\qquad (4)$$

Applying Poisson's equation and combining terms (2) and (4) of (12.13), we obtain

$$(2) + (4) = \frac{\partial}{\partial p}\left(\frac{p^\kappa - \bar{p}_{s_0}^\kappa}{p^\kappa}\omega T\right);$$

and combining (1) and (3) yields

$$(1) + (3) = \nabla\bullet\left(\frac{p^\kappa - \bar{p}_{s_0}^\kappa}{p^\kappa}\vec{V}T\right).$$

Defining the so-called *efficiency factor* as $N \equiv (p^\kappa - \bar{p}_{s_0}^{-\kappa})/p^\kappa$, (12.13) can be written as

$$\frac{\partial A_{s_ol}}{\partial t} = \frac{1}{gs_0}\int_{s_0}\int_0^{p_0}\alpha\omega\,dp\,ds + \frac{1}{gs_0}\int_{s_0}\int_0^{p_0}N\dot{Q}\,dp\,ds$$
$$\qquad\qquad C(A,K)\qquad\qquad\qquad G(A)$$

$$- \frac{1}{gs_0}\int_{s_0}\int_0^{p_0}\nabla\bullet(c_p N\vec{V}T)\,dp\,ds + \frac{1}{gs_0}\int_{s_0}\int_0^{p_0}\frac{c_p T}{p^\kappa}\frac{d\bar{p}_{s_0}^\kappa}{dt}\,dp$$
$$\qquad\qquad B(A)\qquad\qquad\qquad\qquad DP$$

$$ds. \qquad (12.14)$$

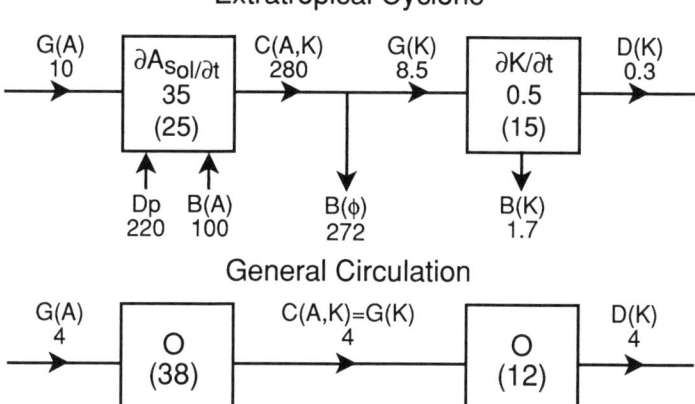

Fig. 12.1: Energy budgets of extratropical cyclones and the hemispheric general circulation. After Smith, 1980. Energy contents are shown by parenthetical numbers with units of 10^5 J m^{-2}. Energy conversions have units of W m^{-2}.

Equation (12.14) is the available potential energy budget equation of an open domain. $G(A)$ is the generation of available potential energy by diabatic heating, $B(A)$ the horizontal flux divergence of available potential energy, and DP the integrated change of the area-mean pressure within an open domain. DP vanishes when integrating over the entire atmosphere (Boer, 1976). The averaged extratropical cyclone available potential energy budget and the corresponding hemispheric values evaluated by Smith (1980) are displayed in Table 12.2. The energy budget analysis of extratropical cyclones presented in Tables 12.1 and 12.2 can be summarized by Fig. 12.1; the parenthetical values are energy contents. The numerical sum of all of the energetics terms shown in Table 12.2 results in a residual of 15 W m^{-2}, which may be the result of computational uncertainty.

Table 12.2: The units of A_{sol} are 10^5 J m^{-2} and that of energy conversion is W m^{-2}.

	A_{sol}	$\partial A_{sol}/\partial t$	$G(A)$	$C(A,K)$	$B(A)$	DP
Extratropical cyclones	25	35	10	-280	100	220
Hemispheric general circulation	38	0	4	-4	0	0

According to the Lorenz energy cycle and the hemispheric energetics analysis, $G(A)$ is the only energy source driving the atmospheric circulation. However, this argument does not seem to hold for the energetics budget of open domains shown in Fig. 12.1. The significant values of $B(A)$ and DP indicate that $G(A)$ is insignificant in developing extratropical cyclones. In fact, the pronounced contrast between $G(A)$ and either $B(A)$ or DP suggests that a substantial amount of available potential energy is exchanged between the open domain accommodating the cyclone and its surrounding environment. Additionally, it is well known that the atmospheric general circulation is a very inefficient thermodynamic engine: the generation of atmospheric circulation energy (4 W m^{-2}) is about 1% of input solar energy ($\sim$ 350 W m^{-2}). For extratropical cyclones, the contrasting numerical values between $C(A, K)$ and $G(K)$ that shown in Fig. 12.1 indicate that only 3% of the released available potential energy is used to generate the kinetic energy of cyclones. Comparing this figure to that of general circulation energetics, one may legitimately claim that the cyclone systems are also rather inefficient thermodynamic systems.

12.2 Quasi-Lagrangian Energy Budget Analysis

The energy budget equations developed in the previous section were formulated in a coordinate system fixed in space – that is, an Eulerian system. Thus, energy budget analyses must be performed in a region sufficiently large to contain the meteorological phenomenon concerned – for example, the extratropical cyclone – during the time period of interest. Because synoptic-scale weather systems generally migrate, one may compute the energy budget over a smaller area containing the storm at any given time in a moving frame of reference – that is, a Lagrangian system.

a. The Kinetic Energy Equation and the Budget Analysis

First, let us deal with the Lagrangian kinetic energy budget of an open domain. The relation between the local rate of change in the Eulerian ($\partial/\partial t$) and Lagrangian ($\delta/\delta t$) systems is described as

$$\frac{\delta(\ )}{\delta t} = \frac{\partial(\ )}{\partial t} + \vec{C} \cdot \nabla(\ ), \qquad (12.15)$$

where $\vec{C} = \partial x/\partial t\ \hat{i} + \partial y/\partial t\ \hat{j} + \partial \omega/\partial t\ \hat{k}$ is the translation velocity of an open system in the Lagrangian system. Applying the continuity equation to (12.15) yields

$$\frac{\delta(\ )}{\delta t} = \frac{\partial(\ )}{\partial t} + \nabla \cdot \left[\vec{C}(\ ) \right]. \qquad (12.16)$$

Also, the time rate of change of kinetic energy K_{s_0} moving with an open system can be written as

$$\frac{\delta k_{s_0}}{\delta t} = \frac{1}{gs_0} \int_{s_0} \int_0^{p_0} \frac{\partial k}{\partial t} \, dp \, ds + \frac{1}{gs_0} \int_s \int_0^{p_0} \nabla \bullet (\vec{C} \, k) \, dp \, ds. \qquad (12.17)$$

Substituting (12.2) into the previous equation, we obtain

$$\begin{aligned}
\frac{\delta K_{s_0}}{\delta t} =& \frac{1}{gs_0} \int_{s_0} \int_0^{p_0} -\vec{V_h} \bullet \nabla \phi \, dp \, ds \\
& + \frac{1}{gs_0} \int_{s_0} \int_0^{p_0} \nabla \bullet [(\vec{C_h} - \vec{V_h})k] \, dp \, ds \\
& \qquad\qquad\qquad B_L(K) \qquad B(K) \\
& + \frac{1}{gs_0} \int_{s_0} \int_0^{p_0} \vec{V_h} \bullet \vec{F_h} \, dp \, ds, \qquad (12.18) \\
& \qquad\qquad\qquad D(K)
\end{aligned}$$

where $(\ )_h$ denotes the horizontal component of $(\ )$ – for example, $\vec{V_h} = u\,\hat{i} + v\,\hat{j}$.

On the right-hand side of (12.18), the first term represents kinetic energy generation due to cross-contour flow, the second term the horizontal divergence of kinetic energy flow across the boundaries of an open, moving system, and the last term the dissipation. The second term can also be split into $B_L(K)$, the divergence of the horizontal system kinetic energy flux, and $B(K)$, the divergence of horizontal Eulearian kinetic energy flux. The comparison between (12.4) and (12.18) shows that the mathematical difference between them is $B_L(k)$, caused by the translation of an open system. Moreover, in the practical budget analysis, both the boundary of s_0 and the translation velocity $\vec{C_h}$ will be subjectively determined using synoptical arguments.

As a result of the subjective constraint of determining the boundary of s_0 and $\vec{C_h}$, few kinetic energy budget analyses of an open system have been performed using the Lagrangian approach. Chen and Bosart (1977) selected four similar cases of well-developed cyclone-anticyclone couplets over North and Central America for this purpose. These winter cases involved cold-air penetration from the North American continent deep into low latitudes behind a major east coast cyclone. A grid spacing of 254 km on a polar stereographic map valid at 60° N was used. Presented in Table 12.3 is the Lagrangian kinetic energy budget of the cyclones integrated from 1000 to 100 mb and averaged over an 11x11 grid area.

For the anticyclone region, $B(K)$ decreases from 37.2 to 5.5 W m^{-2}, while the cross-contour destruction of kinetic energy decreases to 6 W m^{-2}. However, the horizontal system flux increases from 0.6 W m^{-2} during times 3-4 to 15.3 W m^{-2} in the next 12 hours. The opposite situation is observed in the downstream cyclogenetic region: the kinetic energy influx increases

Table 12.3: Unit: W m^{-2}

	Time period	$\delta K_{s_0}/\delta t$	$G(K)$	$B_L(K)$	$B(K)$	$D(K)$
Anticyclone	3-4	−0.8	−9.5	0.6	−37.2	45.4
	4-5	−0.8	−6.0	15.3	−5.5	44.8
Cyclone	3-4	3.7	24.5	−11.8	17.7	-26.9
	4-5	−14.3	64.5	−20.1	32.2	-91.3

from 17.7 to 32.2 W m^{-2} from times 3-4 to times 4-5. The kinetic energy generation also increases from 24.5 to 64.5 W m^{-2} for the same time periods. In contrast, the kinetic energy of the cyclone region is lost by the horizontal system flux 11.8 and 20.1 W m^{-2}, respectively, for the same time periods. Thus, the in-situ generation of kinetic energy by cross-contour flow is apparently the major energy source of developing cyclones, while the kinetic energy exported from the upstream anticyclone region is not a negligible source. The dissipation of kinetic energy, computed as a residual, is 26.9 and 91.3 W m^{-2} in the cyclone region for the two time periods respectively. Peculiarly, kinetic energy is generated by the dissipation term 45.4 and 44.8 W m^{-2}, respectively, during the two time periods over the anticyclone region. In addition to the computational errors, the dissipation computed as a residual may also include other physical processes involving subsynoptic-scale disturbances that cannot be resolved by the operational radiosonde network.

Finally, a comment should be made about comparing the Eulearian (Table 12.1) and Lagrangian (Table 12.3) kinetic energy budget analyses. It is obvious that the intensity of Chen and Bosart's kinetic energy budget is much greater than Kung and Baker's (1975) long-term time averaged result due to the selected cyclones and the size of the computational domain.

b. The Available Potential Energy Equation

The time rate of change of available potential energy A_{s_0} moving with an open system is

$$\frac{\partial A_{s_0}}{\partial t} = \frac{c_p}{g s_0} \frac{\partial}{\partial t} \int_{s_0} \int_0^{p_0} NT \, dp \, ds, \tag{12.19}$$

Following the same procedure as in the formulation of the Lagrangian kinetic energy equation, we obtain

$$\frac{\delta A_{s_0}}{\delta t} = \frac{c_p}{g s_0} \int_{s_0} \int_0^{p_0} \frac{d}{dt}(NT) \, dp \, ds$$

$$+ \frac{c_p}{g s_0} \int_{s_0} \int_0^{p_0} \nabla \cdot \left[NT \left(\vec{C_h} - \vec{V_h} \right) \right] \, dp \, ds. \tag{12.20}$$

ENERGETICS OF AN OPEN DOMAIN

Using the thermodynamic equation, the integrand of the first term on the right-hand side of (12.20) may be written as

$$\frac{d}{dt}(NT) = \frac{N}{c_p}(\dot{Q} + \omega\alpha) - T\left(\frac{\bar{p}_{s0}}{p}\right)^{\kappa-1}\frac{d}{dt}\left(\frac{\bar{p}_{s0}}{p}\right). \tag{12.21}$$

Let us apply the equation of state and $dp/dt = \omega$ to (12.21). We may then express this equation as

$$\frac{d}{dt}(NT) = \frac{1}{c_p}\left[N\dot{Q} + \alpha\omega - \alpha\left(\frac{\bar{p}_{s0}}{p}\right)^{\kappa-1}\frac{dp_{s0}}{dt}\right]. \tag{12.22}$$

Substituting (12.22) into (12.20) yields

$$\frac{\delta A_{s0}}{\delta t} = \underbrace{\frac{1}{g_{s0}}\int_{s0}\int_0^{p_0} N\dot{Q}\, dp\, ds}_{G(A_{s0})} + \underbrace{\frac{1}{g_{s0}}\int_{s0}\omega\alpha\, dp\, ds}_{C(A_{s0}, K_{s0})}$$
$$-\underbrace{\frac{1}{g_{s0}}\int_{s0}\int_0^{p_0} \nabla\bullet\left[NT\left(\vec{C}_h - \vec{V}_h\right)\right]dp\, ds}_{B_L(A_{s0})\quad B(A_{s0})}$$
$$-\frac{1}{g_{s0}}\int_{s0}\int_0^{p_0}\alpha\left(\frac{\bar{p}_{s0}}{p}\right)^{\kappa-1}\frac{dp_{s0}}{dt}dp\, ds. \tag{12.23}$$

This is the Lagrangian available potential energy equation of an open system. $G(A_{s0}), C(A_{s0}, K_{s0})$, and $B(A_{s0})$ have the same physical meaning as before. $B_L(A_{s0})$ represents the available potential energy flux across the boundaries due to the translation of the boundaries. The last term in the right-hand side of (12.23) is the effect of the time change in the reference pressure of the moving system on its available potential energy.

As expressed earlier, $C(A_{s0}, K_{s0})$ can be rewritten as

$$C(A_{s0}, K_{s0}) = \underbrace{-\frac{1}{g_{s0}}\int_{s0}\int_0^{p_0}\vec{V}_h\bullet\nabla\phi\, dp\, ds}_{G(K_{s0})}$$
$$-\underbrace{\frac{1}{g_{s0}}\int_{s0}\int_0^{p_0}\nabla_h\bullet\left(\vec{V}_h\,\phi\right)dp\, ds}_{B(\phi)} \tag{12.24}$$

or

$$C(A_{s0}, K_{s0}) = G(K_{s0}) - B(\phi). \tag{12.25}$$

To illustrate the energy budget of a moving open system, a schematic energy diagram is presented in Fig.12.2. No diagnostic analysis of the Lagrangian available potential energy equation (12.23) has been made.

140 FUNDAMENTALS OF ATMOSPHERIC ENERGETICS

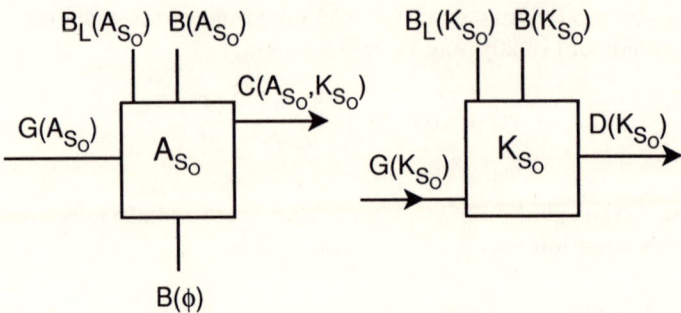

Fig. 12.2: The energy budget diagram of a moving open system; the arrow indicates the direction of the energy conversion.

12.3 The Kinetic Energy Budget of Baroclinic and Barotropic Flow in an Open Domain

The divergence of kinetic energy flux and the generation of kinetic energy by cross-contour flow from (12.4) can be written as

$$-\frac{1}{g}\int_0^{p_0}\int_s \nabla \bullet \left(\vec{V}\, k\right) ds\, dp$$

$$= -\frac{1}{g}\int_0^{p_0}\int_s \nabla \bullet \left(\vec{V}_M\, k\right) ds\, dp$$

$$-\frac{1}{g}\int_0^{p_0}\int_s \nabla \bullet \left(\vec{V}_S\, k\right) ds\, dp \qquad (12.26)$$

$$-\frac{1}{g}\int_0^{p_0}\int_s \vec{V} \bullet \nabla \phi\, ds\, dp$$

$$= -\frac{1}{g}\int_0^{p_0}\int_s \vec{V}_M \bullet \nabla \phi_M\, ds\, dp$$

$$-\frac{1}{g}\int_0^{p_0}\int_s \vec{V}_S \bullet \nabla \phi_S\, ds\, dp. \qquad (12.27)$$

Using (5.14), we can obtain the barotropic kinetic energy equation of an open domain:

$$\frac{\partial K_M}{\partial t} = -\underbrace{\frac{1}{g}\int_0^{p_0}\int_s \nabla \bullet \left(\vec{V}_M\, k\right) ds\, dp}_{B(K_M)} - \underbrace{\frac{1}{g}\int_0^{p_0}\int_s \vec{V}_M \bullet \nabla \phi_M\, ds\, dp}_{G(K_M)}$$

$$-\underbrace{\frac{1}{g}\int_0^{p_0}\int_s \left[\vec{k}\bullet\left(\vec{V}_S \times \vec{V}_M\right) + \left(\vec{V}_S \bullet \vec{V}_M\right)\nabla\bullet\vec{V}_S\right]ds\, dp}_{C(K_S, K_M)}$$

$$+ \frac{1}{g}\int_0^{p_o}\int_s \vec{V}_M \bullet \mathbf{F}_M \, ds \, dp. \tag{12.28}$$
$$D(K_M)$$

Compared to (5.16), two extra terms appear in the barotropic kinetic energy equation of an open domain. These two terms are the divergence of kinetic energy flux by barotropic flow, $B(K_M)$, and the generation of the barotropic kinetic energy, $G(K)$.

Due to our boundary condition, the barotropic atmospheric flow does not have available potential energy to release. As indicated by (5.15) or (5.16), $G(K_M)$ integrates to zero in a closed domain, but not in an open domain. The generation of barotropic kinetic energy can be written as

$$\frac{1}{g}\int_0^{p_o}\int_s \vec{V}_M \bullet \nabla \phi_M \, ds \, dp \;=\; -\frac{1}{g}\int_0^{p_o}\int_s \nabla \bullet \left(\vec{V}_M \, \phi_M\right) ds \, dp.$$
$$B(\phi_M) \tag{12.29}$$

Apparently, the barotropic kinetic energy of an open system can be supplied by the transport of barotropic potential energy through boundaries $B(\phi_M)$. Up to this point, the discussion concerning (12.29) makes clear that the rate of change of barotropic kinetic energy within an open domain can be induced by the external energy reservoir through lateral transport, in addition to the internal conversion and dissipation processes.

The baroclinic kinetic energy equation may be formulated using the following convenient approach:

$$\frac{\partial K_S}{\partial t} = \frac{\partial K}{\partial t} - \frac{\partial K_M}{\partial t}. \tag{12.30}$$

Subtracting (12.28) from (12.4), we find that

$$\frac{\partial K_S}{\partial t} = -\frac{1}{g}\int_0^{p_o}\int_s \nabla \bullet \left(\vec{V}_S \, k\right) ds \, dp - \frac{1}{g}\int_0^{p_o}\int_s \vec{V}_S \bullet \nabla \phi_S \, ds \, dp$$
$$B(K_S) \qquad\qquad G(K_S)$$
$$+ \frac{1}{g}\int_0^{p_o}\int_s \left[\vec{k} \bullet \left(\vec{V}_S \times \vec{V}_M\right)\zeta_S + \left(\vec{V}_S \bullet \vec{V}_M\right)\nabla \bullet \vec{V}_S\right] ds \, dp$$
$$-C(K_S, K_M)$$
$$+ \frac{1}{g}\int_0^{p_o}\int_s \vec{V}_S \bullet \mathbf{F}_S \, ds \, dp. \tag{12.31}$$
$$D(K_S)$$

In order to relate the rate of change of baroclinic kinetic energy to available potential energy, $G(K_S)$ can be written as

$$-\frac{1}{g}\int_0^{p_o}\int_s \vec{V}_S \bullet \nabla \phi_S \, ds \, dp \;=\; -\frac{1}{g}\int_1^{p_o}\int_s \nabla \bullet \left(\vec{V}_S \, \phi_S\right) ds \, dp$$
$$B(\phi)$$

$$+\frac{1}{g}\int_0^{p_0}\underbrace{\int_s w\alpha\,ds\,dp}_{C(A,K_S)}.\qquad(12.32)$$

The generation of baroclinic kinetic energy is a sum of the internal conversion from available potential energy into baroclinic kinetic energy $C(A, K_S)$ and the transport of baroclinic potential energy through boundaries. It was shown in (5.17) that the baroclinic kinetic energy of a closed domain is maintained by the release of available potential energy, the conversion from baroclinic into barotropic kinetic energy, and the dissipation of baroclinic kinetic energy. As with the open domain, the rate of change of baroclinic kinetic energy is also affected by the external energy reservoir through the transport of baroclinic potential and kinetic energy across its lateral boundaries.

Equations (12.28) and (12.31) can be written symbolically as

$$\frac{\partial K_M}{\partial t} = B(K_M) + \underbrace{B(\phi_M)}_{G(K_M)} + C(K_S, K_M) + D(K_M),\qquad(12.33)$$

$$\frac{\partial K_S}{\partial t} = B(K_S) + \underbrace{B(\phi_S) + C(A, K_S)}_{G(K_S)} - C(K_S, K_M) + D(K_S).\qquad(12.34)$$

These two equations can be illustrated by the schematic energy diagram shown in Fig. 12.3. Note that $B(\phi_S)$ and $B(\phi_M)$ have not been evaluated in any practical energetics analyses. Instead, $G(K_S)$ and $G(K_M)$ have been computed. The development and structure of all the weather systems within a limited area may not always be the same. It is therefore extremely difficult to draw any universal conclusion from the energetics analysis of an open domain. Nevertheless, for illuminating the dynamic processes involved in the development of any synoptic-scale weather system, the kinetic energy budget analysis of an open domain is certainly informative. Some effort has been made along this line to explore the kinetic energy budget of baroclinic and barotropic flow over North America.

Petterssen and Smebye (1971) classified two types of cyclones. Type A contains a frontal wave and is initiated as a result of baroclinic instability. Type B is developed by an incipient upper trough over a region where fronts may or may not be present and where baroclinicity in the lower troposphere is relatively weak. Two consecutive cyclones over North America were developed from the cyclone in the vicinity of such an initial baroclinic contrast during March 1973. One evolved between 0000 GMT 13 March and 0000 GMT 16 March, and the other between 0000 GMT 16 March and 0000 GMT 18 March. The first cyclone exhibited Type B characteristics, whereas the second cyclone was Type A. The kinetic energy budget analysis of baroclinic and barotropic flow would therefore be a good dynamic

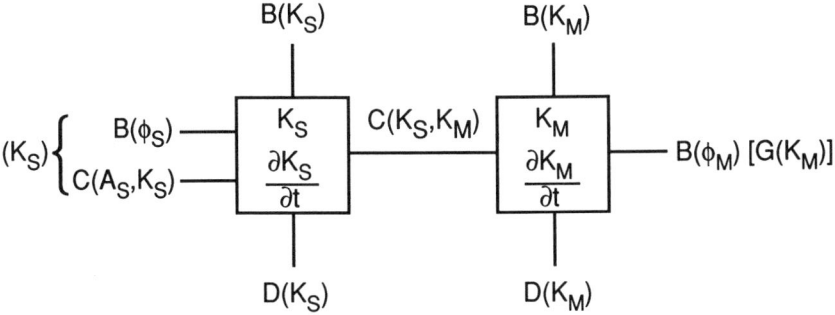

Fig. 12.3: The energy budget diagram of baroclinic (subscript S) and barotropic (subscript M) flows in an open system.

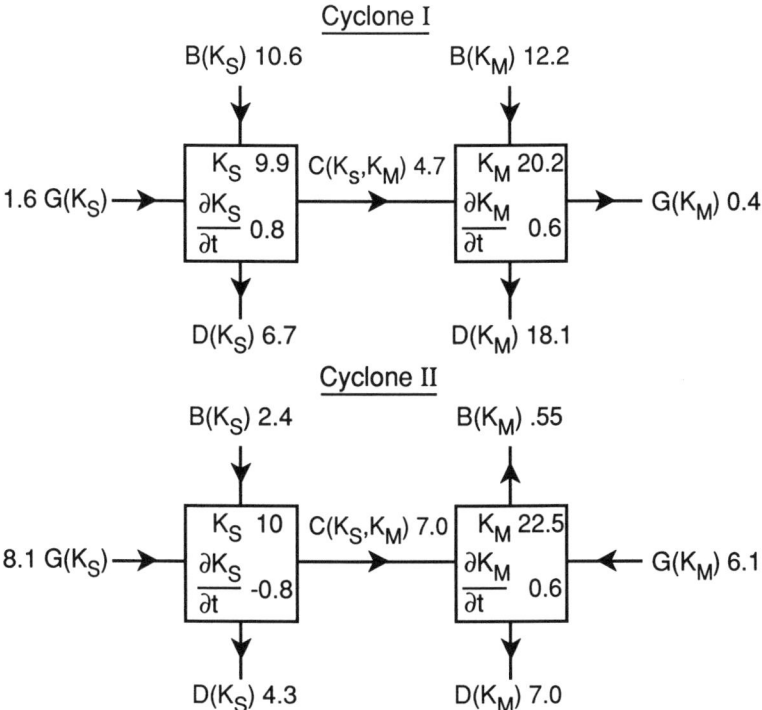

Fig. 12.4: Barotropic and baroclinic kinetic energy budget diagrams for Cyclone I (upper) and Cyclone II (lower) averaged over their entire life cycles (after Alpert, 1981). Arrows indicate the direction of energy conversion. Units: kinetic energy (10^5 J m^{-2}); energy conversion (W m^{-2}).

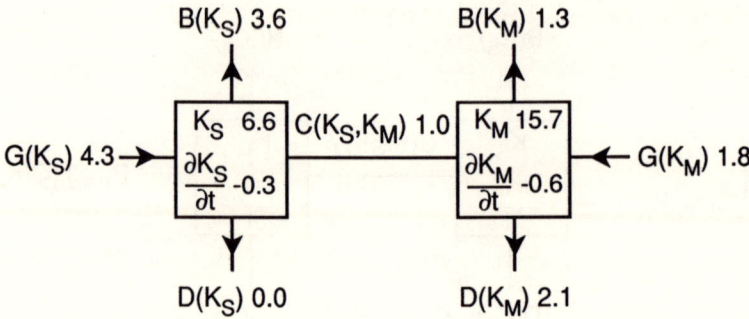

Fig. 12.5: Barotropic and baroclinic kinetic energy budget diagrams averaged for 111 medium- and short-scale waves. Units: kinetic energy (10^5 J m^{-2}); energy conversion (W m^{-2}).

diagnosis to illustrate the difference between the development of these two cyclones.

Such an energetics analysis task was undertaken by Alpert (1981). The results taken over the life cycle of these two cyclones are summarized in Fig. 12.4. The top energy diagram of Fig. 12.4 shows that the horizontal kinetic energy fluxes, $B(K_S)$ and $B(K_M)$, are much larger than the generations of kinetic energy, $G(K_S)$ and $G(K_M)$. It seems that Cyclone I developed from the input of external kinetic energy instead of from internal baroclinic processes. In contrast, the horizontal kinetic energy flux by the barotropic flow of Cyclone II, shown in the lower energy diagram, exports energy out of the cyclone vicinity. Thus, the major energy source developing this cyclone comes from generations of kinetic energy by both baroclinic and barotropic flow. Although the schematic energy diagram of Fig. 12.3 indicates that $G(K_M)$ and part of $G(K_S)$ are attributed to the transport of external baroclinic and barotropic potential energy across boundaries, the energetics of these two cyclones' development shown in Fig. 12.4 are consistent with Petterssen and Smebye's classification.

Sheu and Smith (1981) also applied the same energy analyses to 111 medium- and short-scale waves of various 500–mb flow regimes over North America during the 1967 and 1969 winters and the 1970 spring. Their composite energy diagram is shown in Fig. 12.5. Sheu and Smith's results differ in some ways from those of Alpert's case study or Wiin-Nielsen and Drake's (1965, 1966) hemispheric analysis. The dissipation of baroclinic kinetic energy and $C(K_S, K_M)$ are much smaller in Sheu and Smith's composite results, although their baroclinic energy transfers are on average more vigorous than such transfers usually are. They argued that their results might differ from others because of their exclusion of longer waves.

12.4 The Kinetic Energy Budget Of Rotational and Divergent Flow in an Open Domain

The contributions of rotational and divergent winds to the divergence of the kinetic energy flux and the generation of kinetic energy by cross-contour flow from (12.4) can be written as

$$-\frac{1}{g}\int_0^{p_0}\int_s \nabla\bullet\left(\vec{V}\,k\right)\,ds\,dp = -\frac{1}{g}\int_0^{p_0}\int_s \nabla\bullet\left(\vec{V}_2\,k\right)\,ds\,dp$$
$$-\frac{1}{g}\int_0^{p_0}\int_s \nabla\bullet\left(\vec{V}_3\,k\right)\,ds\,dp, \quad (12.35)$$

$$-\frac{1}{g}\int_0^{p_0}\int_s \vec{V}\bullet\nabla\phi\,ds\,dp = -\frac{1}{g}\int_0^{p_0}\int_s \vec{V}_2\bullet\nabla\phi_1\,ds\,dp$$
$$-\frac{1}{g}\int_0^{p_0}\int_s \vec{V}_3\bullet\nabla\phi_1\,ds\,dp. \quad (12.36)$$

Multiplying (8.13) with $\vec{V}_2$, we obtain the rotational kinetic energy equation of an open domain:

$$\frac{\partial K_2}{\partial t} =$$

$$\underbrace{-\frac{1}{g}\int_0^{p_0}\int_s \vec{V}_2\bullet\frac{d\vec{V}_3}{dt}\,ds\,dp}_{INTR} \underbrace{-\frac{1}{g}\int_0^{p_0}\int_s \nabla\bullet\left(\vec{V}_2\,k\right)\,ds\,dp}_{B(K_2)}$$

$$\underbrace{-\frac{1}{g}\int_0^{p_0}\int_s \left[(f+\zeta_2)\vec{V}_2\bullet\left(\vec{k}\times\vec{V}_3\right) + k_3\nabla\bullet\vec{V}_3 + \omega_3\vec{V}_2\bullet\frac{\partial\vec{V}_3}{\partial p}\right]}_{C(K_3,K_2)}$$

$ds\,dp$

$$\underbrace{-\frac{1}{g}\int_0^{p_0}\int_s \vec{V}_2\bullet\nabla\phi_1\,ds\,dp}_{G(K_2)} + \underbrace{\frac{1}{g}\int_0^{p_0}\int_s \vec{V}_2\bullet\vec{F}\,ds\,dp}_{D(K_2)}. \quad (12.37)$$

The time variation of K_2 within an open domain can be induced in some way by $\vec{V}_3$ through $INTR$, which is physically difficult to interpret. $B(K_2)$ is the divergence of the kinetic energy flux by the rotational flow, and $G(K_2)$ is the generation of rotational kinetic energy by cross-contour rotational flow. According to (8.14), (8.15), and (8.18), $INTR$, $B(K_2)$, and $G(K_2)$ integrate to zero in a closed domain. Since it vanishes in a closed domain, $G(K_2)$ is considered by Pearce (1974) and Chen and Wiin-Nielsen (1976)

146 FUNDAMENTALS OF ATMOSPHERIC ENERGETICS

as the generation of kinetic energy due to the barotropic process. Alternatively, (8.18) implies that the rotational kinetic energy of an open domain can be changed through the transport of potential energy by rotational flow through boundaries:

$$G(K_2) = -\frac{1}{g}\int_0^{p_o}\int_s \nabla \bullet \left(\vec{V}_2\, \phi_1\right)\, ds\, dp = BR\,(\phi_1). \qquad (12.38)$$

Both $B(K_2)$ and $BR(\phi_1)$ indicate that the time variation of the rotational kinetic energy inside an open domain may be due to the external energy reservoir through lateral transports, as well as to the internal conversion and dissipation processes.

Substituting (12.35) and (12.36) into (12.4) and subtracting (12.38) from the resulting equation, we can obtain the divergent kinetic energy equation:

$$\frac{\partial K_3}{\partial t} = -\underbrace{\frac{1}{g}\int_0^{p_o}\int_s \vec{V}_3 \bullet \frac{\partial \vec{V}_2}{\partial t}\, ds\, dp}_{INTD} - \underbrace{\frac{1}{g}\int_0^{p_o}\int_s \nabla \bullet \left(\vec{V}_3\, k\right)\, ds\, dp}_{B(K_3)}$$

$$+ \underbrace{\frac{1}{g}\int_0^{p_o}\int_s \left[(\zeta_2 + f)\vec{V}_3 \bullet \left(\vec{k} \times \vec{V}_2\right) + k_2 \nabla \bullet \vec{V}_3\right.}_{-C(K_3, K_2)}$$

$$+ \left. \omega_3 \vec{V}_2 \bullet \frac{\partial \vec{V}_3}{\partial p}\right]\, ds\, dp - \underbrace{\frac{1}{g}\int_0^{p_o}\int_s \vec{V}_3 \bullet \nabla \phi_1\, ds\, dp}_{G(K_3)}$$

$$+ \underbrace{\frac{1}{g}\int_0^{p_o}\int_s \vec{V}_3 \bullet \vec{F}\, ds\, dp.}_{D(K_3)} \qquad (12.39)$$

The time variation of K_3 may be caused by $\vec{V}_2$ through $INTD$. As $INTR$, it is also difficult to interpret physically $INTD$. $G(K_3)$ does not integrate to zero in a closed domain and is regarded by Pearce (1974) and Chen and Wiin-Nielsen (1976) as the generation of kinetic energy due to baroclinic processes. To relate the time variation of divergent kinetic energy inside an open domain to available potential energy, $G(K_3)$ may be written as

$$G(K_3) = -\frac{1}{g}\int_0^{p_o}\int_s \vec{V}_3 \bullet \nabla \phi_1\, ds\, dp$$

$$= \underbrace{-\frac{1}{g}\int_0^{p_o}\int_s \nabla \bullet \left(\vec{V}_3\, \phi_1 dp_1\right)\, ds\, dp}_{BD(\phi_1)} + \underbrace{\frac{1}{g}\int_0^{p_o}\int_s \alpha\omega\, ds\, dp.}_{-C(A, K_3)} \quad (12.40)$$

$G(K_3)$ is influenced by the internal conversion from available potential to divergent kinetic energy, $C(A, K_3)$, and by the potential energy transport of

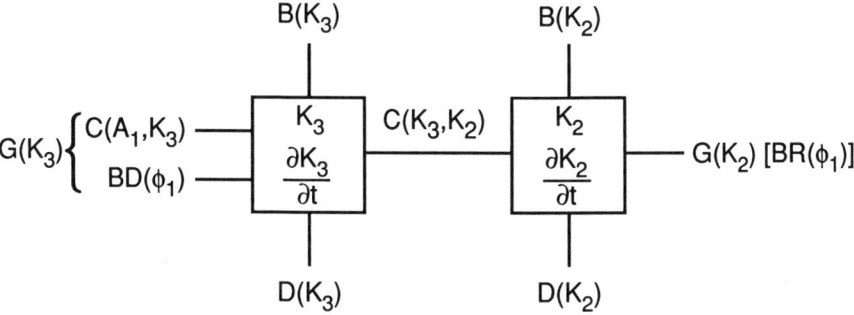

Fig. 12.6: Divergent (subscript 3) and rotational (subscript 2) kinetic energy budget diagrams in an open system.

the divergent flow through the boundaries, $BD(\phi_1)$. For a closed domain, (8.23) indicates that divergent kinetic energy is maintained by the release of available potential energy, the conversion from divergent to rotational kinetic energy, and the dissipation of divergent kinetic energy. In addition to these energy transformation processes, (12.35) and (12.40) show that the transport of kinetic and potential energy by divergent flow from the external energy reservoir across lateral boundaries also affects the time variation of divergent kinetic energy inside an open domain.

The symbolic expressions of (12.37) and (12.39) can be written as

$$\frac{\partial K_2}{\partial t} = -INTR + B(K_2) + C(K_3, K_2) + \underbrace{BR(\phi)}_{G(K_2)} + D(K_2), \qquad (12.41)$$

$$\frac{\partial K_3}{\partial t} = -INTD + B(K_3) - C(K_3, K_2) + \underbrace{BD(\phi) + C(A, K_3)}_{G(K_3)}$$
$$+ \; D(K_3). \qquad (12.42)$$

The schematic energy diagram representing these two equations is illustrated in Fig. 12.6: A short-wave disturbance (13–16 March 1973) developing into a cutoff low aloft and resulting in an intense surface cycle was classified by Chen et al. (1978) as the Type B cyclone of Petterssen and Smebye (1973). The cyclone system was analyzed by Alpert (1982) for the kinetic energy budget of baroclinic and barotropic flows. The kinetic energy budget of this cyclone was characterized as distinctively different from a Type A cyclone of Petterssen and Smebye. The import of kinetic energy by a strong and persistent jet stream in the upper troposphere provided most of the energy for development. In contrast, the generation of kinetic

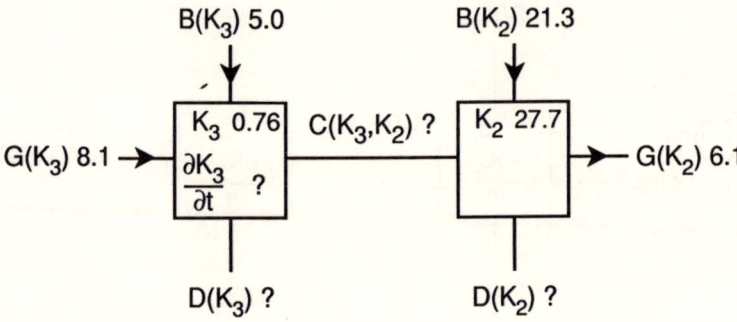

Fig. 12.7: Divergent and rotational kinetic energy budgets of a shortwave disturbance during 13–16 March 1973. Units: kinetic energy (K_2 and K_3) (10^5 J m^{-2}); others (W m^{-2}).

energy by cross-contour flow was a secondary energy source. Chen et al. (1978) analyzed the kinetic energy budget of divergent and rotational flow of this cyclone system. Their results are shown in Fig. 12.7

Chen et al. found that $\vec{V}_3$ contributes 12% of the total kinetic energy but nearly 25% of the convergence of the total kinetic energy flux during the lifetime of the cyclone. The generation of kinetic energy by divergent cross-contour flow, $G(K_3)$, is comparable to the rotational cross-contour flow, $G(K_2)$, averaged over the lifetime of the cyclone. Thus, one may safely replace the total wind by the divergent wind when calculating the total kinetic energy. In contrast, the convergence of kinetic energy flux and the generation of kinetic energy are sensitive to the magnitude of the divergent wind. Any significant reduction of the divergent wind component affected by smoothing or filtering the analyzed wind fields will cause significant errors in the time variation of the kinetic energy of the cyclone. Later, the conclusion by Chen et al. was further confirmed by Tibaldi et al. (1980), DiMego and Bosart (1982), and Boyle and Bosart (1986) for various mid-latitude cyclones. Nevertheless, these studies never included the computation of $C(K_3, K_2)$ and could not estimate dissipation of both K_3 and K_2 – that is, of $D(K_3)$ and $D(K_2)$.

To explore the roles of the divergent and rotational flows in the kinetic energy budget of a limited open domain associated with intense convection activity, Fuelberg and Browning (1983) and Buecheler and Fuelberg (1989) analyzed the kinetic energy budgets of two intense convective periods: (1) the fourth Atmospheric Variability Experiment (AVE IV) and the first Atmospheric Variability Experiment – Severe Environment Storm and Mesoscale Experiment (AVE–SESAME I). The AVE IV period extended from 0000 GMT 14 April to 1200 GMT 25 April, 1975. Soundings were taken at 6-hour intervals and covered the eastern United States. The AVE IV period contained the development of two mesoscale convective complexes (MCC) (Maddox, 1980) in the absence of major cyclone activity.

ENERGETICS OF AN OPEN DOMAIN 149

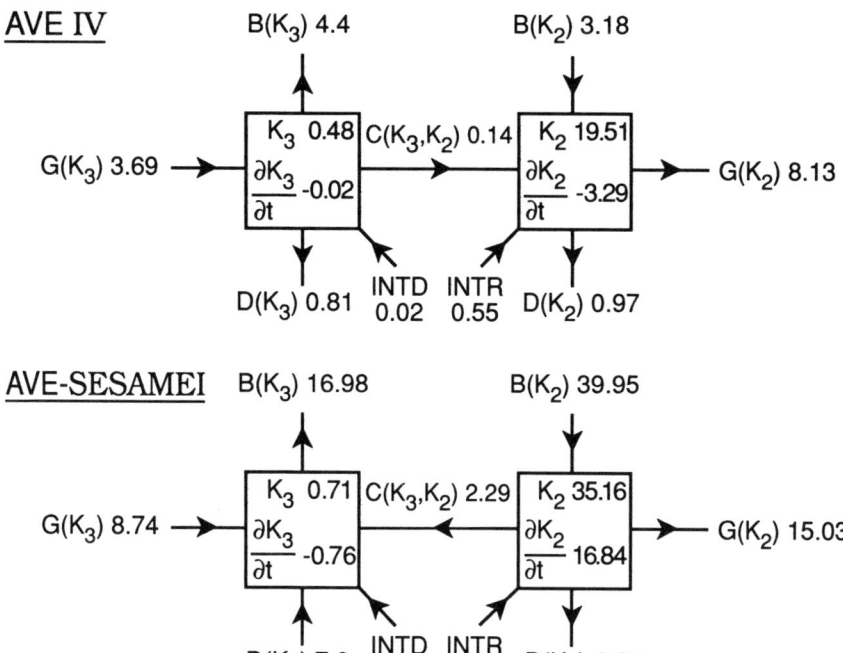

Fig. 12.8: The divergent and rotational kinetic energy budgets averaged over the periods of AVE IV and AVE-SESAME I. Units: kinetic energy (K_2 and K_3) (10^5 J m^{-2}); others (W m^{-2}).

The first convective outbreak attained its maximum spatial extent over the Midwest near 0060 GMT 24 April 1975, and the second convective outbreak peaked over eastern Kansas near 0060 GMT 25 April, 1975. The AVE IV data were objectively analyzed onto a grid with a spacing of 158 km by Buechler and Fuelberg (1986) using the Barnes (1964) technique. The AVE–SESAME I period included 1200 GMT 10 April to 1200 GMT 11 April 1979. The soundings provided meso α-scale resolution every 3 hours and covered the southern–central United States. The data were objectively analyzed onto a grid with 127 km spacing (Buecheler and Fuelberg, 1986). AVE–SESAME I was characterized by a synoptic condition common to severe storm outbreaks, namely strong lower- and upper-level jets and an intense cyclone with an associated front. During AVE-SESAME I, the Red River Valley tornado occurred between 2100 GMT and 0200 GMT and the Wichita Falls tornado touched down near 0000 GMT. The kinetic energy budget analysis of the divergent and rotational flows of these two intense convection periods are shown in Fig. 12.8.

The intensity of the AVE–SESAME I period is much stronger than that of the AVE IV period because of the stormy synoptic conditions of the former. Regardless of this difference, $G(K_3)$ and $B(K_2)$ are energy sources, and $G(K_2)$ and $B(K_2)$ are energy sinks during both intense convection periods. As exhibited in Fig. 12.8, $C(K_3, K_2)$ does not seem to be a crucial energy conversion. In fact, according to Buecheler and Fuelberg (1986), the smallness of $C(K_3, K_2)$ is due to the cancellation of contributions from physical processes involved in this energy conversion. Moreover, the magnitudes of both $D(K_3)$ and $D(K_2)$ are, in contrast, comparable to the global model atmosphere case (Chen and Wiin-Nielsen, 1976), in which $D(K_3)$ is about 10% of $D(K_2)$. Apparently, the energetic characteristics of convectively active regions differ substantially from those of large-scale motions. In view of its contribution to various energy conversions and transports of limited open domains associated with intense convection, the divergent wind is an indispensable ingredient of the kinetic energy budget. Finally, it should be pointed out that although K_3 is much smaller than K_2, it still acts as a "catalyst" during the intense convection periods, as can be observed in major cyclone systems and in the global atmosphere.

13

ENERGETICS OF SOME SPECIAL PHENOMENA

13.1 Subtropical Jet Streams

The most conspicuous elements of the atmospheric general circulation in the extratropics are the subtropical jet streams. Since Rossby and his collaborators (1947) initiated a pioneer study of several aspects of subtropical jet streams, numerous efforts have been made to explore their structure, their relation to other weather systems and the mechanisms that create and maintain them. However, we still lack a satisfactory mechanism explaining the existence of subtropical jet streams. In contrast, the maintenance of these jet streams, to a great extent has been successfully explained in several studies.

Namias and Clapp (1947) proposed the *confluence theory* to illustrate the maintenance of subtropical jet streams. They showed that a thermally direct cross-jet circulation exists in the jet entrance regions with the rising warm air to the south of the jet and the sinking cold air to the north. The former regions are located over the Gulf Stream and the Kuroshio current, while the latter regions are located over the cold continents. Conversely, a thermally indirect cross-jet circulation exists in the jet exit regions with the rising cold air to the north of the jet over the Aleutian and Icelandic lows, and the sinking warm air to the south of jets where the subtropic oceanic anticyclones are located. Analyzing various physical variables with periods of 2.5–6 days and 10–90 days, Blackmon et al. (1977) confirmed the confluence theory that jet streams depicted by a time-mean flow are maintained upstream by the thermally direct cross-jet circulations. In contrast, the thermally indirect cross-jet circulation maintained by baroclinic waves slows down jet streams downstream.

Later, Holopainen (1978) and Lau (1979) analyzed the kinetic energy budgets of time-mean and transient modes. Their results indicated that the kinetic energy generated by ageostrophic flow in the upstream of jet streams is transported downstream and destroyed by ageostrophic flows. The time-mean ageostrophic winds in the upper troposphere are geographically consistent with cross-jet circulations. The kinetic energy budget analysis of Holopainen and Lau is therefore consistent with Namias and Clapp's confluence theory and Blackmon et al.'s illustration for the generation or destruction of jet-stream kinetic energy by the ageostrophic cross-jet circulation.

It was illustrated by Krishnamurti (1961) and Blackmon et al. (1977) that the circumpolar maximum westerlies in the mid-latitudes of the Northern Hemisphere possess a three-jet structure. These three subtropical jet streams are located at the northern end of the Hadley circulation (Palmén and Newton, 1969). Constructing the planetary-scale divergent circulation in the upper troposphere (Krishnamurti, 1971; Krishnamurti et al., 1973), Krishnamurti (1979) pointed out that divergent centers of the upper-level planetary-scale divergent circulation over the three tropical continents correspond well to the three subtropical jet streams primarily depicted by the rotational flow. Thus, we may infer that the planetary-scale divergent circulations maintained by the tropical diabatic heating (Chen and Yen, 1990) contribute in some way to the maintenance of the subtropical jet streams. Seemingly, the interhemispheric interaction between the divergent and rotational circulations may be an important factor in maintaining the subtropical jet streams. Chen et al. (1988) used the energetics scheme of divergent and rotational flows developed in Chapter 8 and Section 12.3 to illustrate the maintenance of the three subtropical jet streams based on interhemispheric interaction.

This brief historical account concerning the maintenance of subtropical jet streams by energetics analyses shows the two approaches that have been taken: regional ageostrophic circulation mechanisms and interhemispheric interactions between planetary-scale divergent circulation and midlatitude rotational flow. The energetics analyses for the maintenance of subtropical jet streams will be developed and illustrated in this section, based upon these two approaches. And a discussion of the relationship between these two approaches will follow the energetics analyses.

13.1.1 Regional Ageostrophic Circulation Mechanism

The equations of time-mean and transient motions can be written as

$$\frac{\partial \vec{\overline{V}}}{\partial t} + \vec{\overline{V}} \cdot \nabla \vec{\overline{V}} + \overline{\omega}\frac{\partial \overline{V}}{\partial p} - f\hat{k} \times \vec{\overline{V}} = \nabla\overline{\phi} + \overline{\mathbf{F}} + \overline{\mathbf{V'} \cdot \nabla \mathbf{V'}} + \overline{\omega'\frac{\partial \mathbf{V'}}{\partial p}} \quad (13.1)$$

and

$$\frac{\partial \mathbf{V'}}{\partial t} + \overline{\mathbf{V}} \cdot \nabla \mathbf{V'} + \mathbf{V'} \cdot \nabla \overline{\mathbf{V}} - \overline{\mathbf{V'} \cdot \nabla \mathbf{V'}} + \overline{\omega}\frac{\partial \mathbf{V'}}{\partial p} + \omega'\frac{\partial \overline{\mathbf{V}}}{\partial p} - \overline{\omega'\frac{\partial \mathbf{V'}}{\partial p}} = -\nabla\phi + \mathbf{F'}. \quad (13.2)$$

Multiplying (13.1) with $\overline{\mathbf{V}}$ and (13.2) with $\mathbf{V'}$ and applying the continuity equation, we obtain the time-mean ($\overline{k}$) and transient (k') kinetic energy equations as follows:

$$\frac{\partial \overline{k}}{\partial t} = \underbrace{- \nabla \cdot (\overline{\mathbf{V}}\overline{k})}_{B_1(\overline{k})} \underbrace{- \nabla \cdot \overline{(\overline{u}u' + \overline{v}v')\mathbf{V'}}}_{B_2(\overline{k})}$$

ENERGETICS OF SOME SPECIAL PHENOMENA

$$-\frac{\partial}{\partial p}\left[(\overline{\omega}\overline{k}) + \overline{(\overline{u}u' + \overline{v}v')\,\omega'}\right]$$
$$-\,\overline{\mathbf{V}}\bullet\nabla\overline{\phi} + \overline{\mathbf{V}}\bullet\overline{\mathbf{F}} - C\left(\overline{k}, k'\right), \qquad (13.3)$$
$$G(\overline{k}) \qquad D(\overline{k})$$

and

$$\frac{\partial k'}{\partial t} = -\nabla\bullet\mathbf{V}k' - \nabla\bullet\overline{(\mathbf{V}'k')} - \frac{\partial}{\partial p}\overline{(\omega' k')} - \overline{\mathbf{V}'\bullet\nabla\phi'} + \overline{\mathbf{V}'\bullet\mathbf{F}'} + C\left(\overline{k}, \overline{k}'\right) \qquad (13.4)$$
$$B_1(k') \qquad B_2(k') \qquad\qquad\qquad G(k') \qquad D(k')$$

The combination of (13.3) and (13.4) yields the conventional kinetic energy equation

$$\frac{\partial k}{\partial t} + \nabla\bullet\overline{(\mathbf{V}k)} - \frac{\partial}{\partial p}\overline{(\omega k)} - \overline{\mathbf{V}\bullet\nabla\phi} + \overline{\mathbf{V}\bullet\mathbf{F}}, \qquad (13.5)$$
$$B(k) \qquad\qquad\qquad G(k) \qquad D(k)$$

where $k = \left(u^2 + v^2\right)/2$, $\overline{k} = \left(\overline{u}^2 + \overline{v}^2\right)/2$, and $k' = \left(\overline{u'^2} + \overline{v'^2}\right)/2$. The time-mean transient components of any variable are defined as $\overline{(\ )} = (1/T)\int_0^T (\ )\,dt$ and $(\ )' = (\ ) - \overline{(\ )}$. Note also $-\overline{\mathbf{V}\bullet\nabla\phi} = -\overline{\mathbf{V}}\bullet\nabla\overline{\phi} - \overline{\mathbf{V}'\bullet\nabla\phi'}$, and

$$C(\overline{k}, k') = -\frac{\overline{u'u'}}{a\cos\varphi}\frac{\partial\overline{u}}{\partial\lambda} - \frac{\overline{u'v'}\cos\varphi}{a}\frac{\partial}{\partial\varphi}\left(\frac{\overline{u}}{\cos\varphi}\right) - \frac{\overline{u'v'}}{a\cos\varphi}\frac{\partial\overline{v}}{\partial\lambda}$$
$$-\frac{\overline{v'v'}}{a}\frac{\partial\overline{v}}{\partial\varphi} + \overline{v'v'v}\frac{\tan\varphi}{a} - \overline{u'\omega'}\frac{\partial\overline{u}}{\partial p} - \overline{v'\omega'}\frac{\partial\overline{v}}{\partial p}. \qquad (13.6)$$

Integrating (13.3)–(13.6) vertically and averaging them over a long period of time, we obtain

$$-\langle B_1(\overline{k}) + B_2(\overline{k})\rangle - \langle G(\overline{k})\rangle = \langle D(\overline{k})\rangle + \langle C(\overline{k}, k')\rangle \qquad (13.7)$$

and

$$-\langle B_1(k') + B_2(k')\rangle - \langle G(k')\rangle = \langle D(k')\rangle - \langle C(\overline{k}, k')\rangle, \qquad (13.8)$$

where $\langle\ \rangle = (1/p_0)\int_0^{p_0}(\ )\,dp$. The assumptions $\omega = 0$ at $p = p_0$ and $p = 0$ were applied to (13.7) and (13.8). We can combine (13.7) and (13.8) to form

$$-\langle B(k)\rangle - \langle G(k)\rangle = \langle D(k)\rangle. \qquad (13.9)$$

The computation of the kinetic energy generation by cross-isobaric flow

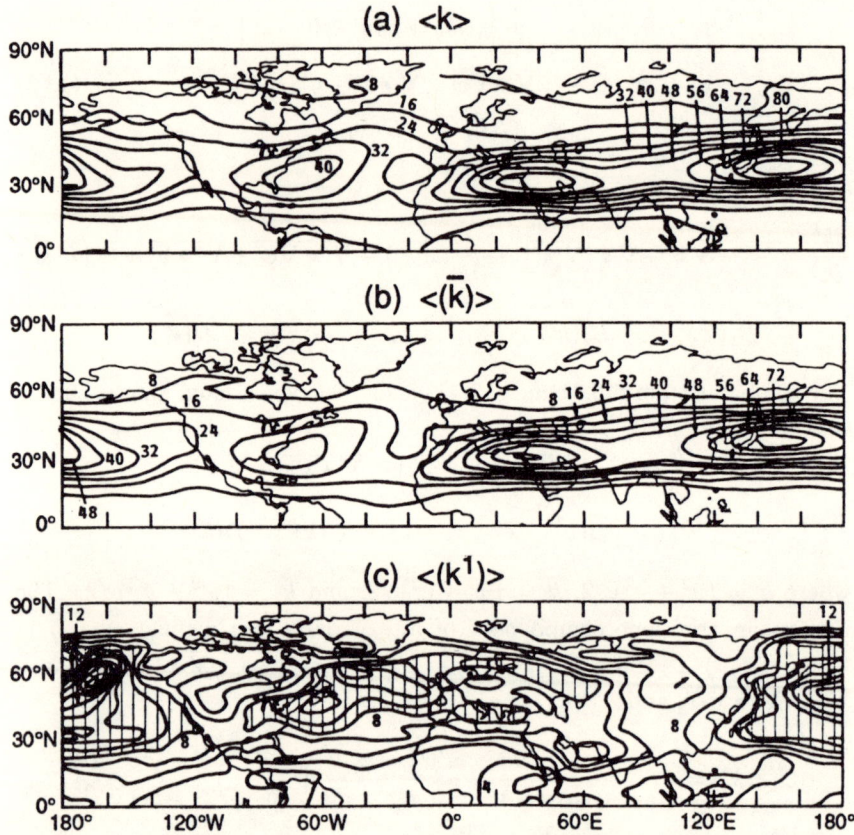

Fig. 13.1: Horizontal distribution of (a) total kinetic energy $\langle k \rangle$, (b) kinetic energy of time-mean flow $\langle \bar{k} \rangle$, and (c) kinetic energy of transient flow $\langle k' \rangle$ in the GLAS climate model. The values of $\langle k' \rangle$ larger than 8×10^5 J m^{-2} are hatched.

is sensitive to the ageostrophic wind contained in the analyzed data. Generally speaking, the evaluations of $G(k)$, $G(\bar{k})$ and $G(k')$ are subject to a bias from the objective analysis scheme used in preparing the observational data on a certain grid system. The bias makes it difficult to assess a quantitative spatial relationship between the generation and transportation of kinetic energy. The history data generated by the simulation of a general circulation model are, of course, not real, but provide an alternative data source for this purpose. Chen and Lee (1983) made an attempt to use the history data generated by the GLAS climate model for the winter simulation to illustrate the maintenance of the Northern Hemisphere subtropical jet streams in terms of the kinetic energy budget analysis (13.7)-(13.8). Their results are presented in Fig. 13.1.

The spatial distribution of $\langle k \rangle$ (Fig. 13.1a) exhibits three maxima as-

ENERGETICS OF SOME SPECIAL PHENOMENA 155

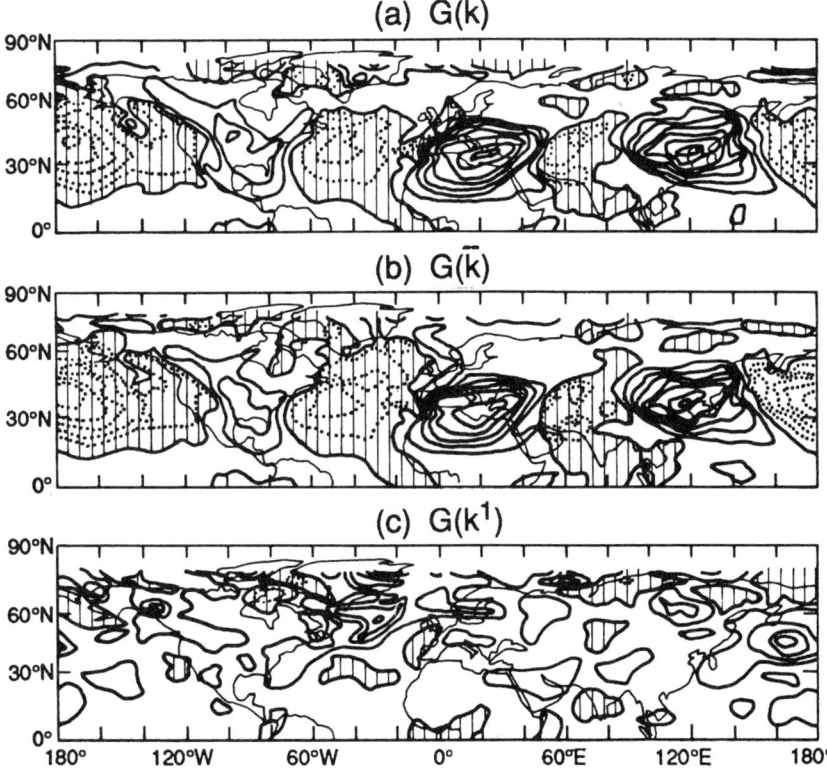

Fig. 13.2: Horizontal distribution of (a) $\langle G(k) \rangle$, (b) $\langle G(\bar{k}) \rangle$, and (c) $\langle G(k') \rangle$ in the GLAS climate model. The negative values are hatched. The contour interval is 10 W m^{-2} for (a) and (b), and 5 W m^{-2} for (c).

sociated with the subtropical jet streams over the east coasts of North America, East Asia, and North Africa. The resemblance between the $\langle k \rangle$ and $\langle \bar{k} \rangle$ (Fig. 13.1b) distributions reflect the fact that the major part of $\langle k \rangle$ is contributed by the time-mean flow. The maxima of $\langle k' \rangle$ appear northeast of either maximum $\langle k \rangle$ or maximum $\langle \bar{k} \rangle$ where the maximum baroclinicity exists, but these are absent northeast of the North African jet. The 2.5–6 day bandpassed statistics of some general circulation variables shown by Blackmon et al. (1977) are significant in association with this jet stream. They seem to suggest that the cyclogensis frequency northeast of the North African jet is much less in the GLAS model.

The k, $\bar{k}$, and k' generations are displayed in Fig. 13.2. Surprisingly, spatial distributions of $\langle G(k) \rangle$ and $\langle G(\bar{k}) \rangle$ resemble each other and the magnitudes of these two variables are also comparable. Obviously, $\langle G(k) \rangle$ is mainly explained by the time-mean ageostrophic effect. In comparing with $\langle k \rangle$ and $\langle \bar{k} \rangle$, both $\langle G(k) \rangle$ and $\langle G(\bar{k}) \rangle$ are positive (negative) in the upstream (downstream) side of jets. That is, the kinetic energy of jet streams is generated (destroyed) in the upstream (downstream) side. In contrast, the spatial distribution of $\langle G(k')\rangle$ is less organized than either $\langle G(k)\rangle$ or $\langle G(\overline{k})\rangle$, and the numerical values of $\langle G(k')\rangle$ are also much smaller than those of the other two forms of kinetic energy generation. Even so, it is still discernible that $\langle G(k')\rangle$ has some relatively notable values over regions where $\langle k'\rangle$ is significant – that is, northeast of jet streams. Because $\langle G(k)\rangle$ and $\langle G(\overline{k})\rangle$ alternate between positive and negative values in the longitudinal direction, the area averages of both $\langle G(k)\rangle$ and $\langle G(\overline{k})\rangle$ over the Northern Hemisphere result in a numerical value comparable to that of $\langle G(k')\rangle$ as shown in Table 13.1.

Table 13.1: Unit: W m^{-2}

Variable	Value
$\langle G(k)\rangle$	2.10
$\langle G(\overline{k})\rangle$	0.23
$\langle G(k')\rangle$	1.87

The function of the kinetic energy flux is to redistribute the kinetic energy of atmospheric motion. The area of import (export) of kinetic energy is denoted by the convergence (divergence) of that energy. The comparisons between Figs. 13.3 and 13.2 show that $\langle B(k)\rangle$ and $\langle G(k)\rangle$ possess similar spatial distributions and are comparable in magnitude, but opposite in sign. In other words, these two energetical processes counterbalance each other. Also, the contrast between $\langle B(k)\rangle$ and $\langle G(k)\rangle$ confirms Blackmon et al.'s (1977) proposal that the kinetic energy of jet streams is generated by an ageostrophic process in the upstream side. The generated kinetic energy is transported downstream and destroyed by ageostrophic processes.

Recall from (13.3)–(13.4) that $\langle B(k)\rangle = \langle B_1(\overline{k})\rangle + \langle B_2(\overline{k})\rangle + \langle B_1(k')\rangle + \langle B_2(k')\rangle$. Additionally, $\langle B_1(\overline{k})\rangle + \langle B_2(\overline{k})\rangle = \langle B(\overline{k})\rangle$ and $\langle B_1(k')\rangle + \langle B_2(k')\rangle = \langle B(k')\rangle$ belong to the $\overline{k}$ and k' equations, respectively. Although $\langle B(k)\rangle$, $\langle B(\overline{k})\rangle$, and $\langle B(k')\rangle$ are relatively consistent in their spatial structure, it is observed that $\langle B(k)\rangle \simeq \langle B_1(k)\rangle$. Apparently, the time-mean flow is responsible for the major part of the local transport of the atmospheric kinetic energy. Moreover, the contrast between $\langle G(\overline{k})\rangle$ and $\langle B_1(\overline{k})\rangle$ resembles that between $\langle G(k)\rangle$ and $\langle B(k)\rangle$, which suggests that the time-mean flow redistributes the kinetic energy generated by the ageostrophic process of time-mean flow. In other words, the counterbalance between $\langle G(\overline{k})\rangle$ and $\langle B(\overline{k})\rangle$ confirms the confluence theory of Namias and Clapp (1947).

Finally, let us discuss $\langle C(\overline{k}, k')\rangle$. The magnitude of this energy variable is very small compared to other energy variables in the kinetic energy equations. The area-mean value of $\langle C(\overline{k}, k')\rangle$ over the Northern Hemisphere is -0.04 W m^2. It does not seem that $\langle C(\overline{k}, k')\rangle$ plays any vital role in the local maintenance of subtropical jet streams, although some studies

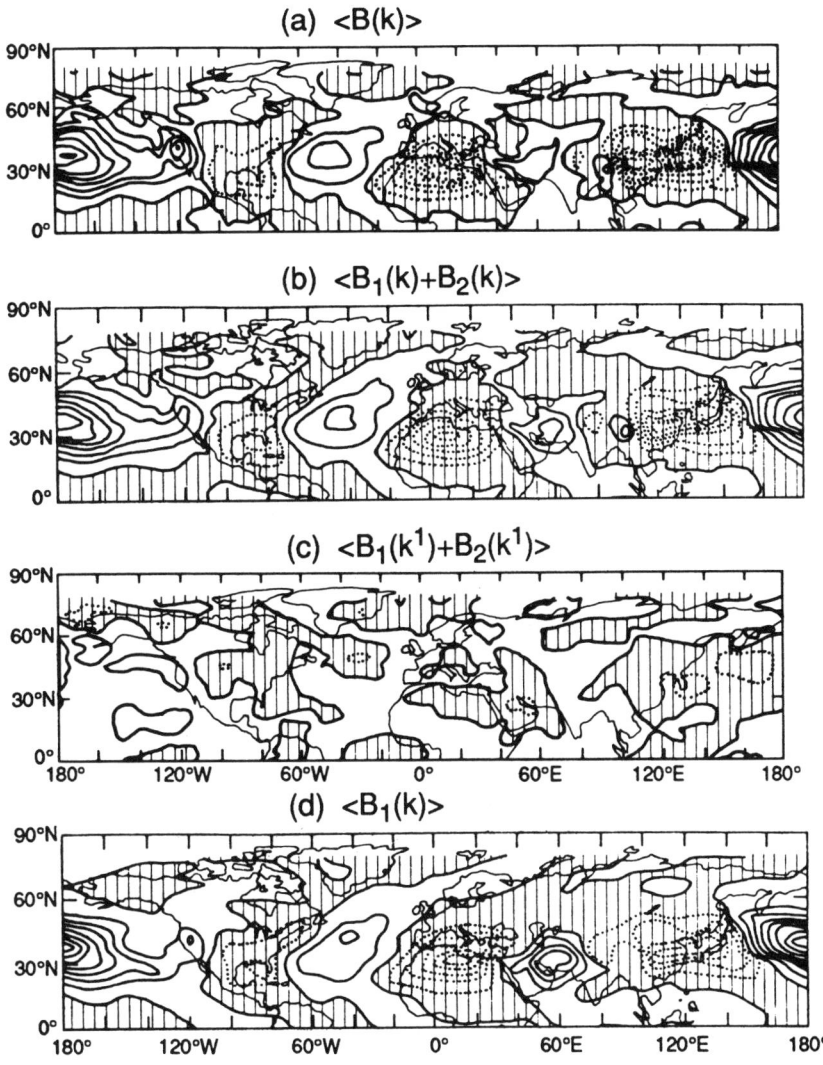

Fig. 13.3: Horizontal distribution of (a) $\langle B(k) \rangle$, (b) $\langle B_1(k) + B_2(k) \rangle$, (c) $\langle B_1(k') + B_2(k') \rangle$, and (d) $\langle B_1(k) \rangle$ in the GLAS climate model. The negative values are hatched. The contour interval is 10 W m^{-2} for (a), (b) and (c), and 5 W m^{-2} for (d).

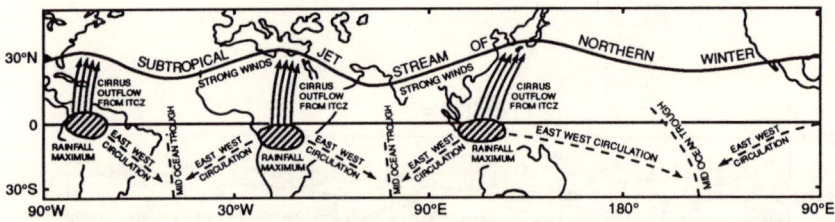

Fig. 13.4: A schematic illustration of the winter subtropical jet streams in the Northern Hemisphere and some associated features of the upper-level circulation. After Krishnamurti, 1979.

– Holopainen (1978) for example – claim that the transient eddies may be important to the maintenance of the long-term mean flow.

13.1.2 Hemispheric Interaction Mechanism

Using the schematic diagram shown in Fig. 13.4, Krishnamurti (1979) suggested that the formation of the three subtropical jet streams may be related to the rainfall over the three tropical continents through the planetary-scale divergent circulation. To illustrate this possibility, we shall use (12.37) and (12.39) and neglect terms involving vertical motion because these terms are generally small in magnitude. Thus, we may write these two equations in the following approximate form:

$$\frac{\partial k_3}{\partial t} \simeq -\nabla \bullet (\mathbf{V}_3 k) - (\zeta + f)(u_3 v_2 - u_2 v_3) - \mathbf{V}_3 \bullet \nabla \phi, \quad (13.10)$$

$$\qquad B(k_3) \qquad\qquad -C(k_3, k_2) \qquad\qquad G(k_3)$$

$$\frac{\partial k_2}{\partial t} \simeq -\nabla \bullet (\mathbf{V}_2 k) + (\zeta + f)(u_3 v_2 - u_2 v_3) - \mathbf{V}_2 \bullet \nabla \phi \quad (13.11)$$

$$\qquad B(k_2) \qquad\qquad C(k_3, k_2) \qquad\qquad G(k_2)$$

Chen et al. (1989) perform the k_2 and k_3 budget analysis at 200 mb with the wind fields generated by the FGGE III-b analyses of the European Centre for Medium Range Weather Forecasts for the 1978/79 winter (December-February). The three northern subtropical jet streams described previously are well depicted by $\mathbf{V}_2$ in Fig. 13.5a. The planetary-scale divergent circulation (Fig. 13.5b) exhibits three convergence centers in the Northern Hemisphere connected to the three subtropical jet streams. The comparison between the rotational circulation $\mathbf{V}_2$ and the generation of rotational kinetic energy $G(k_2)$ reveals that k_2 is generated in the upstream side of jet streams by the downgradient cross-contour flow and is destroyed in the downstream side of the upgradient cross-contour flow. On

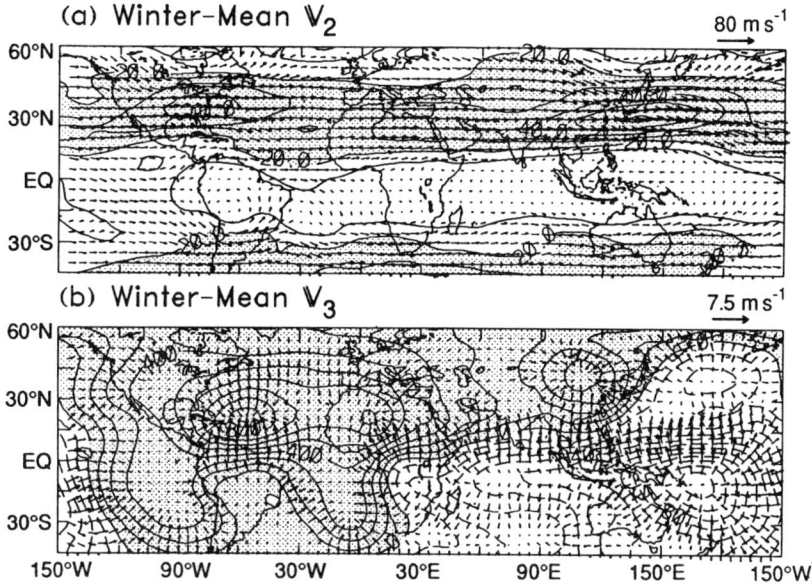

Fig. 13.5: The winter-mean (December-Feburary) 200 mb (a) rotational (V_2) and (b) divergent (V_D) circulation. Thin solid lines of (a) are isotachs and contour intervals are 10 m s^{-1}. The contours of (b) are velocity potential (χ) whose positive values are shaded. The contour intervals of χ are 2×10^6 m^2 s^{-1}.

the other hand, the generated k_2 is transported from upstream to downstream as indicated by $B(k_2)$ (Fig. 13.5b). The counterbalance between $G(k_2)$ and $B(k_2)$ seems to maintain the subtropical jet streams in a similar manner as that between $G(k)$ and $B(k)$, illustrated by the ageostrophic circulation mechanism.

The comparison between the k_3 generation $G(k_3)$ (Fig. 13.6c) and planetary-scale divergent circulation reveals that k_3 is generated by the downgradient divergent flow between the equator and the northern subtropic jet streams. Moreover, the contrast between $G(k_3)$ and the interaction between divergent and rotational flows $C(k_3, k_2)$ (Fig. 13.6d) shows that these two energy variables possess a similar distribution: positive and significant along the equatorward side of subtropical jet streams. The divergence of kinetic energy flux by divergent flow is generally small compared to the other two energetics processes. Because k_3 is always much smaller than k_2, the resemblence between the $G(k_3)$ and the $C(k_3, k_2)$ distributions confirms Chen and Wiin-Nielsen's (1976) suggestion that most of the generated k_3 is converted to maintain the rotational flow, especially the three subtropical jet streams.

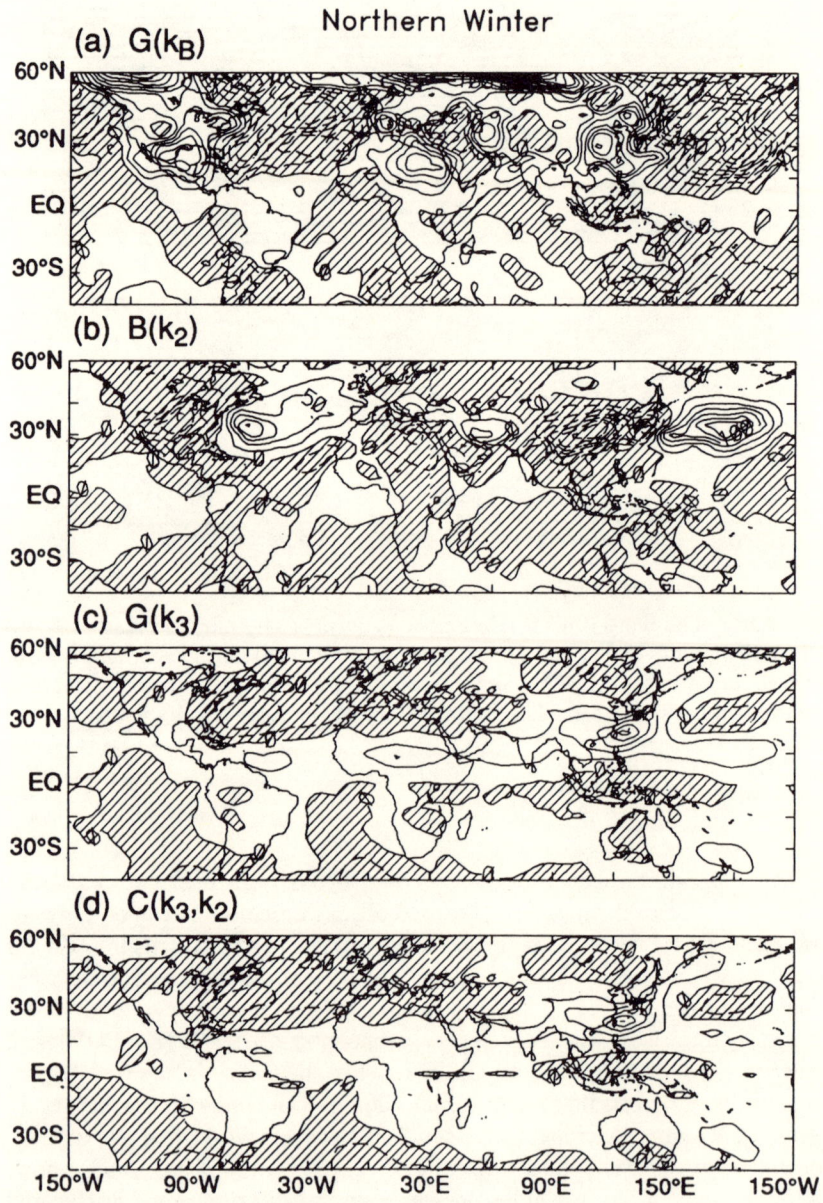

Fig. 13.6: Various energy conversions of (13.10) and (13.11) at 200 mb averaged over the entire northern winter: (a) $G(k_2)$, (b) $B(k_2)$, (c) $G(k_3)$, and (d) $C(k_3, k_2)$. Contour intervals are 2.5×10^{-3} m^2 s^{-3}.

13.1.3 Relation Between the Two Mechanisms

Blackmon et al. (1977) illustrated the maintenance of subtropical jet streams in terms of the zonal momentum budget analysis

$$\frac{du}{dt} = fv_a, \qquad (13.12)$$

where $v_a = (v - v_g)$ is the ageostrophic meridional wind. According to (13.12), the zonal flow of a subtropical jet stream is accelerated by the Coriolis force on the upstream side of jets where v_a flows poleward at upper levels – that is, thermally direct cross-jet circulation. The opposite situation occurs on the downstream side of jets. One may illustate this argument using energetics analyses. Multiplying (13.12) by u, one can obtain the total time variation of the zonal kinetic energy caused by the work done by the Coriolis force (fuv_a), which is equivalent to the work done by the cross-contour flow along the x-direction – that is, $-u\, \partial\phi/\partial x$. This simple argument brings out the essence of the ageostrophic circulation mechanism maintaining subtropical jet streams.

From the viewpoint of interaction between rotational and divergent flows, Chen and Wiin-Nielsen demonstrated that for a *closed* domain

$$G(k_3) \simeq C(k_3, k_2). \qquad (13.13)$$

Because k_3 is always small and $B(k_3)$ is insignificant, the diagnostic result of the k_3 budget analyses displayed in Fig. 13.6 shows that (13.12) is also *locally* true. As suggested by Fig. 13.4, the planetary-scale divergent circulation is correlated with the atmospheric circulation between the tropics and mid-latitudes and contributes to the local maintenance of the subtropical jet streams. This suggestion is demonstrated by (13.12).

Regardless of the difference between them, both mechanisms discussed previously explain the maintenance of the same subtropical jet streams. It is believed that these two mechanisms are related in a certain way. Because (13.3) is true for the maintenance of subtropical jet streams locally, the total generation of kinetic energy by cross-contour flow can be written as

$$G(k) = -\mathbf{V}\bullet\nabla\phi = -\mathbf{V}_2\bullet\nabla\phi - \mathbf{V}_3\bullet\nabla\phi = -\mathbf{V}_2\bullet\nabla\phi + C(k_3, k_2). \quad (13.14)$$

Based upon (13.14), the maintenance of subtropical jet streams by the ageostrophic circulation mechanism described by (13.9) can be illustrated by (13.11). On the other hand, the hemispheric interaction between divergent and rotational flows maintaining subtropical jet streams can be delineated by (13.10) or (13.13). Thus, it becomes clear that the kinetic energy budgets of rotational and divergent flows can be used to depict the two mechanisms.

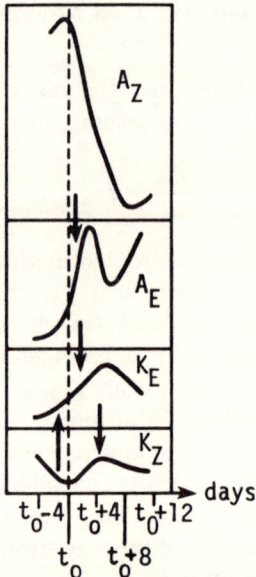

Fig. 13.7: Schematic picture of the energy changes and energy conversions during a blocking situation. Day $t = t_0$ is the onset of the blocking. After Lejenäs, 1977.

13.2 Spectral Energetics of Blocking

Summarizing the hemispheric energetics analyses of blocking that have been conducted in several previous studies (Winston and Krueger 1961; Miyakoda 1963; Paulin 1970), Lejenäs (1977) presented a schematic diagram of blocking energetics (Fig. 13.7) over the life cycle of the phenomenon. The time evolution of blocking energetics is characterized by energy conversions, indicated by arrows, during various phases of blocking. The energetics characteristics of blocking observed by Lejenäs have also emerged in more recent studies by Hansen and Chen (1982), Chen and Shukla (1983), and Fischer (1984). Of course, Lejenäs' schematic diagram of blocking energetics sheds some light on the physical processes involved in blocking development. The hemispheric energetics analysis of blocking episodes contains contributions not only from blocking itself, however, but also from other effects possibly unrelated to blocking. Thus, the former may be masked to some extent by the latter.

Compiling the geographic distributions of blocking, Rex (1950) found two major blocking regions during the Northern Hemisphere winter: one in the eastern Pacific, centered around 140° E; and one in the Atlantic, between 0° and 40° W (Fig. 13.8). Thus, Rex's blocking climatology supports

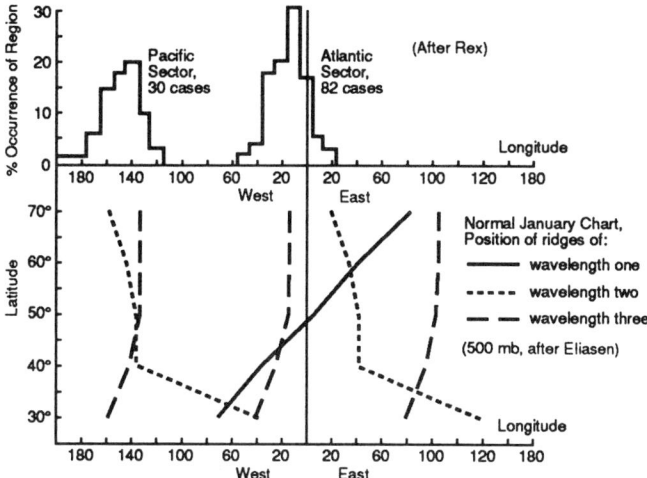

Fig. 13.8: Comparison between the longitude of the initial splitting of the jet (Rex, 1950) and the normal phases of the planetary waves at 500 mb (Eliasen, 1958). After Austin, 1980

the argument that it may be improper to use the hemispheric energetics analysis to illustrate physical processes in developing blocks. Eliasen (1958) analyzed the structure of stationary long waves (wavenumbers 1-3) with the 500 mb height field. The latitudinal variations of the longitudinal positions of ridges are also displayed in the bottom of Fig. 13.8. By contrasting Rex's geographic locations of maximum blocking occurrence and Eliasen's structure of stationary long waves, one can easily observe that stationary long waves interfere constructively between 50° and 60° N at the following locations: (1) wavenumbers 2-3, at about 140° W and (2) wavenumbers 1 and 2 at about 20 ° E. Based upon the studies of Rex and Eliasen, Austin (1980) drew the conclusion that "blocking in the two (Rex's) geographic locations is associated with planetary waves of very large amplitude but normal phase." The wavenumber *signatures* observed by Austin are listed in the following table:

Wavenumber	Pacific	Atlantic	Double
1	small	large	large
2	large	large	large
3	large	small	large

The "very large amplitude" of stationary long waves means that the development of blocking is caused by the amplification of certain quasi-stationary waves. In view of Austin's observation, the spectral energetics analyses over the life cycle of blocking events would be a powerful tool to gain an understanding of the mechanism amplifying stationary long waves associated with blockings. Keeping this scenario in mind, Chen and his collaborators (Chen and Shukla, 1983; Hansen and Chen, 1982) employed the spectral energetics scheme proposed by Saltzman (1957) to perform a number of case studies of blocking. This approach was later adopted by Fischer (1984) and Kung and Baker (1986) as well.

Shutts (1983) suggested that blocking can be maintained by the straining mechanism of time-mean flow caused by transient eddies. This theory was widely accepted. Furthermore, numerous diagnostic studies have been performed in the past several years to support or to extend Shutts proposal (e.g., Illari, 1984; Holopainen and Fortelius, 1987). In fact, most diagnostic studies following Shutts' theory essentially deal with the maintenance of time-mean blocking. It was pointed out by Holopainen and Fortelius that the *Reynolds* expression for the eddy effects does not apply when the "mean" flow is not clearly separated by a spectral gap from the eddy part of the flow. The role of high-frequency eddies in the onset and breakdown of blocking – that is, the time evolution of blocking – cannot, therefore, be studied in terms of the Reynolds approach. Chen (1982a) has shown that the stationary and transient modes are the primary contributors to the energy amounts of the long-wave and short-wave regimes, respectively. It is therefore not surprising that the nonlinear interaction between long waves and short waves revealed in the spectral energetics study should represent the nonlinear interaction between stationary and transient eddies rather well.

13.2.1 Spectral Energetics Analyses of Some Blocking Events

The wavenumber representation of a meteorological variable in the longitudinal direction, as determined by the Fourier analyses, has been presented in Chapter 9. Following this approach, Saltzman (1957) formulated the Lorenz energy cycle (Lorenz 1955) in the one-dimensional spectral domain:

$$\frac{dK_Z}{dt} = \sum_{n=1}^{N} C(K_n, K_Z) + C(A_Z, K_Z) + B(K_Z) - D(K_Z), \tag{13.15}$$

$$\begin{aligned}\frac{dK_n}{dt} = &- C(K_n, K_Z) + CK(n|m, \ell) + C(A_n, K_n) \\ &+ B(K_n) - D(K_n),\end{aligned} \tag{13.16}$$

$$\frac{dA_Z}{dt} = - \sum_{n=1}^{N} C(A_Z, A_n) - C(A_Z, K_Z) + B(A_Z) + G(A_Z), \tag{13.17}$$

$$\frac{dA_n}{dt} = C(A_Z, A_n) - C(A_n, K_n) + CA(n|m, \ell) \tag{13.18}$$

ENERGETICS OF SOME SPECIAL PHENOMENA

$$+ \quad B(A_n) + G(A_n).$$

Note that

$$\sum_{n=1}^{N} C(K_n, K_Z) = C(K_E, K_Z),$$

$$\sum_{n=1}^{N} C(A_Z, A_n) = C(A_Z, A_E).$$

Equations of eddy energies

$$K_E = \sum_{n=1}^{N} K_n \quad \text{and} \quad A_E = \sum_{n=1}^{N} A_n$$

can be obtained by summing (13.16) and (13.18) from $n = 1$ to N, that is,

$$\frac{dK_E}{dt} = -C(K_E, K_Z) + C(A_E, K_E) + B(K_E) - D(K_E) \qquad (13.19)$$

and

$$\frac{dA_E}{dt} = C(A_Z, A_E) - C(A_E, K_E) + B(K_E) + G(A_E). \qquad (13.20)$$

The conventional Lorenz energy cycle consists of (13.15), (13.17), (13.19), and (13.20). Presumably, Austin's *constructive interference* argument describes the basic mechanism forming blockings. However, we still need to understand how each wave involved in blocking development amplifies and decays. To accomplish this goal, spectral energetics analyses using (13.16) and (13.20) for individual waves becomes necessary. The case studies performed by Chen and Shukla (1983) and Hansen and Chen (1983) will be used here for illustration.

A. A Case Study of Constructive Interference

To test the sensitivity of quasi-stationary waves to the north Pacific sea surface temperature (SST) anomalies of the 1976/77 winter compiled by Namias (1978), Shukla and Bangaru (1979) integrated the climate model of the Goddard Laboratory for Atmospheric Sciences (GLAS) for 60 days with the observed initial conditions for 1 January 1975 and with the anomalous SST. Two persistent ridge-trough systems over western North America and western Europe were produced. The spectrally filtered (SF) Hovmöller diagrams at 500 mb and 50° N are shown in Fig. 13.9 for three wave groups:

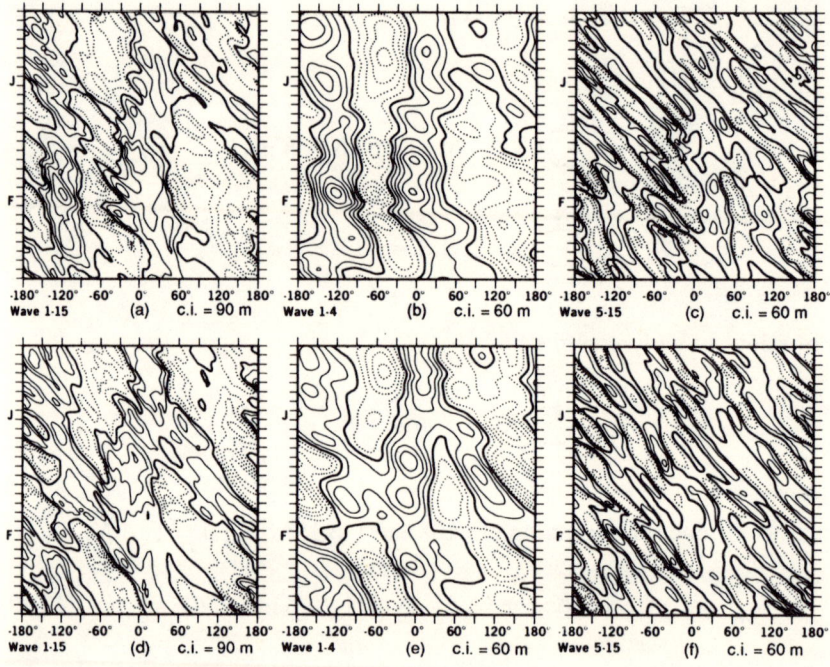

Fig. 13.9: Spectrally filtered Hovmöller diagrams of the 500-mb height (m) at 50° N for three wave groups (indicated in each panel) of both anomaly [(a)-(c)] and control experiments [(d)-(f)]. Dotted contours are negative values.

(1) wavenumbers 1-15, (2) wavenumbers 1-4, and (3) wavenumbers 5-15. A comparison between the first two wave groups clearly shows that the large-scale pattern of the trough-ridge system is primarily explained by the long-wave regime. The persistent blocking ridges simulated by the model appeared in the first two weeks of February and were a result of the long-waves becoming stationary and amplifying. Moreover, the combination of wavenumbers 2 and 3 forms the two persistent blocking ridges well.

The height and temperature structure of wavenumbers 2 and 3, and the combination of these two waves at the mature stage of blocking, 1200 GMT 4 February, is exhibited in Fig. 13.10. The high centers of wavenumber 2 are located over the Alaska Bay and northern central Europe, whereas those of wavenumber 3 are anchored at the west coast of Canada, in the North Sea, and in Siberia. The combination of these two waves forms two centers on the west coast of Canada and the North Sea respectively, with a deep low situated at Labrador and another broad low extending from North Africa to the northwestern Pacific. These two high centers form the persistent blocking ridges. However, the thermal centers of wavenumber 2, as opposed to those of wavenumber 3, have significant phase lags behind height. This contrast of thermal structure reveals that wavenumber 2 is

ENERGETICS OF SOME SPECIAL PHENOMENA

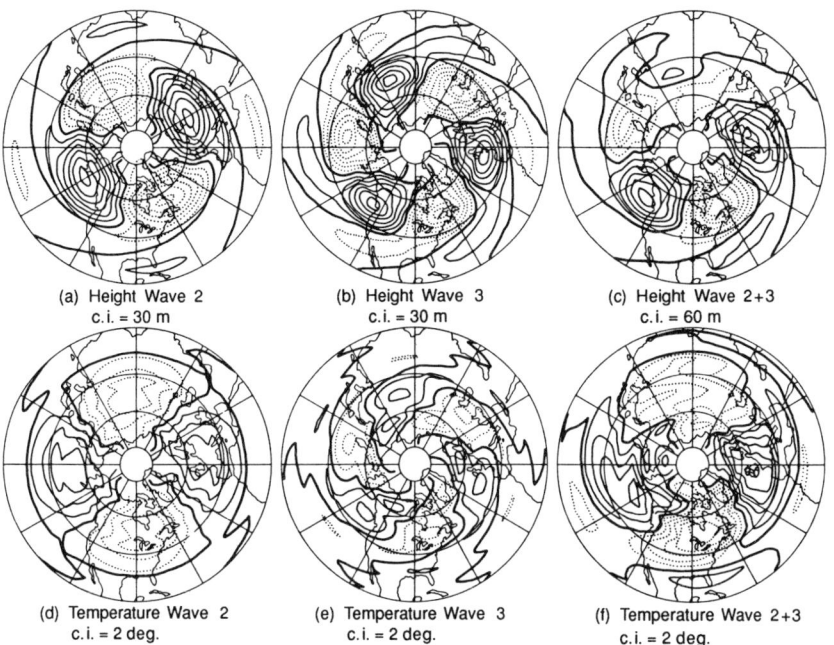

Fig. 13.10: Synoptic charts of the spectrally filtered height (m) and temperature (°C) of the anomaly experiment for wavenumbers 2 and 3 and combinations of these two waves at 500 mb on Feburary 4.

baroclinic and wavenumber 3 *barotropic*. Such a synoptic analysis of these two waves should ease our understanding of their respective spectral energetics analyses.

The time evolutions of various energies and energy conversions averaged between 26° and 86° N for wavenumbers 1-4 are shown in Fig. 13.11 and 13.12. The inclusion of wavenumber 1 and 4 energetics is for the purpose of comparison. Compared to other waves, A_2 and K_2 are enhanced systematically after the end of January when the persistent blocking ridges start to develop. According to (13.18), A_2 is maintained by $C(A_Z, A_2)$, $C(A_Z, K_2)$, and $CA(2|m, \ell)$. Note that A_2, $C(A_Z, A_2)$, and $C(A_2, K_2)$ all vary in phase and in time. Moreover, both $C(A_Z, A_2)$ and $C(A_Z, K_2)$ are positive over the entire period of analysis, while $CA(2|m, \ell)$ extracts energy from A_2 over most of the blocking period. K_2 is maintained by $C(A_2, K_2)$, $C(K_2, K_2)$, and $CK(2|m, \ell)$. The sign of $C(A_2, K_2)$ indicates that A_2 is always converted to K_2, which is then converted to K_Z, except in the second week of February, as shown in Fig. 13.12. The nonlinear interaction $CK(2|m, \ell)$ essentially drains K_2 during the persistent blocking period. $C(A_Z, A_2)$ and $C(A_2, K_2)$ involve the baroclinic processes of horizontal and vertical sensible heat transport, respectively, whereas $C(K_2, K_Z)$ is caused by the

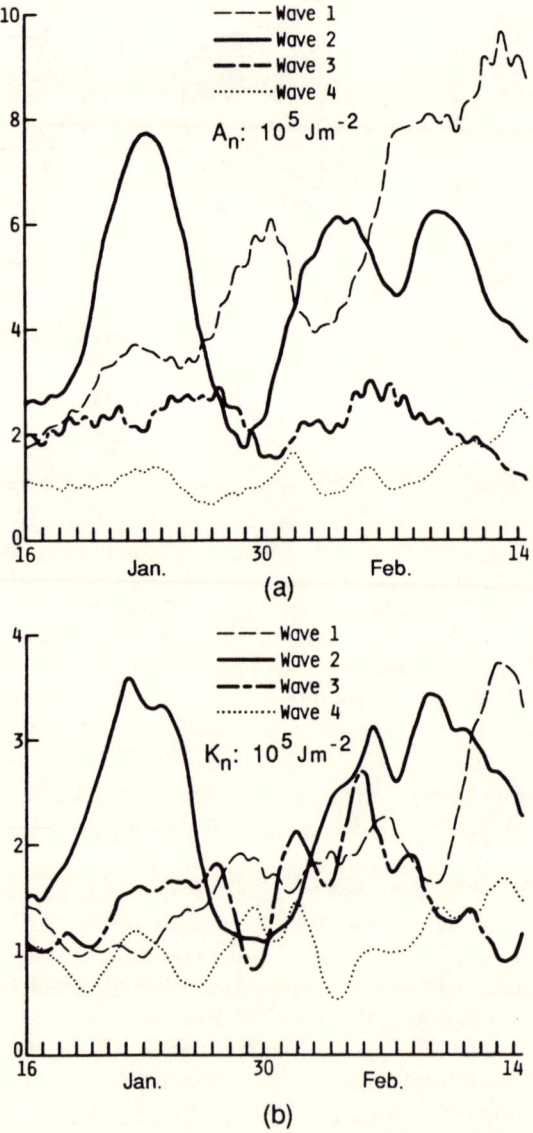

Fig. 13.11: Time variations of (a) available potential energy and (b) kinetic energy for wavenumbers 1, 2, 3, and 4 in the anomaly experiment.

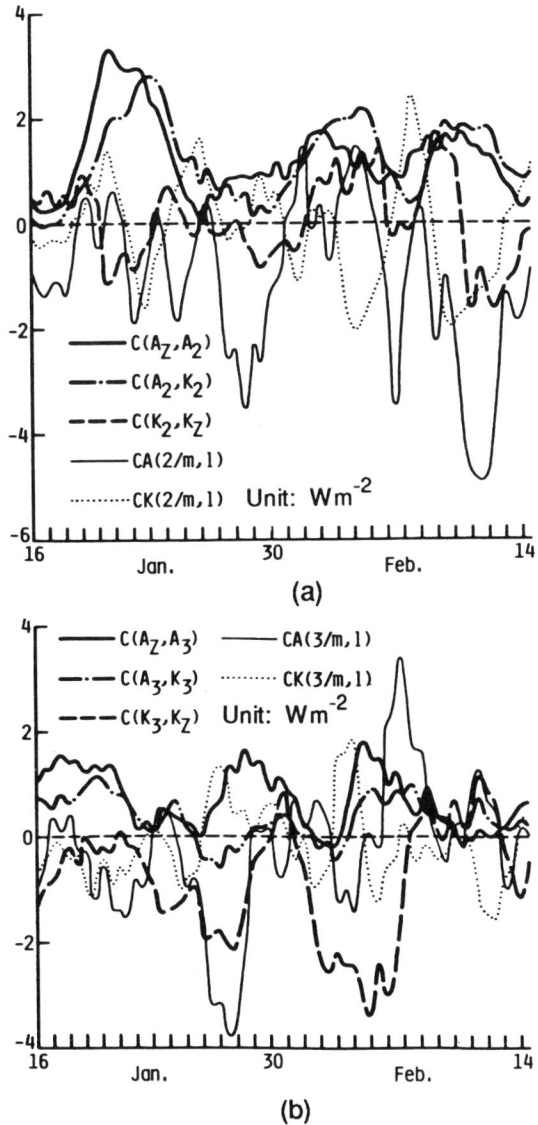

Fig. 13.12: Time variations of energy conversions of (a) wavenumber 2 and (b) wavenumber 3 in the anomaly experiment.

barotropic process of momentum transport. The maintenance of A_2 and K_2 by $C(A_Z, A_2)$ and $C(A_Z, K_2)$ is consistent with the synoptic condition of wavenumber 2 in that its temperature structure has a phase lag behind its height. Apparently, the development and maintenance of wavenumber 2 is attributable to the baroclinic process.

Compared with A_1 and A_2, A_3 does not change significantly during the persistent blocking. Athough $C(A_Z, A_3)$ and $CA(3|m, \ell)$ are major energy sources for the maintainance of A_3, neither of these two energy conversions is augmented during blocking. Furthermore, $C(A_3, K_3)$ does not seem to play a vital role in blocking developments. In contrast, K_3 increases to a certain extent and is maintained by $C(K_3, K_Z)$, the barotropic conversion releasing the available zonal kinetic energy. Clearly the physical processes maintaining K_3 differ from those maintaining K_2. As analysed previously, wavenumber 3 height and temperature do not possess a phase lag, as does the wavenumber 2 synoptic structure. Thus, the synoptic condition and energetic analysis indicate that wavenumber 3 is largely developed and maintained barotropically during the existence of the simulated persistence of blocking ridges.

B. A Case Study of Nonlinear Forcing

During the second half of December 1978, a major blocking ridge developed over the North Atlantic Ocean, located around 30° W. A shortwave 500 mb trough appeared over the western Great Lakes on 17 December. By 19 December, this trough had deepened considerably and extended southeastward over the Atlantic. In the meantime, a ridge was building south of Greenland ahead of this trough. The block attained maturity on 22 December, with a closed high centered at 68° N over the Greenland shore of the Denmark Strait. The block remained at about the same location until 28 December. Strong 500 mb easterlies along 60° N persisted throughout the mature period of this Atlantic block, and the 500-mb contour pattern exhibited a classic Rex blocking structure (Fig. 13.13). However, beginning on about 28 December, the westward extension and retrogression of the blocking ridge started. By 1 January 1979, the Atlantic block had completely decayed. The synoptic evolution of this Atlantic blocking ridge resembled that depicted by Berggren et al. (1947). In summary, the block developed from 19 to 22 December, reached maturity from 22 to 29 December, and decayed rapidly from 29 December to 1 January.

Hansen and Chen (1982) used the Atlantic block case just described to illustrate the blocking development forced by cyclone-scale disturbances. To accomplish this goal, the spectral energetics delineated by (13.16) and (13.18) are grouped into long-wave (wavenumbers 1-4) and cyclone-wave (wavenumbers 5-10) regimes. An energy variable of the long (cyclone)-wave regime is obtained by summing the contributions of all wave components contained in this wave regime as expressed by the following mathematical

ENERGETICS OF SOME SPECIAL PHENOMENA 171

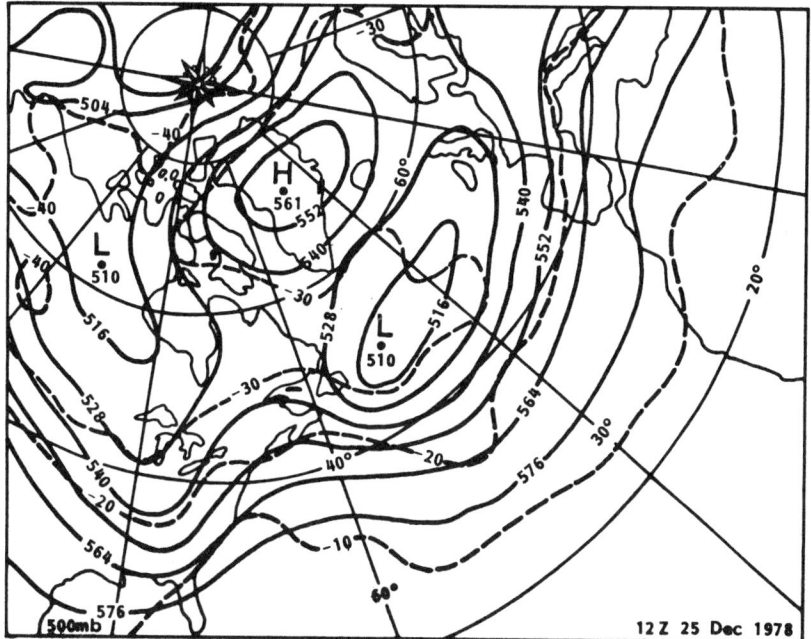

Fig. 13.13: The 500-mb contour charts at 1200 GMT 25 December 1978. The contour interval is 2 decameters.

form:

$$(\ )_L [(\ )_c)] = \sum_{n=1}^{4}(\ )_n \left[\sum_{n=5}^{10}(\ )_n\right].$$

The time evolutions of the spectral energetics for both long-wave and cyclone-wave regimes, averaged between 30° and 80° N over the second half of December 1978, are shown in Figs. 13.14 and 13.15. The values of A_L and K_L peaked when the Atlantic block reached its mature phase, but both A_C and K_C attained their maximum values before the Atlantic block matured. The large energy conversions of the long-wave regime before 20 December (not shown) were primarily attributable to a deepening ultralong-wave trough over East Asia. In contrast, the significant energy contents and conversions of the cyclone-wave regime during the third week of December were essentially located over eastern North America and the western Atlantic.

The physical processes involving the development of blocking may be inferred from the time evolution of spectral energetics in the long-wave and cyclone-wave regimes. The spectral energetics analysis, however, is inadequate to illustrate contributions from local physical processes. The synoptic and geographic delineation of these possible physical processes should be employed to supplement the spectral energetics analysis. In the

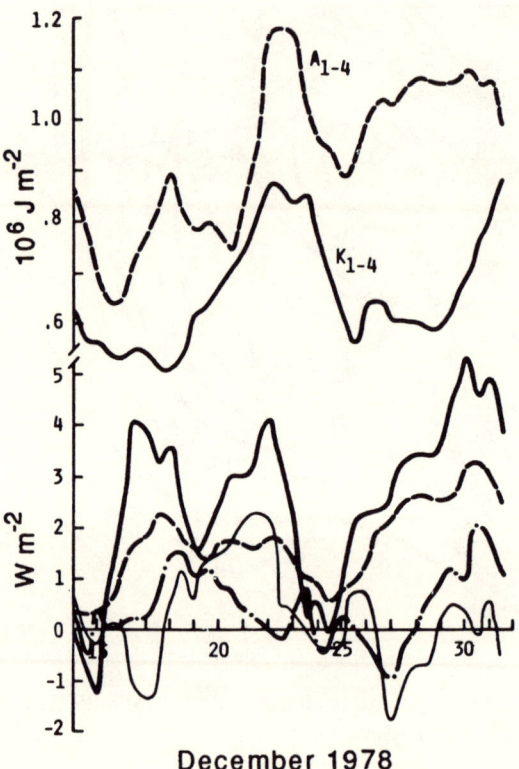

Fig. 13.14: Upper part: time variations of K_{1-4} (solid line) and A_{1-4} (dashed line); lower part: time variations of $C(A_Z, A_{1-4})$ (heavy solid line), $C(A_{1-4}, K_{1-4})$ (dashed line), $C(K_Z, K_{1-4})$ (dot-dashed line), and $CK(n/m, l)_{1-4}$ (thin solid line).

developing phase of the Atlantic blocking – for example, at 1200 GMT 18 December – the cyclone-scale 500 mb height field (Fig. 13.16a) exhibited a deep NW-SE -tilted trough extending from northeastern America into the western Atlantic, with strong cyclone-scale ridges both upstream and down. The cyclone-scale trough corresponded to the trough located over northeastern America in the synoptic chart of this date. Also, the maximum center of $C(A_C, K_C)$ appeared at this time and was located around the northeastern American cyclone system (not shown). The corresponding cyclone-scale heat transports and $C(A_Z, A_C)$ were also centered around the same trough-ridge system. Heat transport increased during this stage, especially the northward warm air transport on the eastern side of the trough in the lower level. The 850-mb cyclone-scale heat transport at 1200 GMT 19 December is shown in Fig. 13.16b.

The time evolutions of the spectral energetics for both long-wave and

ENERGETICS OF SOME SPECIAL PHENOMENA 173

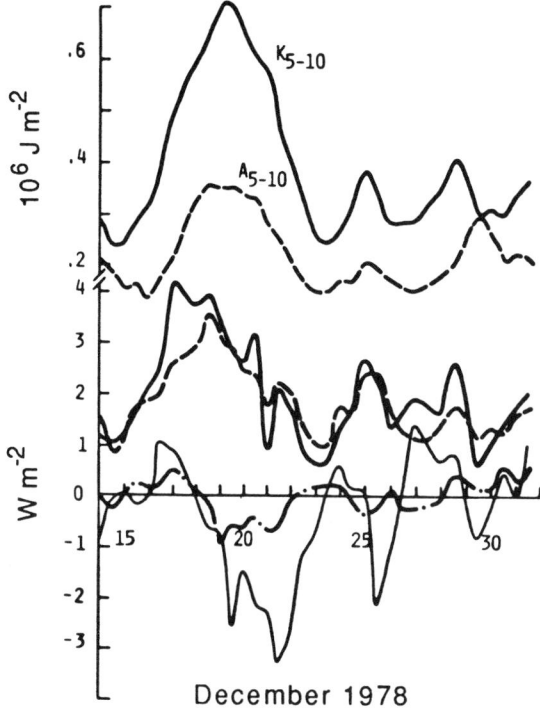

Fig. 13.15: Same as Fig. 13.14, except for wavenumbers 5-10.

cyclone-wave regimes, averaged between 30° and 80° N over the second half of December 1978, are shown in Figs. 13.14 and 13.15. The values of A_L and K_L peaked when the Atlantic block reached its mature phase, but both A_C and K_C attained their maximum values before the Atlantic block matured. The large energy conversions of the long-wave regime before 20 December (not shown) were primarily attributable to a deepening ultralong-wave trough over East Asia. In contrast, the significant energy contents and conversions of the cyclone-wave regime during the third week of December were essentially located over eastern North America and the western Atlantic.

The physical processes involving the development of blocking may be inferred from the time evolution of spectral energetics in the long-wave and cyclone-wave regimes. The spectral energetics analysis, however, is inadequate to illustrate contributions from local physical processes. The synoptic and geographic delineation of these possible physical processes should be employed to supplement the spectral energetics analysis. In the developing phase of the Atlantic blocking – for example, at 1200 GMT 18 December – the cyclone-scale 500 mb height field (Fig. 13.16a) exhibited

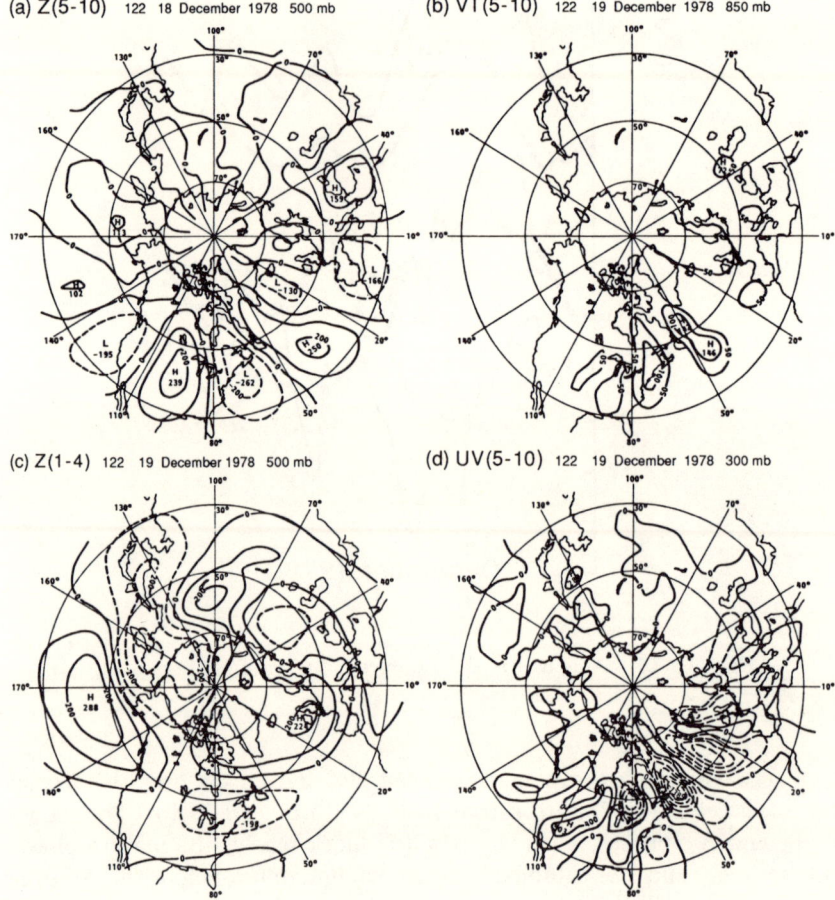

Fig. 13.16: (a) The 500 mb height field of wavenumbers 5-10 (Z_{5-10}) at 1200 GMT 18 December 1978; the contour interval is 100 m, solid (dashed) contours indicate positive (negative) values in this and the other three panels, (b) the 300 mb momentum transport (uv_{5-10}) at 1200 GMT 19 December 1978, (c) Same as (a), except for wavenumbers 1-4 (Z_{1-4}) at 1200 GMT 19 December 1978, and (d) the 850 mb sensible heat transport of wavenumbers 5-10 (vT_{5-10}) at 1200 GMT 19 December 1978.

a deep NW-SE -tilted trough extending from northeastern America into the western Atlantic, with strong cyclone-scale ridges both upstream and down. The cyclone-scale trough corresponded to the trough located over northeastern America in the synoptic chart of this date. Also, the maximum center of $C(A_C, K_C)$ appeared at this time and was located around the northeastern American cyclone system (not shown). The corresponding cyclone-scale heat transports and $C(A_Z, A_C)$ were also centered around the same trough-ridge system. Heat transport increased during this stage, especially the northward warm air transport on the eastern side of the trough in the lower level. The 850-mb cyclone-scale heat transport at 1200 GMT 19 December is shown in Fig. 13.16b.

During the developing phase of the Atlantic block, the long-wave 500 mb height field showed a diffuse ridge over the North Atlantic, with easterly flow extending from south of Greenland to the Hudson Bay (Fig. 13.16c). The NW-SE tilting of the cyclone-scale trough-ridge system causes considerable negative momentum transport. Let us use 1200 GMT 19 December to illustrate this claim. Momentum transports at this time (Fig. 13.16d) were particularly pronounced over Newfoundland and south of Greenland. In fact, the cyclone-scale easterly momentum converged towards the long wave easterly flow from 18 to 22 December. As indicated by the contrast between Figs. 13.14 and 13.15, a negative correlation existed between K_L and K_c. The rapid development of the Atlantic blocking coincided with both the reduction of K_c and the growth of K_L. Moreover, a large nonlinear interaction between the cyclone and long waves also appeared in the same period. Thus, the argument up to this point suggests that the growth of K_L associated with the Atlantic blocking may be caused by this nonlinear interaction.

On 19 December, the K_C maxima existed over northeastern America and the North Atlantic (Fig. 13.17a), but K_L (Fig. 13.17b) was insignificant. In contrast, this situation was reversed after the Atlantic block reached maturity. On 23 December, the K_C center over the North Atlantic disappeared, and the K_L maximum associated with the blocking high was located over Greenland. A comparison of K_C and K_L distributions between Figs. 13.17a and b supports the suggestion that K_c was converted nonlinearly to enhance K_L over the North Atlantic. In view of both the large $CK(n|m,\ell)_L$ and $CK(n|m,\ell)_C$ shown in Figs. 13.14 and 13.15 and the existence of easterly flow south of Greenland during the developing phase of the Atlantic block, the mechanism responsible for this large energy conversion was the northward transport of easterly momentum by cyclone waves into the region of long-wave easterly flow over the North Atlantic, resulting in the acceleration of the easterly flow. Although $C(A_L, K_L)$ (Fig. 13.14) was also as large as or larger than $C(n|m,\ell)_L$, the former quantity was scattered throughout various regions during this time period. We may conclude therefore that the nonlinear interaction provides the main kinetic energy source for the Atlantic block.

A comment should be made here concerning the enhancement of A_L

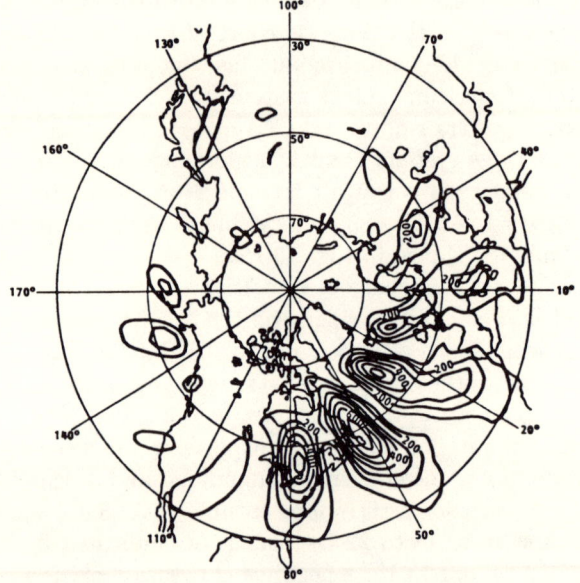

Fig. 13.17: (a) K_{5-10} and (b) K_{1-4} at 500 mb of 1200 GMT December 1978. The contour interval is 100 m^2 s^{-2}.

during the development phase of the Atlantic block. The value of A_L peaked on 22 and 23 December. The maximum A_L of this time period corresponded to the maximum amplitude of the Atlantic thermal ridge associated with the block. The nonlinear interaction $CA(n|m,\ell)_L$ was not a significant quantity, and $C(A_L, K_L)$ was not concentrated in the vicinity of the Atlantic block. The major energy source enhancing A_L came from the excess of $C(A_Z, A_L)$ over $C(A_L, K_L)$. These A_L energetics were consistent with the synoptic structure of long waves associated with the Atlantic block. The height and temperature of this wave regime remained almost in phase throughout the development and mature phases of this block.

All energy conversions supporting K_L had diminished to almost zero when K_L reached its maximum value on 23 December. Although amplitudes of the long-wave regime around the Atlantic block remained pronounced from 22 to 28 December, no significant energy sources existed in the vicinity of the block throughout its mature phase, and K_L declined rapidly (Fig. 13.14) after 23 December. By 1200 GMT 28 December, the Atlantic block started to retrograde, and this retrogression eventually led to the disappearance of the block.

13.3 Energetics of Stationary Eddies

The importance of stationary eddies to atmospheric general circulation is suggested by two examples. As reported by numerous studies (e.g., Hollingsworth et al., 1980), the most common systematic errors of operational forecast models are underforecasts of stationary waves and the excessive zonality of subtropical jet streams. The El Niño/ Southern Oscillation (ENSO) is the most pronounced short-term climate variability. During the ENSO warm event, the effect of anomalously warm sea-surface temperatures over the tropical eastern Pacific exerts on the extratropical atmospheric circulation through the energy propagation from the tropics to the mid-latitudes by stationary eddies (Hoskins and Koraly, 1981). In fact, generation and maintenance of these eddies have been prominent subjects in dynamic meteorology (e.g., Charney and Eliassen, 1949; Bolin, 1950; Smagorinsky, 1953) since the inauguration of numerical weather prediction in the middle of this century.

To understand the maintenance of stationary eddies, several diagnostic approaches have been introduced. Lau (1979) examined time mean local vorticity and heat budgets in the Northern Hemisphere. He found that the vorticity of stationary eddies is maintained primarily by the counterbalance between the local horizontal advection of relative vorticity by the stationary flow and the vortex stretching term. Additionally, he pointed out that heat transports by transient eddies in the lower troposphere have a strong tendency to destroy the zonally asymmetrical component of the stationary temperature field. Note that stationary eddies can be well depicted by time-mean eddy streamfunctions. Moreover, vorticity can be expressed in terms of the Laplacian of this streamfunction; Chen and Chen (1990) have suggested that the Laplacian inverse of the vorticity equation – that is, the *streamfunction budget equation* – may be a highly informative diagnostic tool for illustrating the maintenance of stationary eddies. Additionally, Chen and Yen (1991a) formulated a *velocity-potential maintenance equation* by combining the thermodynamic and the continuity equations. These two equations were adapted by Chen and Yen (1991c) to establish a chain relationship between diabatic heating and divergent and rotational circulations, and to explore the maintenance of stationary eddies.

In addition to diagnostic approaches proposed by both Lau and Chen, the Lorenz energy cycle was adopted by Holopainen (1970) to examine the maintenance of stationary eddies. He found that stationary eddies in the Northern Hemisphere winter derive their available potential energy from zonally averaged flow and convert it to support the kinetic energy dissipated by friction and transient eddies. The Lorenz energy cycle was formulated in terms of the zonally averaged form of the energy variables. To shed light on local contributions to the global or zonal averages of various energy variables of stationary eddies, Lee and Chen (1986) displayed geographically all these energy variables before zonal averaging took place. To illustrate how stationary eddies are maintained, the energetics approach and results of a study will be summarized in this section.

13.3.1 Energetics Scheme

A variable X in the mixed space-time domain may be expressed as

$$X = \overline{X}_Z + \overline{X}_E + X'. \tag{13.21}$$

The kinetic energy of horizontal motion and the available potential energy integrated over the whole atmosphere and averaged over a period of time may be written as

$$K = \frac{1}{g} \int_s \int_0^{p_0} \frac{1}{2} \left(\overline{u^2} + \overline{v^2} \right) dp\,ds = K_Z + K_S + K_T \tag{13.22}$$

and

$$A = \frac{1}{g} \int_s \int_0^{p_0} \frac{1}{2} C_p \gamma \left(T - \widetilde{\overline{T}} \right)^2 dp\,ds = A_Z + A_S + A_T, \tag{13.23}$$

where

$$K = \frac{1}{g} \int_s \int_s \int_0^{p_0} \frac{1}{2} \left(\overline{u_Z^2} + \overline{v_Z^2} \right) dp\,ds, \tag{13.24}$$

ENERGETICS OF SOME SPECIAL PHENOMENA

$$K_S = \frac{1}{g} \int_s \int_0^{p_0} \frac{1}{2} \left(\bar{u}_E^2 + \bar{v}_E^2 \right) \, dp \, ds,$$

$$K_T = \frac{1}{g} \int_s \int_0^{p_0} \frac{1}{2} \left(\overline{u'^2} + \overline{v'^2} \right) \, dp \, ds, \qquad (13.25)$$

and

$$A_Z = \frac{1}{g} \int_s \int_s^{p_0} \frac{1}{2} c_p \gamma \overline{T''}_Z^2 \, dp \, ds, \qquad (13.26)$$

$$A_S = \frac{1}{g} \int_s \int_s^{p_0} \frac{1}{2} c_p \gamma \bar{T}_E^2 \, dp \, ds,$$

$$A_T = \frac{1}{g} \int_s \int_s^{p_0} \frac{1}{2} c_p \gamma \overline{T'^2}_E \, dp \, ds. \qquad (13.27)$$

Note that $(\sim) = 1/s \int_s (\ ) \, ds$ and $(\ )'' = (\ ) - (\sim)$ are departures from the area mean. γ is a stability factor:

$$\gamma = \frac{R}{c_p p} \left(\frac{p_0}{p} \right)^{R/c_p} \left(-\frac{\partial \tilde{\theta}}{\partial p} \right)^{-1}.$$

The energy equations of stationary eddies can be derived from both the equations of motion and the thermodynamic equation:

$$\frac{\partial K_S}{\partial t} = C(A_S, K_S) - C(K_S, K_Z) - C(K_S, K_T) \qquad (13.28)$$
$$\quad - D(K_S) + B(K_S),$$

$$\frac{\partial K_T}{\partial t} = C(A_T, K_T) - C(K_Z, K_T) + C(K_S, K_T) \qquad (13.29)$$
$$\quad - D(K_T) - B(K_T),$$

$$\frac{\partial A_S}{\partial t} = C(A_Z, A_S) - C(A_S, K_S) - C(A_S, A_T) \qquad (13.30)$$
$$\quad + G(A_S) + B(A_S),$$

$$\frac{\partial A_T}{\partial t} = C(A_Z, A_T) - C(A_T, K_T) + C(A_S, A_T) \qquad (13.31)$$
$$\quad + G(A_T) + B(A_T).$$

The detailed mathematical expressions of energy conversions shown in (13.28)-(13.31) are

$$C(A_Z, A_S) = \frac{1}{g} \int_s \int_s^{p_0} \left\{ -c_p \gamma \left(\bar{v}_E \bar{T}_E \right)_Z \frac{\partial \bar{T}_Z}{a \, \partial \phi} \right.$$
$$\left. - c_p \gamma \left(\frac{p}{1000} \right)^{R/c_p} \left(\bar{\omega}_E \bar{T}_E \right)_Z \frac{\partial \theta_Z}{\partial p} \right\} dp \, ds,$$

$$C(A_Z, A_T) = \frac{1}{g}\int\int_s\int_0^{p_0}\left\{-c_p\gamma\,\overline{(v'T')}_Z\frac{\partial \overline{T}_Z}{a\partial\phi}\right.$$
$$\left.-c_p\gamma\left(\frac{p}{1000}\right)^{R/c_p}\overline{(\omega'T')}_Z\frac{\partial\theta''_Z}{\partial p}\right\}dp\,ds,$$

$$C(A_S, A_T) = \frac{1}{g}\int\int_s\int_0^{p_0} c_p\gamma\left\{\left[\overline{T}_E\frac{\partial\overline{(u'T')}_E}{a\cos\phi\,\partial\lambda}\right]_Z + \left[\overline{T}_E\frac{\partial\overline{(v'T')}_E}{a\,\partial\phi}\right]\right.$$
$$\left. + \left[\overline{T}_E\frac{\partial\overline{(\omega'T')}_E}{\partial p}\right]_Z\right\}dp\,ds,$$

$$C(A_S, K_S) = \frac{1}{g}\int\int_s\int_0^{p_0} -(\overline{\omega}_E\overline{\alpha}_E)_Z\,dp\,ds = \frac{1}{g}\int\int_s\int_0^{p_0} -\frac{R}{p}(\overline{\omega}_E\overline{T}_E)_Z\,dp\,ds,$$

$$C(A_T, K_T) = \frac{1}{g}\int\int_s\int_0^{p_0} -\overline{(\omega'\alpha')}_Z\,dp\,ds = \frac{1}{g}\int\int_s\int_0^{p_0} -\frac{R}{p}\overline{(\omega'T')}_Z\,dp\,ds,$$

$$C(K_S, K_Z) = \frac{1}{g}\int\int_s\int_0^{p_0}\left\{(\overline{u}_E\overline{v}_E)_Z\frac{\partial\overline{u}_Z}{a\,\partial\phi} + (\overline{u}_E\overline{\omega}_E)\frac{\partial\overline{u}_Z}{\partial p} + \overline{v}_E\frac{\partial[\overline{v}]}{a\,\partial\phi}\right.$$
$$\left. + (\overline{v}_E\overline{\omega}_E)_Z\frac{\partial\overline{v}_Z}{\partial p} - \frac{\tan\phi}{a}\left[\overline{v}_Z(\overline{u}_E^2)_Z - \overline{u}_Z(\overline{u}_E\overline{v}_E)_Z\right]\right\}dp\,ds,$$

$$C(K_T, K_Z) = \frac{1}{g}\int\int_s\int_0^{p_0}\left\{\overline{(u'v')}_Z\frac{\partial\overline{u}_Z}{a\,\partial\phi} + \overline{(v'\omega')}_Z\frac{\partial\overline{u}_Z}{\partial p} + \overline{(v'^2)}_Z\frac{\partial\overline{v}_Z}{\partial p\phi}\right.$$
$$\left. + \overline{(v'\omega')}_Z\frac{\partial\overline{u}_Z}{\partial p} - \frac{\tan\phi}{a}\left[\overline{v}_Z\overline{(u'^2)}_Z - \overline{u}\,\overline{(u'v')}_Z\right]\right\}dp\,ds,$$

$$C(K_S, K_T) = \frac{1}{g}\int\int_s\int_0^{p_0}\left\{\left[\overline{u}_E\left(\frac{\partial\overline{(u'^2)}_E}{a\cos\phi\,\partial\lambda} + \frac{\partial\overline{(u'v')}_E}{a\,\partial\phi} + \frac{\partial\overline{(u'v')}_E}{\partial p}\right)\right]_Z\right.$$
$$+ \left[\overline{v}_E\left(\frac{\partial\overline{(u'v')}_E}{a\cos\phi\partial\lambda} + \frac{\partial\overline{(v'^2)}_E}{a\partial\phi} + \frac{\partial\overline{(v'\omega')}_E}{\partial p}\right)\right]_Z$$
$$\left. - \frac{\tan\phi}{a}\left[\overline{u}_E\overline{(u'v')}_E\right]_Z - \left[\overline{v}_E\overline{(u'^2)}_E\right]_Z\right\}dp\,ds,$$

$$C(A_Z, K_Z) = \frac{1}{g}\int\int_s\int_0^{p_0} -\frac{R}{p}\overline{\omega}_Z\overline{T}_Z''\,dp\,ds. \tag{13.32}$$

The energetics schemes of the stationary and transient eddies shown in (13.28)-(13.31) are illustrated by the following schematic diagram (Fig.

ENERGETICS OF SOME SPECIAL PHENOMENA

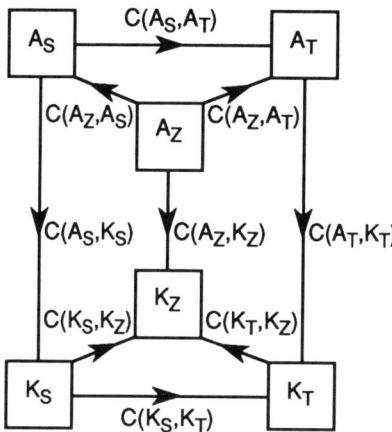

Fig. 13.18: Schematic energy diagram of the stationary (subscribt s) and transient (subscribt T) eddies.

13.18); of course, only an overview of the energetics of stationary and transient eddies can be obtained from such a diagram. Nevertheless, our main concern in this section is the local maintenance of stationary eddies. Various physical processes related to the conversions between energy reservoirs are therefore to be examined. As shown in (13.32), the vertical velocity ω is an important variable in several energy conversions and is usually computed in terms of simplified indirect methods employing certain assumptions. Therefore, the indirectly computed ω field may not be dynamically consistent with other variables used in (13.32). To circumvent this possible inconsistency among data, the history data generated by a general circulation model (GCM) will be used. As Pitcher et al. (1983) have pointed out, the atmospheric circulation simulated by the Community Climate Model (CCM) of the National Center for Atmospheric Research (NCAR) is relatively *realistic*, although not real. Lee and Chen (1986) used the history data generated by the NCAR CCM in the energetics analysis of simulated stationary eddies. These results will be used to illustrate the maintenance of stationary eddies.

13.3.2 The Three-Dimensional Structure of Stationary Eddies

The horizontal structure of stationary eddies is illustrated in terms of seasonal-mean height ($\overline{Z}_E$) and temperature ($\overline{T}_E$) eddy anomalies. The upper troposphere $\overline{Z}_E$ anomalies (Fig 13.19a) exhibit distinct low- and high-latitude regimes, with a reversal of the $\overline{Z}_E$ sign at about 30° N, as observed by Holopainen (1970) and Lau (1979). The upper troposphere $\overline{Z}_E$

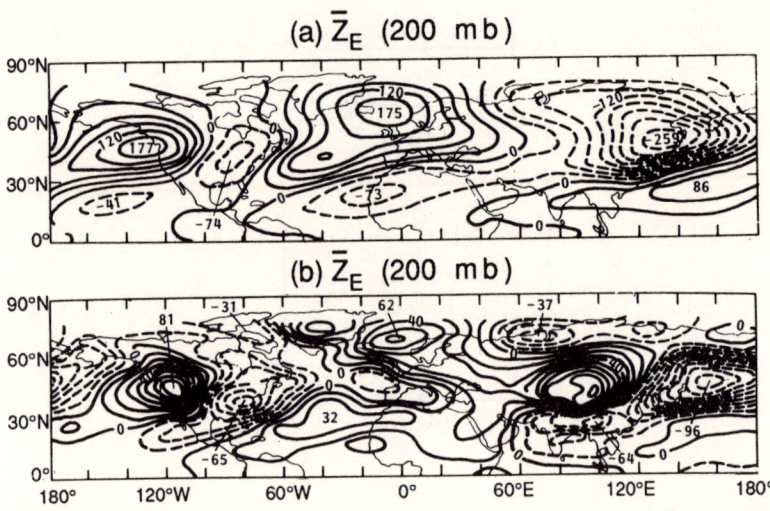

Fig. 13.19: Geopotential heights of stationary eddies ($\bar{Z}_E$) at (a) 200 mb and (b) 850 mb simulated by the NCAR Community Climate Model. The contour intervals of (a) and (b) are 30 and 10 m, respectively.

anomalies of the high-latitude regime possess negative centers over the east coasts of East Asia and North America and positive centers over the eastern Pacific and Atlantic. These centers correspond to troughs and ridges, respectively; the simulation of $\bar{Z}_E$ anomalies of all positive and negative centers are somewhat smaller in their amplitudes than indicated by observations. Pitcher et al. (1983) suggested that this simulation deficiency is a result of a colder-than-observed tropospheric temperature. Lower-troposphere $\bar{Z}_E$ anomalies (Fig. 13.19b) exhibit a structure similar to that of upper-troposphere $\bar{Z}_E$ anomalies.

The longitude-pressure cross sections of the $\bar{Z}_E$ vertical structure (Fig. 13.20) tilt slightly westward, a typical baroclinic structure of stationary eddies. In fact, this feature can be seen by comparing $\bar{Z}_E$ (200 mb) with $\bar{Z}_E$ (850 mb), especially over the east coast of Asia. The $\bar{Z}_E$ maximum amplitudes at middle latitudes exist in the vicinity of the tropopause, whereas those at higher latitudes extend into the lower stratosphere. In contrast, a phase change of the $\bar{Z}_E$ vertical structure between different latitudes indicates that stationary eddies are trapped in the low latitudes, and that the stationary eddies' upward propagation of energy from troposphere to stratosphere takes place in the high latitudes.

$\bar{T}_E$ and $\bar{Z}_E$ can be related to each other through the hydrostatic relation. It is expected that the $\bar{T}_E$ maximum amplitude exists at the level where the maximum $\bar{Z}_E$ vertical gradient is located. According to the $\bar{Z}_E$ anomalies shown in Fig. 13.20, the $\bar{T}_E$ maximum amplitude may appear in the middle troposphere of the middle latitudes. We therefore select $\bar{T}_E$(500

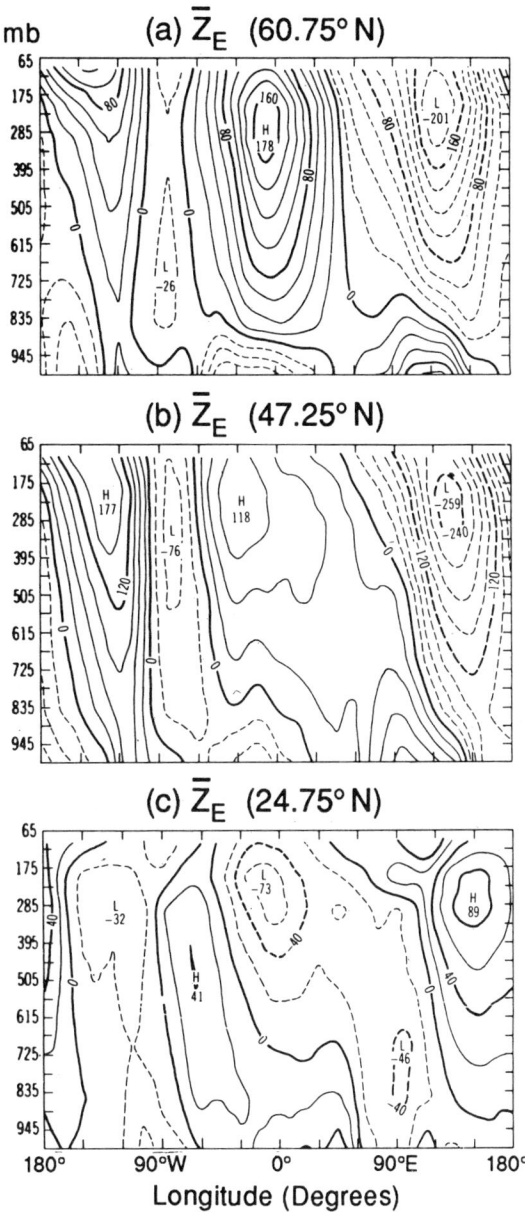

Fig. 13.20: Longitude-pressure cross sections of winter $\bar{Z}_E$ of the NCAR CCM at (a) 60.75°N, (b) 47.25° N and (c) 24.75° N. The contour intervals of (a)-(c) are 20 m, 30 m and 20 m, respectively.

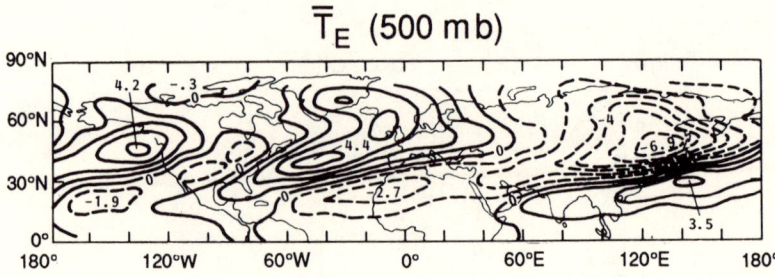

Fig. 13.21: The winter temperature of stationary eddies ($\overline{T}_E$) at 500 mb simulated by the NCAR CCM. The contour interval is 1 K.

mb) to show the horizontal thermal structure of the stationary eddies (Fig. 13.21). Like upper-troposphere $\overline{Z}_E$ anomalies (Fig. 13.20a), $\overline{T}_E$ (500 mb) undergoes a spatial phase change at about 30° N. Also, positive and negative $\overline{T}_E$ anomalies are associated with oceans and continents, respectively, and correspond well to positive and negative $\overline{Z}_E$ anomalies.

The longitude-pressure cross sections of the $\overline{T}_E$ anomalies at three latitudes are shown in Fig. 13.22. Although minor maximum amplitudes appear in the middle troposphere of the mid-latitudes, $\overline{T}_E$'s major maximum amplitudes generally exist in the lower troposphere. A vertical phase change of $\overline{T}_E$ occurs in the upper troposphere where the $\overline{Z}_E$ maximum amplitudes are located. As discussed by Lau (1979), the $\overline{T}_E$ vertical structure does not exhibit a significant westward tilting in the troposphere as the $\overline{Z}_E$ vertical structure does. Comparing Figs. 13.20 and 13.22, one finds that, in the lower troposphere, the $\overline{T}_E$ vertical structure may be closely related to the underlying orography, with the cold air lying over the east coasts of East Asia and North America, and the warm air over the eastern oceans. Furthermore, a comparison of the $\overline{T}_E$ and $\overline{Z}_E$ vertical structures suggests that in the lower troposphere, thermal lows are located over warm oceans and thermal highs over continents, and that the opposite occurs in the upper troposphere. Additionally, based upon both the westward tilting of $\overline{Z}_E$ anomalies and the fact that the positions of maxima and minima $\overline{T}_E$ are about a quarter-wavelength to the west of those of the $\overline{Z}_E$ anomalies, it can be assumed that the stationary eddies are baroclinic.

13.3.3 Energetics of Stationary Eddies

According to the schematic energy diagram (Fig. 13.18), stationary eddies can be maintained by three energetic processes: (a) interactions between zonal-mean states and stationary eddies, (b) release of thermal energy through baroclinic processes, and (c) interactions between stationary

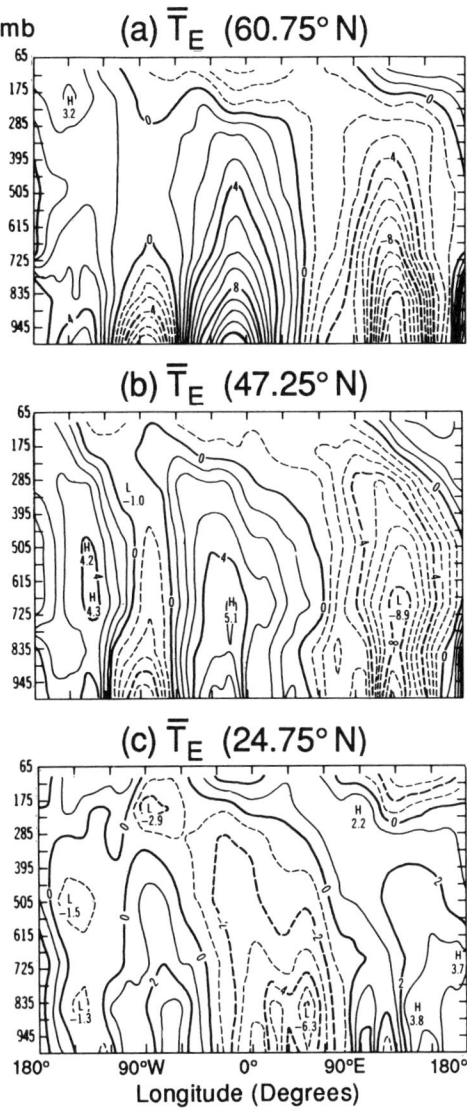

Fig. 13.22: Longitude-pressure cross sections of winter $\overline{T}_E$ of the NCAR CCM at (a) 60.75° N, (b) 47.25° N and (c) 24.75° N. The contour intervals of (a)-(c) are 1 K.

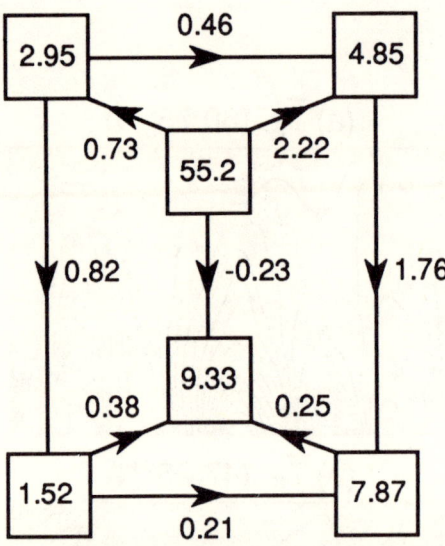

Fig. 13.23: Energy diagrams of the stationary and transient eddies in the winter Northern Hemisphere of the NCAR CCM, following Fig. 13.18. Units of energy and energy conversions are 10^5 J m^{-2} and W m^{-2}, respectively.

and transient eddies. The energetics analysis of stationary eddies simulated by the model is displayed in Fig. 13.23 and summarized in this section. The thermal field ($\overline{T}_E$) of stationary eddies, indicated by A_s, is maintained by thermal interaction (a) – that is, $C(A_Z, A_s) > 0$. Furthermore, the dynamic fields ($\overline{u}_E$ and $\overline{v}_E$), indicated by K_s, are supported through release of the available potential energy (A_s) of stationary eddies by baroclinic processes. In turn, the stationary eddies maintain zonal-mean flow ($\overline{u}_Z$) through dynamic interactions (a). Transient eddies essentially extract energy out of stationary eddies by both thermal and dynamic interactions (c) between these two eddy components of the atmospheric circulation. The energetics analysis of the model stationary eddies is therefore consistent with that of Holopainen's (1970).

The Lorenz energy cycle of stationary and transient eddies provides an overview of the maintenance of stationary eddies discussed in Section 13.3.2. The horizontal distribution of various energy variables and the physical processes involved in these energy variables should be pursued.

(A) *Energies*

In view of the vertical structure of $\overline{T}_E$ (Fig. 13.22), it follows that large A_s exists in the middle and lower troposheres (Fig. 13.24a). Compared with Chen's (1982) observational analysis for the 1976/77 winter, the model simulates smaller (larger) A_s in the middle (lower) troposphere. Recall that thermal ridges and troughs are located over the eastern oceans and continents and are reflected in the horizontal distribution of vertically integrated A_s (Fig. 13.24c).

The maximum K_s is associated with the maximum westerlies, as can be seen in the latitude-pressure cross section of K_s (Fig. 13.24c). However, the model K_s is smaller than K_s in Chen's (1982) observational analysis. The upper-level zonal wind anomalies ($\overline{u}_E$) of stationary eddies can be estimated by the north-south gradient of $\overline{Z}_E$ (200 mb) (Fig. 13.19a). Thus, we can expect negative $\overline{u}_E$ (200mb) anomalies to exist in the eastern Pacific and in the Atlantic, and positive $\overline{u}_E$ (200 mb) to be associated with the East Asian jet. This is shown in Fig. 13.24e. Moreover, the horizontal distribution of the vertically integrated K_s of the model generally corresponds well with that of $\overline{u}_E$ (200 mb).

(B) *Interaction between the Zonal-Mean State and Stationary Eddies*

Fig. 13.25 illustates how two types of interactions – *thermal* and *dynamic* – exist between the zonal-mean state and stationary eddies. The former is measured in terms of the covariance between $(\overline{v}_E\overline{T}_E)_Z$ and $\partial T_Z/a\partial\varphi$ – that is, $C(A_Z, A_E)$ – whereas the latter is measured in terms of the covariance between $(\overline{u}_E\overline{v}_E)_Z$ and $\partial u_Z/a\,\partial\varphi$ – that is, $C(K_E, K_Z)$.

Significant $(\overline{v}_E\overline{T}_E)_Z$ (Fig. 13.25b) and zonal-mean temperature gradients appear in the mid-latitudes of the middle and low tropospheres. Thus, over this region, the thermal interaction field converts A_Z to support the thermal field of stationary eddies denoted by A_s. Note that the maximum value of $C(A_Z, A_s)$ is about half that of Chen's observational result. Lee (1983) has pointed out that $\partial \overline{T}_Z/a\,\partial\varphi$ was properly simulated by the model; the small value of the $C(A_Z, A_E)$ is thus attributed to the small $(\overline{v}_E\overline{T}_E)_Z$ of the model.

In the troposphere, $\overline{T}_Z$ generally decreases towards the poles. Therefore, positive $\overline{v}_E\overline{T}_E$ represents the local contribution to positive $C(A_Z, A_E)$. Several positive $\overline{v}_E\overline{T}_E$ (850 mb) centers (Fig. 13.25e) are located over East Asia, the west coast of North America, and most of the Atlantic. Despite smaller magnitudes, these $\overline{v}_E\overline{T}_E$ (850 mb) centers are relatively consistent with those found by Blackmon et al.'s (1977) observational analysis. The contrast between $\overline{v}_E\overline{T}_E$ (850 mb) and A_s (Fig. 13.24c) indicates that significant $C(A_Z, A_s)$ corresponds to large A_s. Obviously, A_s is maintained locally by this energy conversion.

Momentum transports by stationary eddies (Fig. 13.25d) occur primarily in the upper troposphere. A maximum northward transport is located at 30° N, whereas a maximum southward transport is located at 60° N. The separation between northward and southward transport appears around 50° N. However, the northward (southward) $(\overline{u}_E\overline{v}_E)_Z$ of the model is somewhat (much) smaller than the observational findings of Chen (1982). Note

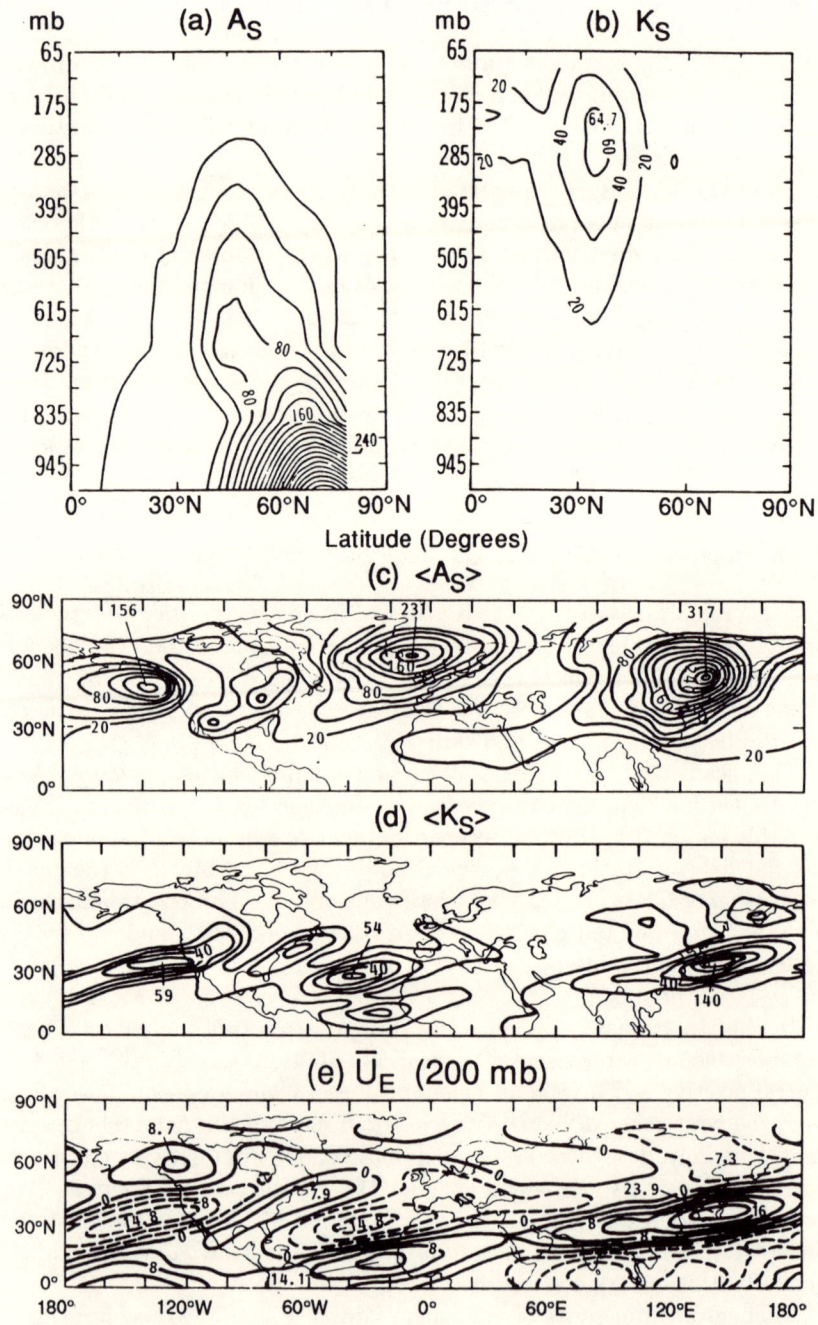

Fig. 13.24: Latitude-pressure cross sections of (a) A_S and (b) K_S, and horizontal distributions of (c) $\langle A_S \rangle$, (d) $\langle K_S \rangle$ and (e) $\overline{u}_E$ (200 mb) in the model. The contour intervals of (a)–(b), (c)–(d), and (e) are 20 J m^{-2} mb^{-1}, 10^4 J m^{-2} and 4 m s^{-1}, respectively.

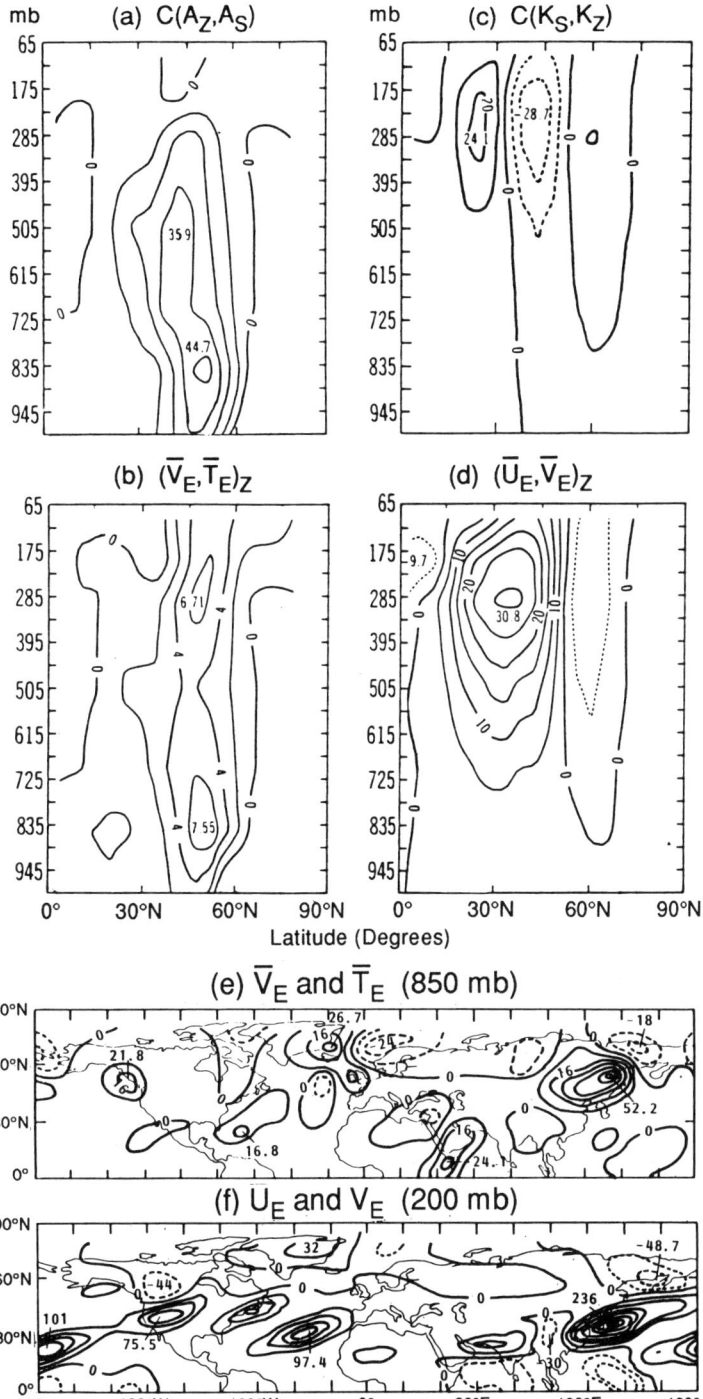

Fig. 13.25: Latitude-pressure cross sections of (a) $C(A_Z, A_S)$, (b) $(\overline{v}_E \overline{T}_E)_Z$, (c) $C(K_S, K_Z)$, and (d) $(\overline{u}_E v_E)_Z$, and horizontal distributions of (e) $\overline{v}_E \overline{T}_E$ (850 mb) and (f) $\overline{u}_E \overline{v}_E$ (200 mb). The contour intervals of (a) and (c), (b), (d), (e), and (f) are 10^{-3} W m^{-2} mb^{-1}, 2 K° m s^{-1}, 10 m^2 s^{-2}, 10 K° m s^{-1}, and 20 m^2 s^{-2}, respectively.

that the maximum $\bar{u}_Z$ (not shown) appears around 35° N. As illustrated by Fig. 13.25c, the interation between $\bar{u}_Z$ and stationary eddies results in the acceleration (deceleration) of the mean-zonal flow south (north) of 30° N.

The north-south gradient of $\bar{u}_Z$ (200 mb) changes sign at about 35° N. Fig. 13.25f indicates that south of the jets three $\bar{u}_E \bar{v}_E$ (200 mb) positive centers over the western and central Pacific and the Atlantic account for the K_Z maintenance by positive $C(K_s, K_Z)$. In contrast, around 45° N the positive $\bar{u}_E \bar{v}_E$ (200 mb) centers around Japan and the west and east coasts of North America contribute to the major conversion from K_Z to K_s. Note that these centers correspond with large K_s (Fig. 13.24d). Such a correspondence indicates that stationary eddies are barotropically maintained by mean-zonal flow over these locations.

(C) *Release of Available Potential Energy by Stationary Eddies*

The release of A_s is accomplished by the thermally direct overturning of stationary eddies and is evaluated by the covariance $(\bar{\omega}_E \bar{T}_E)_Z$. This thermally direct overturning in the midlatitudes is clearly induced by the horizontal sensible heat transport of stationary eddies. As inferred from the maximum $(\bar{v}_E \bar{T}_E)_Z$ (Fig. 13.25b) in the lower troposphere of the midlatitudes, it is expected that the maximum $(\bar{\omega}_E \bar{T}_E)_Z$ will appear in the middle troposphere of the mid-latitudes. This inference is confirmed by Fig. 13.26a, whose maximum value for $C(A_s, K_s)$ is comparable to that of the observational analyses of Chen et al. (1981) and Chen (1982).

In view of the existing relation between $(\bar{v}_E \bar{T})_Z$ and $(\bar{\omega}_E \bar{T}_E)_Z$, $\bar{\omega}_E \bar{T}_E$ (500 mb) [or $C(A_s, K_s)$ (500 mb)] centers in the horizontal domain should correspond to $\bar{v}_E \bar{T}_E$ (850 mb) centers. In fact, this does occur over the east coasts of Asia and North America and over the west coast of North America. By comparing the synoptic charts of $\bar{\omega}_E$ (500 mb) (Fig. 13.26b) and $\bar{T}_E$ (500 mb) (Fig. 13.21a), one finds that $\bar{\omega}_E \bar{T}_E$ (500 mb) centers over the former two regions can be attributed to the sinking of cold air and that over the latter region can be attributed to the rising of warm air. Finally, the relation between energy reservoirs A_s and K_s and $C(A_s, K_s)$ (Fig. 13.26c) can be inferred from the contrast between the horizontal distributions of these energy variables. Significant $C(A_s, K_s)$ is located over regions where A_s and K_s are large. Seemingly, the release of A_s is a vital energy source maintaining K_s.

(D) *Interaction Between Stationary and Transient Eddies*

The interations between stationary and transient eddies are of two types: *thermal* and *dynamic*. The former interaction is primarily contributed by the determined covariance between $\bar{T}_E$ and the divergence of transient eddy heat flux, whereas the latter is largely evaluated by the

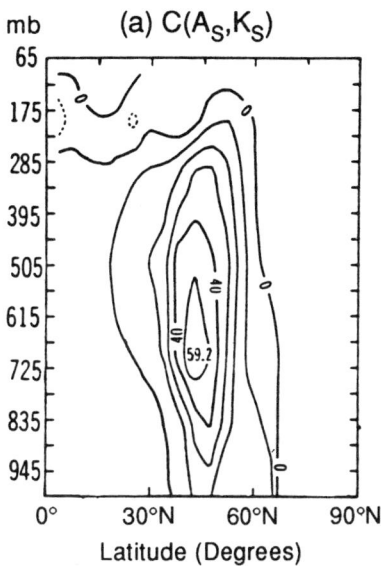

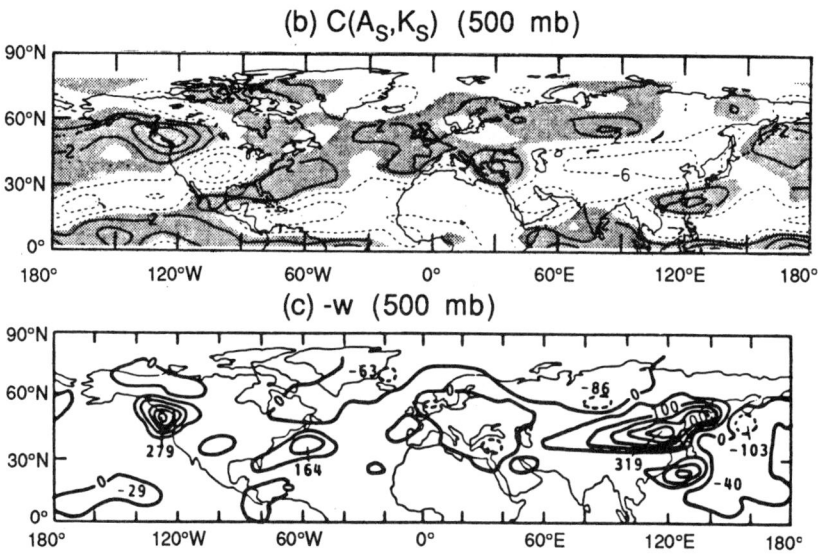

Fig. 13.26: (a) Latitude-pressure cross sections of $C(A_S, K_S)$ (500 mb) and horizontal distributions of (b) the winter-mean pressure velocity [ω (500 mb)] and (c) $C(A_S, K_S)$ (500 mb). Contour intervals of (a), (b), and (c) are 10^{-3} W m^{-2}, 2×10^{-4} mb s^{-1}, and 5×10^{-4} m^2 s^{-3}, respectively.

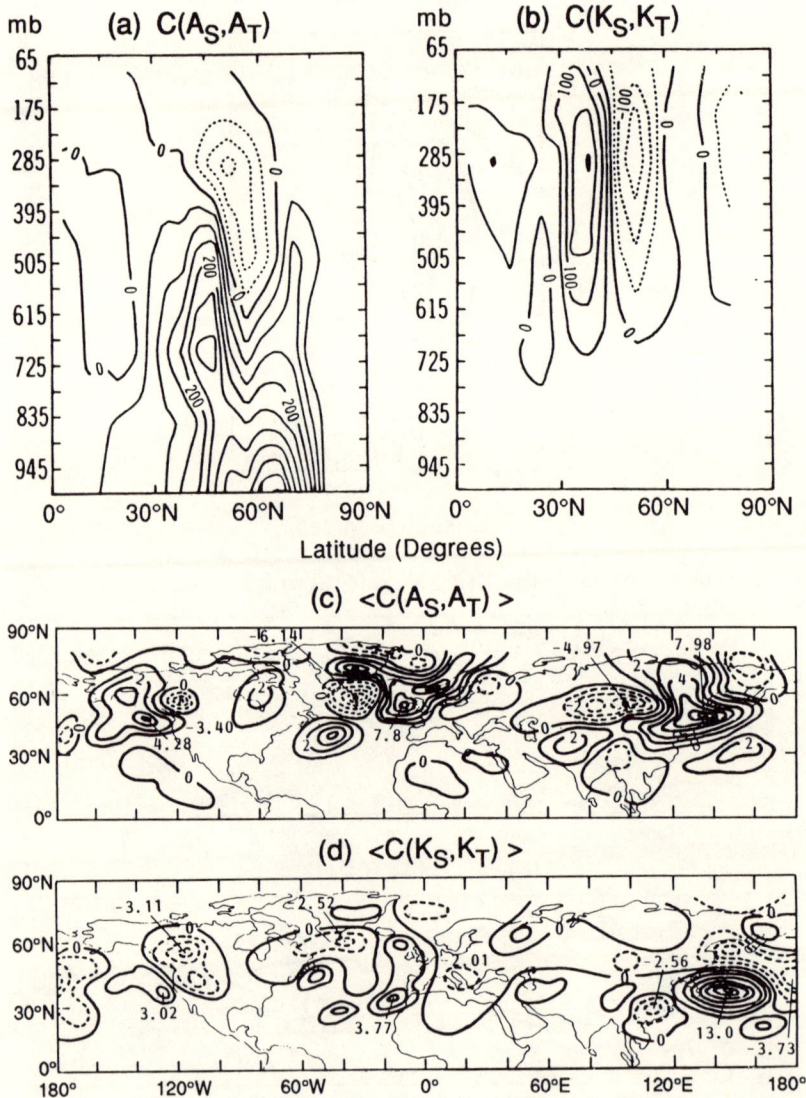

Fig. 13.27: Latitude-pressure cross sections of (a) $C(A_S, A_T)$ and (b) $C(K_S, K_T)$ and horizontal distributions of (c) $\langle C(A_S, A_T) \rangle$ and (d) $\langle C(K_S, K_T) \rangle$. Contour intervals of (a)–(b) and (c)–(d) are 5×10^{-4} W m^{-2} mb^{-1} and 1 W m^{-2}, respectively.

covariance between $\overline{\mathbf{V}}_E$ and the divergence of transient eddy momentum transport.

The hemispheric-average Lorenz energy cycle of stationary and transient modes (Fig. 13.23) shows that the interaction between thermal stationary eddies and transient eddy heat transport consumes A_s. As shown in Fig. 13.27a, this consumption of A_s by transient eddies occurs primarily in the middle and lower tropospheres of the mid-latitudes. Lau (1979) has pointed out that transient eddy heat flux convergence and thermal troughs at 700 mb correspond to each other. Thus, the contrast between geographical distributions of $C(A_s, A_T)$ (Fig. 13.27c) and $\overline{T}_E$ (500 mb) (Fig. 13.21) indicates that thermal troughs (ridges) are associated with the convergence (divergence) of transient eddy heat transport. The role played by transient eddies is therefore to dissipate A_s – that is, to weaken stationary thermal eddies.

The maximum northward momentum transport by transient eddies (not shown) appears near 30° N in the upper troposphere. As can be inferred from the K_S cross section (Fig. 13.24b), the maximum amplitude of $\overline{\mathbf{V}}_E$ is located around 35° N. The dissipation of K_S by transient eddies, as indicated by Fig. 13.27b, occurs primarily around the maximum zonal-mean westerly. In contrast, significant convergence of transient eddy momentum transport exists north of the maximum zonal-mean westerly, and K_S is supported by transient eddies. The contrast between positive and negative $C(K_S, K_T)$ shown here is therefore consistent with Murakami's (1963) analysis, in which $C(K_S, K_T)$ is positive in the mid-latitudes and negative in the high latitudes. Geographically, postive $C(K_S, K_T)$ centers (Fig. 13.27b) exist over the eastern oceans, over the eastern part of the United States, and in the vicinity of Japan, particularly in the latter region. Large K_S also appears over these regions. Evidently, K_S is locally dissipated by transient eddies through the transient eddy momentum transport. Conversely, $C(K_S, K_T)$ is negative; that is, K_S is supported by transient eddies south of the Aleutian Islands, Iceland and the west coast of North America, where convergence of transient eddy momentum transport takes place.

14

QUASI-PERIODIC VARIATION OF ATMOSPHERIC ENERGETICS

An understanding of the physical processes and mechanisms creating or maintaining the quasi-periodic variation and atmospheric circulation properties would benefit both weather and climate prediction. The two quasi-periodic variations of atmospheric energetics appearing most frequently in the meteorological literature are *annual variation* and *vacillation*.

The most significant short-term climatic cycle is the annual variation of the atmospheric circulation. Rasmusson and Carpenter (1982) observed that the life cycle of the Southern Oscillation and the annual cycle of many atmospheric circulation elements over the Pacific Ocean have a phase-lock relation. To predict the interannual variability of atmospheric circulation, we should have a better understanding of the annual cycle. After Krueger et al. (1965) initiated an analysis of the annual variation of atmospheric energetics of the lower half of the troposphere, Wiin-Nielsen (1967) extended this effort to cover the entire troposphere. To identify the phase relationships between annual cycle components of different energy variables, Wiin-Nielsen (1967) performed a spectral analysis of various energy variables in time. This approach was applied by Chen and Buja (1983) to the global data generated by the FGGE III-b analyses of the European Centre for Medium Range Weather Forecasts. However, Krueger et al. and Wiin-Nielsen used the data north of 20° N produced by the National Meteorological Centre. Oort and Peixóto (1974) merged the daily data of the MIT general circulation library and R.E. Newell's (MIT) tropical data between 25° S and 25° N, and with this new data set carried out a more extensive energy analysis for the annual cycle.

The annual variation of atmospheric energetics analyzed by the studies just cited was presented in a hemispheric-averaged format. As revealed from the kinetic energy budget analysis of subtropical jet streams in the winter Northern Hemsiphere (Section 13.1), the annual variation of atmospheric energetics must possess a pronounced spatial variation. Kung (1966,1967) evaluated extensively the source and sink of atmospheric kinetic energy over the North American continent where the kinetic energy of the North American jet is generated. Kung and Soong (1969) later adopted Wiin-Nielsen's (1967) spectral analysis in time to explore the annual cycle component of the regional kinetic energy budget. Holopainen (1973) analyzed the kinetic energy budget over the British Isles, where the downstream region of the North American jet is located, but did not successfully come up with a proper kinetic energy dissipation. Neither was the annual variation of the regional kinetic energy budget pursued in the downstream region of jets.

Since the 1940s, the quasi-periodic variation of the zonal circulation intensity has been a topic in long-range weather forecasting. To measure the intensity of the zonal circulation, one uses zonal-mean surface pressure differences between 35° N and 55° N or the 700 mb westerlies averaged over this latitudinal zone as a *zonal index*. It has been documented by numerous investigators – for example, Petterssen (1956) and Palmén and Newton (1969) – that this zonal index undergoes a quasi-periodic (three-six week) variation, called the index cycle and often referred to as *vacillations*. The alternation between high and low zonal indices is reflected in the quasi-periodic variation of zonal kinetic energy (K_Z), which is contributed primarily by the zonal-mean westerlies. For this reason, the Lorenz energy cycle has been used by many studies to illustrate the physical processes involved in the index cycle.

14.1 Annual Variation in the Northern Hemisphere

14.1.1 Annual Variation of the Lorenz Energy Cycle

The Lorenz energy cycle was developed in Chapter 7, (7.28) and (7.29), based upon the quasi-geostrophic system. If the daily upper-air wind, temperature, and height fields are available, a diagnostic analysis of the Lorenz energy cycle can be performed. In their examination of the long-term variability of atmospheric energetics, Oort and Peixóto (1976) adopted this approach using the daily upper-air data archived in the Massachusetts Institute of Technology General Circulation Library for the five-year period of May 1958 – April 1963. Temperature can be used directly in computations, however. Thus, expressions of energy variables involving specific volume (α) in (7.3) and (7.29) are modified as follows:

$$A_Z = \int_m c_p \gamma \frac{1}{2} T'^2_Z dm, \qquad A_E \int_m c_p \gamma \frac{1}{2} T'^2_E \, dm, \qquad (14.1)$$

$$C(A_Z, A_E) = \int_m -c_p \gamma \, (v_E T_E)_Z \frac{\partial T_Z}{a \, \partial \varphi},$$

$$G(A_Z) = \int_m \frac{\gamma}{c_p} T'_Z H'_Z \, dm,$$

$$G(A_E) = \int_m \frac{\gamma}{c_p} T_E' H_E' \, dm. \qquad (14.2)$$

The conversion between A and K is involved in the ω field. Because of

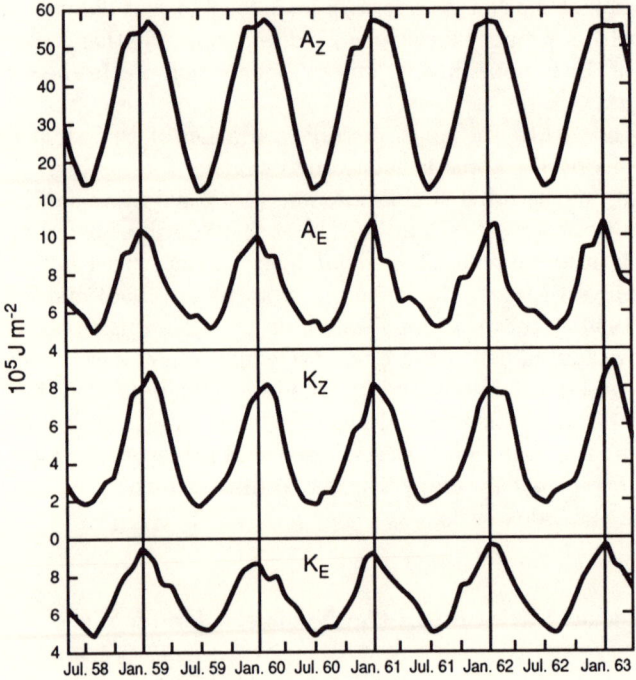

Fig. 14.1: Time varitation of the monthly mean energies (A_Z, A_E, K_Z, K_E) over a five-year period (May 1958-April 1963).

the difficulty of evaluating ω, Oort and Peixóto suggested calculating the work done by ageostrophic flow, thus using instead

$$C(A_Z, K_Z) = \int_m -v_Z \frac{\partial \phi_Z}{a \partial \varphi} \, dm,$$

$$C(A_E, K_E) = \int_m -\mathbf{V}_E \bullet \nabla \phi_E \, dm. \quad (14.3)$$

The atmospheric general circulation attains maximum and minimum strength in winter and summer, respectively, and so do K_Z and K_E. Additionally, the geostrophic thermal wind suggests that the annual variations of A_Z and A_E should behave in manners similar to those of K_Z and K_E. Pronounced annual variations, with a maximum in winter and a minimum in summer, clearly emerge in the five-year time series of monthly mean A_Z, A_E, K_Z, and K_E (Fig. 14.1). The A_Z time series possesses a rather flat maximum around February and a sharp minimum in July. In contrast, the K_Z time series has a wide minimum in July and a sharp maximum in February. Moreover, the A_Z time series is almost symmetric with respect to both maximum and minimum, but the K_Z time series is not, especially

QUASI-PERIODIC VARIATION

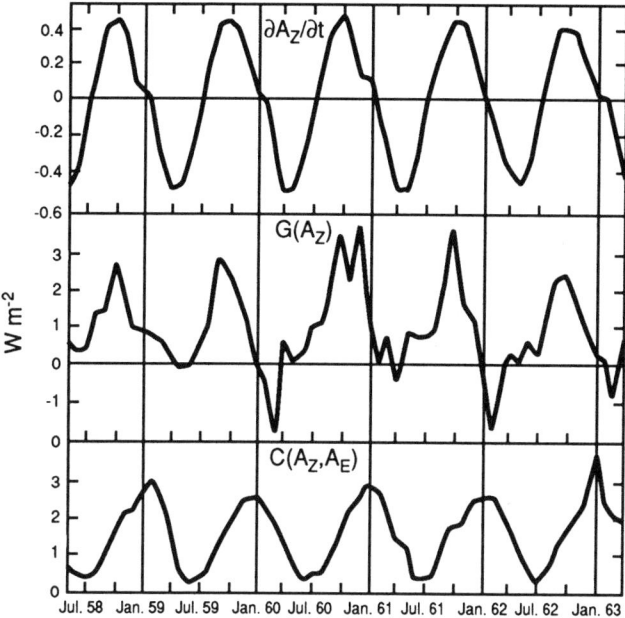

Fig. 14.2: Time variation of the monthly-mean values of various terms in the A_Z equation (7.12) over a five-year period (May 1958–April 1963).

in terms of the summer minima, as observed by Krueger et al. (1965) and Wiin-Nielsen (1967). They found that A_E and K_E annual variations are highly correlated with a fast autumn buildup and a slow spring decline.

The nearly sinusoidal time series of A_Z, A_E, K_Z, and K_E suggests that their time derivatives are about one-quarter cycle out of phase with respect to the energy time series. That is to say, time derivatives of these energies reach their maxima during autumn buildup and their minima during spring decline. The argument here is substantiated by the time series of $\partial A_Z/\partial t$, $\partial A_E/\partial t$, $\partial K_Z/\partial t$, and $\partial K_E/\partial t$ shown in Figs 14.2-14.5. The magnitude of A_Z is approximately an order greater than the magnitudes of A_E, K_Z, and K_E; the magnitude of $\partial A_Z/\partial t$ (with an amplitude of about 0.4 W m^{-2}) is also approximately an order greater than the magnitudes of $\partial A_E/\partial t$, $\partial K_Z/\partial t$, and $\partial K_E/\partial t$. According to (7.28), a time derivative of any energy indicates the combined effect of several physical processes represented by the energy conversions indicated in the equation. To illustrate the physical processes related to annual variations of energy time derivatives, the five-year time series of monthly mean energy conversions is displayed in Figs 14.2-14.5.

As shown in the Lorenz energy cycle (Fig. 7.8), the major energy conversions are $G(A_Z) \to C(A_Z, A_E) \to C(A_E, K_E) \to D(K_E)$. Be-

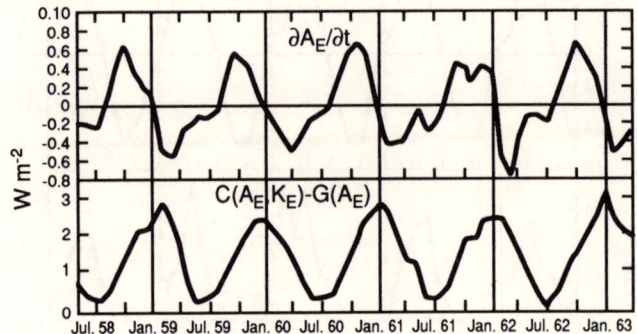

Fig. 14.3: Time variation of the monthly-mean values of various terms in the A_E equation (7.13) over a five-year period (May 1958-April 1963).

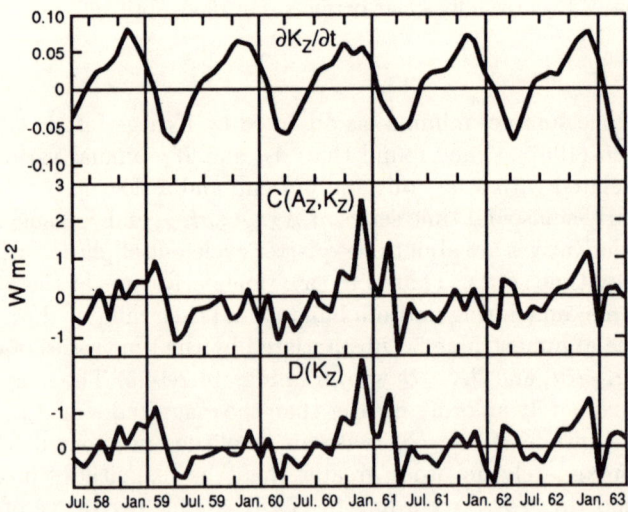

Fig. 14.4: Time variation of the monthly-mean values of various terms in the K_Z equation (7.26) over a five-year period (May 1958-April 1963).

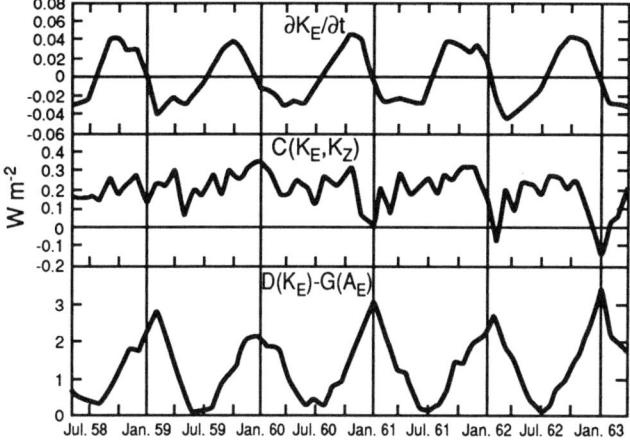

Fig. 14.5: Time variation of the monthly-mean values of various terms in the K_E equation (7.27) over a five-year period (May 1958-April 1963).

cause geostrophic flow had been underestimated in the available data, $C(A_E, K_E)$ was not directly calculated by Oort and Peixóto (1976). Instead, $C(A_E, K_E) - G(A_E)$ and $D(K_E) - G(A_E)$ were computed as residuals. Previous studies reviewed by Wiin-Nielsen (1967) and Oort and Peixóto (1974) concluded that $G(A_E)$ is much smaller than $G(A_Z)$. In essence, $G(A_E)$ plays a minor role in annual variations of $C(A_E, K_E) - G(A_E)$ and $D(K_E)-G(A_E)$. Time series of $G(A_Z)$, $C(A_Z, A_E)$, $C(A_E, K_E) - G(A_E)$, and $D(K_E) - G(A_E)$ (Figs. 14.2-14.5, respectively) reveal that these energy conversions are comparable in terms of amplitudes of their annual variations (2.5~3.0 W m^{-2}) and synchronous in their annual variations, with maxima in winter and minima in summer. In short, annual variations of these energy conversions closely follow those of various energies (Fig. 14.1) – that is, the annual variation of the atmospheric circulation. It is clearly shown in Figs. 14.2-14.5 that $D(K_Z)$, $C(A_Z, K_Z)$, and $C(K_E, K_Z)$ have small numerical values [1]. $D(K_Z)$ is obtained as a residual

---

[1] Because of the small numerical values of $C(A_Z, K_Z)$ and $C(K_E, K_Z)$, caution should be exercised when applying these two quantities to physical explanations of atmospheric general circulation. In the mid-latitudes, T_Z'' changes its sign from positive in the low latitudes to negative in the high latitudes. ω_Z exhibits a three-cell structure: Hadley, Ferrel, and a high-latitude cell. The covariance between T_Z'' and ω_Z alternates between positive and negative values of $C(A_Z, K_Z)$. The small numerical value of this energy conversion is attributed to cancellations caused by the alternation between positive and negative values in the latitude-pressure cross-section (Chen et al. 1981; Chen 1982). Oort and Peixóto (1976) used (14.3) to evaluate $C(A_Z, K_Z)$ and attained an alternation between positive and negative values in the upper troposphere, like Chen and Lee's (1982) calculation of $-(\mathbf{V} \bullet \nabla \phi)_Z$. In another form of (14.3), it still results in a small numerical value.

$C(K_E, K_Z)$ is evaluated with the covariance between $(u_E, v_E)_Z$ and $\partial u_Z/a\, \partial\varphi$. The maximum u_Z exists in the upper troposphere at 30°N, although $(u_E v_E)_Z$ changes its

in the K_Z equation of (7.28) in which $C(A_Z, K_Z)$ is a dominant term. It is not surprising that the variability of $D(K_Z)$ resembles that of $C(A_Z, K_Z)$. The negative values of $D(K_Z)$ are perhaps caused by the uncertainty inherent in the evaluation of $C(A_Z, K_Z)$. Obviously, it is relatively difficult to obtain a clear seasonal cycle of these three energy conversions. This difficulty was pointed out by Wiin-Nielsen (1967) in his pilot study of the annual variation of atmospheric energetics.

The primary energy source driving the atmospheric general circulation is north-south differential heating ($H_Z{'}$). The covariance between this physical quantity and the north-south departure of zonal-mean temperature from its global average provides a measurement of $G(A_Z)$. Many diagnostic studies of atmospheric diabatic heating – for example, Chen and Baker (1987) – come to the conclusion that atmospheric diabatic heating is relatively difficult to compute accurately as a residual term in the heat-budget analysis. Under these circumstances, $G(A_Z)$ is generally evaluated as a residual in the A_Z equation of (7.28). The five-year time series of monthly mean $G(A_Z)$ (Fig. 14.2) surprisingly exhibits maxima in autumn, when maximum A_Z buildup occurs. In view of the close correlation between the $\partial A_Z/\partial t$ and the $G(A_Z)$ time series, it is obvious that $G(A_Z)$ builds up A_Z to attain the latter's maximum in winter. In this season, the atmospheric circulation also peaks, and various physical processes represented by various energy conversions redistribute the built-up available potential energy. This argument is supported by the synchronization of various energies and by the major energy conversions of the Lorenz energy cycle in their annual variations.

14.1.2 Fourier Analysis of Energy Variables

The annual variations of various energy variables are neither exactly sinusoidal in time nor perfectly symmetric with respect to maxima and minima. Although most of the energy variables, shown in Section 14.1.1, possess relatively distinct annual variations, it is unlikely that we can measure the precise temporal relation between these variations. Nevertheless, towards this end, Wiin-Nielsen (1967) proposed the Fourier analysis of energy variables in time with one year as the basic period. Any energy variable can

---

sign from northward to southward transport at 50° N. As show by Chen (1982), the covariance between these quantities offers positive $C(K_E, K_Z)$ south of 32° N, negative $C(K_E, K_Z)$ between 32° and 50° N, and positive $C(K_E, K_Z)$ again north of 50° N. The alternation between positive and negative values in a latitude-pressure cross section results in a small value of the latitudinal integration of $C(K_E, K_Z)$. Consequently, it is relatively difficult to obtain a clear seasonal cycle for these two energy conversions (Wiin-Nielsen, 1967). The small numerical values and unclear seasonal cycles of both $C(A_Z, K_Z)$ and $C(K_E, K_Z)$ can be seen in Figs. 14.4 and 14.5, respectively.

be expressed in such terms as [2]

$$F(t) = \sum_{n=-N}^{N} F_n e^{in(2\pi/T)t} \tag{14.7}$$

where

$$F_n = F_{nr} + iF_{ni}. \tag{14.8}$$

and where T is a one-year period. The amplitude ($|F_n|$) and the phase (γ_n) of the Nth harmonic are defined by

$$F_n^2 = F_{nr}^2 + F_{ni}^2, \quad \text{and} \quad \tan\left(n\frac{2\pi}{T}\gamma_n\right) = -\frac{F_{ni}}{F_{nr}}. \tag{14.9}$$

Using February 1963–January 1964 as the basic period, Wiin-Nielsen (1967) performed a Fourier analysis of monthly mean energy variables. After the global data generated by the FGGE III-b analyses of the European Centre for Medium Range Weather Forecasts became available, Chen and Buja (1983) also analyzed the annual variation of atmospheric energetics for both hemispheres with the scheme of (14.7)–(14.9). The time series of energy variables shown in Figs. 14.1–14.5, however, reveal the existence of interannual and short-term variabilities. To filter out these two variabilities and to obtain a clear climatology of annual variations of energy variables, the Fourier analysis scheme (14.7)–(14.9) is applied to the five-year mean monthly energy variables of Figs. 14.1- 14.2. Fourier analysis results in an energy-variable time series provided by the three aforementioned studies, and this is presented in Table 14.1 for the first three harmonic components: the annual-mean mode ($n = 0$), the annual mode ($n = 1$), and the semiannual mode ($n = 2$).

---

[2] Conventionally, the Fourier expansion (14.7) can be written also as

$$\begin{aligned} F(t) &= F_0 + \sum_{n=1}^{N} \left[FC_n \cos\left(n\frac{2\pi}{T}t\right) + FS_n \sin\left(n\frac{2\pi}{T}t\right)\right] \\ &= F_0 + \sum_{n=1}^{N} \left[\frac{1}{2}(FC_n - iFS_n)e^{in(2\pi/T)t} + \frac{1}{2}(FC_n + iFS_n)e^{-in(2\pi/T)t}\right] \\ &= \sum_{n=-N}^{N} \frac{1}{2}(FC_n - iFS_n)e^{in(2\pi/T)t}. \end{aligned} \tag{14.4}$$

Compared to (14.4) and (14.7), the following identities can be established:

$$F_{nr} = \frac{1}{2}FC_n \quad \text{and} \quad F_{ni} = -\frac{1}{2}FC_n, \tag{14.5}$$

where

$$F_n = F_{nr} + iF_{ni}. \tag{14.6}$$

Table 14.1: Amplitudes and phases of the first three Fourier components of energies and energy conversions. Unit: amplitude of energies is 10^5 J m^{-2}, and that of energy conversions is W m^{-1}; phase is month/day.

N		0			1	
Variable		F_0			F_1	
	OP[1]	WN[2]	CB[3]	OP	WN	CB
A_Z	37.03	31.64	36.91	22.29	11.96	18.20
A_E	7.39	8.14	5.39	2.23	2.78	2.40
K_Z	4.64	8.32	5.60	3.22	6.36	4.12
K_E	7.03	10.17	6.60	1.96	3.46	2.34
$\partial A_Z/\partial t$				0.42		
$G(A_Z)$	0.90			1.23		
$C(A_Z, A_E)$	1.48	2.37	1.45	1.16	1.86	1.18
$C(A_E, K_E) - G(A_E)$			1.17			0.51
$D(K_E) - G(A_E)$	1.31			1.19		
$C(K_E, K_Z)$	0.19	0.48		0.06	0.41	0.09
$C(A_Z, K_Z)$	0.11			0.42		1.00

1. Obtained from Fourier analysis with Oort and Peixóto's (1974) numerical values for various energy variables.
2. Wiin-Nielsen (1967)
3. Chen and Buja (1983)

						2		
	γ_1			F_2			γ_2	
OP	WN	CB	OP	WN	CB	OP	WN	CB
1/10	1/10	1/27	2.28	2.03	2.07	4/1	4/15	2/18
1/3	1/22	1/16	0.54	0.72	0.67	6/14	1/11	6/29
1/14	1/24	1/28	0.68	0.86	0.53	1/29	1/29	2/9
1/5	1/25	1/21	0.18	0.14	0.40	6/1	5/2	5/27
10/10			0.07			2/15		
9/27			0.44			4/13		
12/31	1/18	1/10	0.08	0.01	0.06	3/6	4/30	5/14
		1/15		0.07			3/7	
1/6			0.12			2/15		
	9/13	12/30	0.05	0.04	0.03	4/13	5/17	5/7
		7/22	0.05	0.11		6/16	5/17	

The computational domain of Oort and Peixóto (1976) and Chen and Buja (1983) is the entire Northern Hemisphere, whereas that of Wiin-Nielsen (1967) is only that area north of 15° N. Consequently, departures of zonal-mean temperatures from the domain-average temperature are smaller in Wiin-Nielsen's analysis. On the other hand, the intensity of atmospheric circulation and the magnitudes of atmospheric-disturbance variables – that is, of temperature and wind – are smaller in the tropics than in the midlatitudes. These contrasts, which are caused by differences between the computational domains of the different studies, result in smaller A_Z and larger A_E, K_Z, and K_E in Wiin-Nielsen's annual-mean value than in Oort and Peixóto's (1976) or Chen and Buja's (1983). This contrast in energy magnitudes is also true in the annual-variation mode ($n = 1$).

A pronounced annual-variation component has emerged from the time series of most of the energy variables shown in Figs. 14.1 and 14.2. The amplitudes of all annual-variation modes, F_1, displayed in Table 14.1, are generally larger than a quarter of their annual-mean values, F_0. On the other hand, the amplitudes of semiannual-variation modes, F_2, are much smaller than those of either F_0 or F_1. In view of the magnitude of contrast between the F_1s and F_2s, the seasonal variation of atmospheric energetics is dominated by the annual-variation mode. Except for energy variables with small magnitudes, phases of annual-variation modes γ_1 indicate that the maximum values of the first harmonics obtained by Wiin-Nielsen (1967) and by Chen and Buja (1983) occur in late January, about three to four weeks after the winter solstice. In contrast, the maximum values of the annual-variation components in the Oort and Peixóto (1976) analysis appear about two to three weeks earlier than those of either Wiin-Nielsen or Chen and Buja. The maximum energy input, $G(A_Z)$, and the maximum A_Z tendency, $\partial A_Z/\partial t$, occur in early autumn during the autumn buildup of A_Z.

Shown in Figs 14.6 and 14.7 are the time series of annual ($n = 1$) and semiannual ($n = 2$) harmonics, and a combination of these two harmonics for four energies and several significant energy conversions. It was pointed out in Section 14.1.1 that $A_Z(K_Z)$ has a wide maximum in winter (summer) and a narrow minimum in summer (winter). As revealed by γ_2, the semiannual component of $A_Z(K_Z)$ exhibits its minimum (maximum) in winter (summer). The combination of the first two harmonics leads to a larger (smaller) maximum $A_Z(K_Z)$ in winter and a smaller (larger) minimum $K_Z(A_Z)$ in summer. Additionally, several studies have noted fast autumn buildup and slow spring decline in both A_E and K_E. Fig. 14.6 reveals that the second harmonic exhibits a maximum in the first half of the first harmonic's maximum period but exhibits a minimum in the first half of the first harmonic's minimum period. This phase relation between the first two harmonics of both A_E and K_E results in their fast autumn buildup and slow spring decline. As far as energy conversions are concerned, the second harmonic has negligible amplitude, with the exception of $\partial A_Z/\partial t$ and $G(A_Z)$.

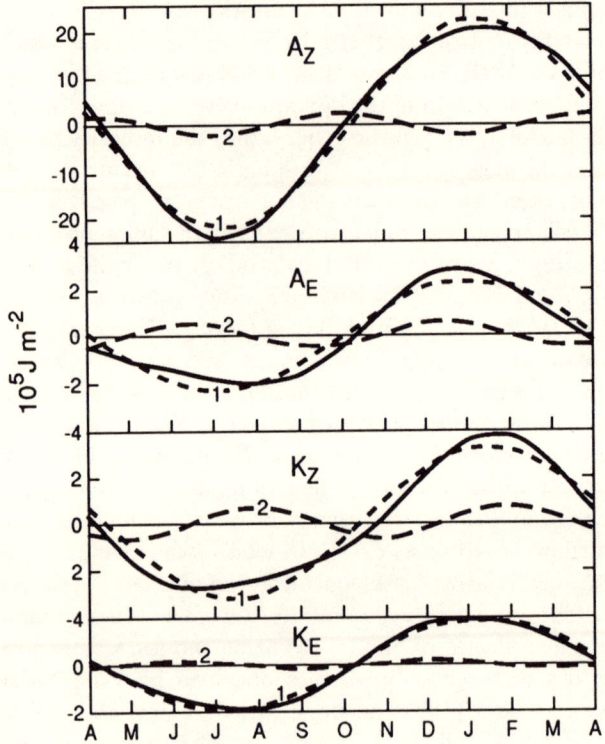

Fig. 14.6: Annual variation of the vertically integrated K, $G(k)$, $B(k)$, and $D(k)$ for the five-year period May 1958–April 1963 over North America. After Kung and Soong, 1969.

14.2 Annual Variation of the Kinetic Energy Budget over North America

The North American continent is located on the upstream side of the North American jet. According to the kinetic energy budget analysis of this jet, kinetic energy is generated over the North American continent and is destroyed over the Atlantic Ocean. To balance sources and sinks of the kinetic energy, the generated kinetic energy is transported by the jet from the source to the sink region. Section 13.1 demonstrated that the hemispheric kinetic energy exhibits a pronounced annual variation. Because the major kinetic energy content of the atmospheric circulation is contributed by jet streams, a distinct annual variation must also exist in the kinetic energy budget of atmospheric circulation over the North American continent. In short, it is of interest to explore the annual variation of the regional kinetic energy budget over the North American continent.

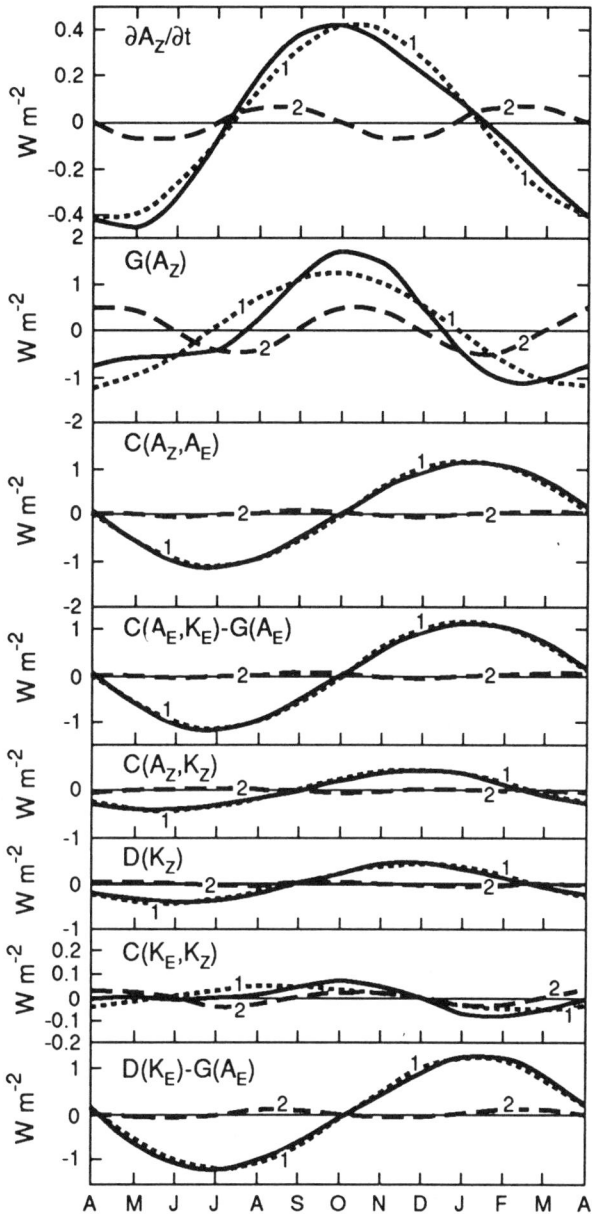

Fig. 14.7: The two largest Fourier components of the mean annual variation of various energy variables shown in Fig. 14.6; generally they are $n = 1$ and 2, except for K, and the combination of these two Fourier components.

Daily wind and geopotential data at 00 and 12 GMT for the five-year period from May 1958 to April 1963 and from surface to 100 mb over the North American continent were compiled and archived in the MIT General Circulation Data Library. Using this data set, Kung (1966a,b; 1967; 1969) has performed extensive analyses of the kinetic energy budget:

$$\underbrace{\frac{\partial K}{\partial t}}_{K_t} + \underbrace{\frac{1}{gs}\int_0^{p_0} \oint \mathbf{V}k \bullet \hat{n}\; dp\; dl}_{B(k)} = \underbrace{\frac{1}{gs}\int_0^{p_0}\int_s -\mathbf{V}\bullet\nabla\phi\; dp\; ds}_{G(k)}$$

$$= \underbrace{\frac{1}{gs}\int_0^{p_0}\int_s \mathbf{V}\bullet\mathbf{F}\; dp\; ds}_{D(k)}, \quad (14.10)$$

where $B(K)$ is changed from an area integral of kinetic energy flux divergence in (12.4) to a line integral of kinetic energy flux along the boundary of the computational domain.

Kung and Soong (1969) extended Kung's studies to examine the annual variation of the North American kinetic energy budget. The monthly-mean-value time series of all energy variables in (14.10) is decomposed into Fourier amplitude and phases. A marked annual-variation component emerges from the time series of K, $G(K)$ and $B(K)$; but none is particulary distinguishable in those of $D(K)$. The lack of a distinguishable annual variation in $D(K)$ may be due to computational errors in association with the residual method. An eyeball examination reveals that the time series of K, $G(K)$, and $B(K)$ shown in Fig. 14.8 are almost in phase with the winter maxima and the summer minima. In view of its regular annual variation, the K autumn buildup and spring decline indicate that the K_t time series is about one season out of phase with respect to the K, $G(K)$, and $B(K)$ time series. The contrast between the K_t, K, and $G(K)$ time series is similar to that in the hemispheric kinetic energy budget discussed in Section 14.1. Interestingly, for all seasons kinetic energy is generated $[G(K) > 0]$ over the North American continent and exported $[B(K) > 0]$ out of the continent, despite annual variations. In short, the active energetics of the North American continent persist throughout the year.

To illustrate quantitatively the phase relation between the annual variations of K, $B(K)$, $G(K)$, and $D(K)$, Kung and Soong applied the Fourier analysis in time (Section 14.1) to the kinetic energy budget over the North American continent. Moreover, the significance of the annual variation component, represented by the Nth harmonic in the Fourier analysis of the kinetic energy budget, is represented by a *fractional variance*, $FV(n)$:

$$FV(n) = \frac{|F_n|^2}{\sum_{n=1}^N |F_n|^2} \times 100, \quad (14.11)$$

where F_n is the amplitude of the Bth harmonic and where N is the maximum harmonic contained in the Fourier analysis. Some short-term, irregular fluctuations appear in the time series of energy variables shown in

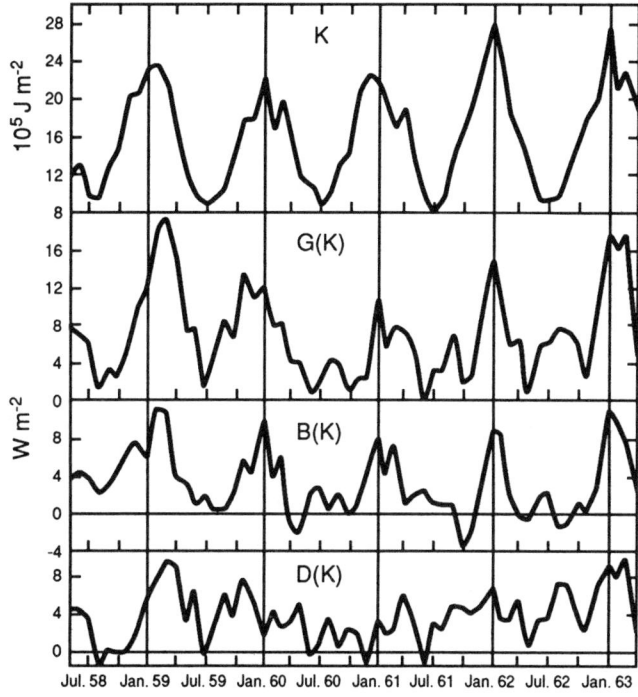

Fig. 14.8: Annual variation of the vertically integrated K, $G(k)$, $B(k)$, and $D(k)$ for a five-year period from May 1958-April 1963 over North America. After Kung and Soong, 1969.

Fig. 14.9. To extract the basic annual-variation components of all energy variables, the multiyear monthly mean process is introduced to the time series of all energy variables in such a way as to filter out the short-term, irregular fluctuations. Fourier analyses in time are performed for all multiyear monthly mean time series of energy variables. The amplitude, phase, and fractional variance of the first three harmonics are displayed in Table 14.2.

Table 14.2 reveals that, compared to Oort and Peixóto's (1974) hemispheric mean values of K (12.4×10^5 J m^{-2}) and $C(A,K)$ (2.0 W m^{-2}), the atmospheric circulation over the North American continent [with $K = 15.93 \times 10^5$ J m^{-2}, and $G(K) = 7.06$ W m^{-2}] is an energetically active region. Moreover, either Fig. 14.9 or Table 14.2 shows that atmospheric kinetic energy is usually exported from this continent. Evidently, as portrayed in Section 13.2, the North American continent is a source region of atmospheric kinetic energy. In view of the fractional variance of the first harmonic, $FV(1)$, the annual variation components of K, $G(K)$, $B(K)$, and $D(K)$ stand out distinctively in Table 14.2, especially that of K. The F_1/F_2 ratios of $G(K)$, $F(K)$, and $D(K)$ are 46%, 62%, and 52%, respec-

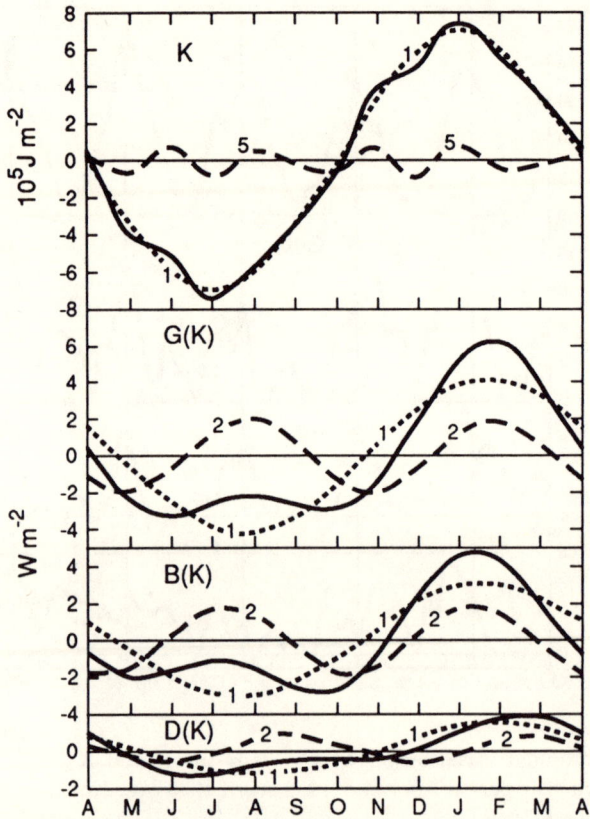

Fig. 14.9: Mean annual variation (solid line with dots) of K, $G(K)$, $B(K)$, and $D(K)$, and the combination of their $n = 0$, 1, and 2 Fourier components, for K.

tively. In fact, the sums of only $FV(1)$ and $FV(2)$ of the four energy variables exceed 90%. Thus, the annual and semiannual variation components seem to be the basic seasonal-variation modes of the regional atmospheric kinetic energy budget over the North American continent. The significance of annual-variations of the four energy variables are implicit in the F_1/F_0 ratio. The F_1/F_0 ratios of K, $G(K)$, $B(K)$, and $D(K)$ are 43%, 59%, 98%, and 34%, respectively. The hemispheric-averaged F_1/F_0 ratios of K and $C(A, K)$ estimated with Oort and Peixóto's (1976) analysis are 22% and 35%, respectively. When Oort and Peixóto's results are compared with those described in Table 14.2, not only does the atmospheric circulation of the North American continent seem energetically more active, but the regional annual variation in the kinetic energy budget of this continent is more pronounced. This argument is relatively consistent with the geographical distribution of the K (300 mb) annual variation component

Table 14.2: Amplitude (F_n), phase (γ_n), and fractional variance [$FV(n)$] of the Fourier components K, $G(K)$, $B(K)$, and $D(K)$ obtained from the five-year-mean monthly time series of these energy variables. Unit: F_n in 10^5 J m^{-2} for K; and W m^{-2} for $G(K)$, $B(K)$, and $D(K)$; γ_n in days from January 1 and $FV(n)$ in percent.

Harmonic (N)		0	1	2
Period (month)			12	6
K	F_n	15.93	6.92	18
	γ_n		18	18
	$FV(n)$		97.6	0
$G(K)$	F_n	7.06	4.20	1.95
	γ		39	42
	$FV(n)$		77.4	16.7
$B(u)$	F_n	3.16	3.11	1.93
	γ_n		36	28
	$FV(n)$		66.6	15.6
$D(K)$	F_n	4.06	1.37	0.71
	γ_n		48	71
	$FV(n)$		74.2	19.7

portrayed by Heddinghaus and Kung (1980) in terms of the empirical orthogonal functions (EOF).

Another important aspect of the annual variation in the regional kinetic energy budget is the phase relation between energy variables. It has been pointed out that the seasonal variation of K is more or less explained by the annual ($n = 1$) component. The phase $\gamma_1(K)$ indicates that K reaches its maximum (minimum) on January 18 (July 18), which is about four weeks after the winter (summer) solstice. This date agrees relatively well with the hemispheric Fourier analyses of Krueger et al. (1965), Wiin-Nielsen (1967), Oort and Peixóto (1976), and Chen and Buja (1983).

Based upon the F_2/F_1 ratios of $G(K)$, $B(K)$, and $D(K)$, the semiannual variations of these energy variables are not negligible. Maxima and minima of $G(K)$, $B(K)$, and $D(K)$ are determined by a combination of annual ($n = 1$) and semiannual ($n = 2$) components. As shown in Fig. 14.10, maximum $G(K)$, $B(K)$, and $D(K)$, generated by combining annual and semiannual components, occur around February 10, 4, and 28, respectively. The seasonal variations of $G(K)$ and $B(K)$ are almost in phase, but the seasonal variation of $D(K)$ is about 2.5 weeks behind the other two energy variables. The composite multiyear-mean monthly time series of K, $G(K)$, $B(K)$, and $D(K)$ shown in Fig. 14.9 attain their maximum values in January. Phase differences between the composite real time series and a combination of $n = 1$ and 2 of these energy variables are attributed to the neglected higher harmonics.

Suppose that $G(K)$ is equivalent to $C(A,K)$ in a relatively large computational domain. The annual variation of the regional kinetic energy budget over the North American continent – that is, the upstream region of the North American jet – behaves in a manner similar to that of a hemispheric example discussed in Section 14.1. Holopainen (1973) and Holopainen and Eerola (1973) pointed out in their kinetic energy budget analysis over the British Isles that $G(K)$ destroys kinetic energy. On the other hand, Holopainen (1978), Lau (1979), and Chen and Lee (1983) showed that characteristics of the kinetic energy budget in the upstream and downstream regions are opposite. The British Isles are located on the downstream side of the North American jet. The contrast between the kinetic energy budget analyses of Kung (1966a,b; 1967; 1969), Holopainen (1973) and Holopainen and Eerola (1979) are not surprising. In view of previous analyses of the kinetic energy budget associated with jet streams, it is to be expected that the annual variation of the destruction of kinetic energy by the ageostrophic effect and the importation of kinetic energy by jets on the downstream side should synchronize with the generation of kinetic energy and the exportation of kinetic energy by jets on the upstream side. An effort along this line is worth pursuing.

14.3 Vacillation of Atmospheric Energetics

The quasi-periodic fluctuation of atmospheric circulation, called the index cycle, is characterized by a relatively zonal (high index) circulation and a relatively wavy (low index) structure. Thus, the alternating of the atmospheric circulation between high and low indices is reflected in quasi-periodic temporal fluctuations of zonal and eddy energies. The Lorenz energy cycle is therefore adopted to examine not only the periodicity of energy variations, but also the physical processes involved in maintaining the vacillation of atmospheric energetics.

Wiin-Nielsen (1961) demonstrated that vacillations can take place in the low-order barotropic model. Analyzing the 200 mb wind field of the Southern Hemisphere mid-latitudes obtained after the EOLE experiment, Webster and Keller (1975) also showed that kinetic energy vacillation is maintained by barotropic processes. In contrast, Kraus and Lorenz (1966) simulated energy vacillation in a two-layer quasi-geostrophic system with a simple thermal forcing. Pfeffer et al. (1974) were able to produce energy vacillation in a rotating annulus with radial differential heating. It seems therefore that the baroclinic process cannot be ruled out in energy vacillation. Numerous spectral energy studies have shown that the long wave regime not only contains eddy energy (e.g., Wiin-Nielsen, 1967; Chen, 1982a), but is also responsible for the major eddy sensible heat and momentum transport (e.g., Chen 1982a). Nevertheless, Chen (1982a) demonstrated that short waves (synoptic waves) are energetically more efficient than long waves. Hunt (1978) showed, based upon the numerical simulation of a general circulation model, that vacillations in the intensity of baroclinic activity essentially occur in the synoptic- wave regime. Moreover, Steinberg et al. (1971), Chen (1982), and others have shown that available potential energy is nonlinearly cascaded from long waves to short waves and that kinetic energy is transferred from synoptic-scale waves to both longer and shorter waves. No doubt, the nonlinear interaction between waves is important to the energetics of an atmospheric system. In contrast, Lorenz (1963) reproduced wave vacillation in a rotating annulus in terms of a spectral model containing only a single wave and the mean-zonal flow. Nonlinear interaction between waves was essentially excluded in his numerical study.

The previous brief review of conventional research regarding energy vacillation suggests that four aspects have been of major concern: (1) the periodicity of vacillation, (2) the primary physical process maintaining vacillation – barotropic or baroclinic, (3) the dominant scales of the eddies, and (4) the effect of nonlinear interaction between waves. As for the last two aspects, the roles played by different wave regimes in the vacillation of atmospheric energy need to be understood.

14.3.1 Vacillation of Eddy Energy

The index cycle was first defined with either zonal-mean surface pressure differences between 30° and 55°N or zonal-mean 700 mb zonal flow averaged over this latitudinal zone. However, the fluctuations of maximum upper-air zonal flow and the eddy activities of atmospheric circulation exist primarily between 20° and 65° N. Using the ECMWF FGGE III-b data of the 1978/79 winter, Chen and Marshall (1984) computed vertically integrated energy variables of the Lorenz energy cycle averaged over the latitudinal zone of 20°-65° N, as revealed from Figs. 14.10 and 14.11. Significant fluctuations appear in the time series of energies and of energy conversions. The winter-mean values and the time-variation ranges of the energy variables are displayed in Table 14.3. Due to the selection of a latitudinal zone (20°-65° N) for the area average, the numerical values of most of the energy variables shown in Table 14.3, except the variable A_Z, are larger than those variables averaged over the entire Northern Hemisphere (Fig. 7.9).

The time-series analysis scheme proposed by Bendat and Piersol (1971) is often used in meteorology. Also, the following criteria are adopted to distinguish noise from significant signals: (1) the consistency of occurrence of the significant signal over a given frequency band and (2) the statistical confidence level (for power spectra) and coherency (for cross spectra). If a significant signal exists, we expect that confidence level to be close to one. The temporal correlation between significant signals of two time series of energy variables is measured in terms of phase and coherency of cross

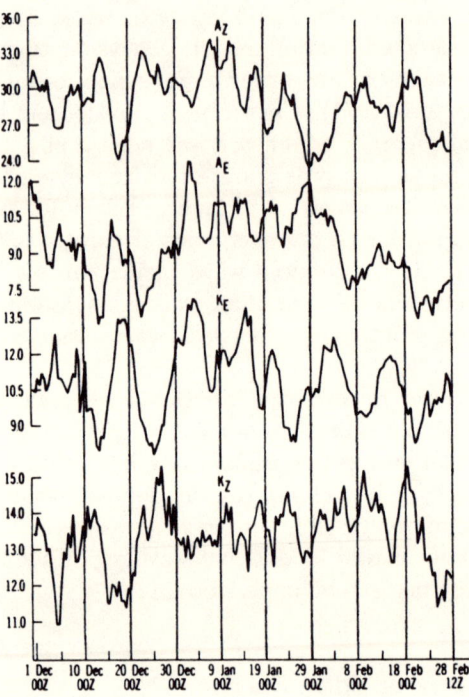

Fig. 14.10: Time series of various energies in the 20°-65° N belt during 1979 summer. Unit: 10^5 J m^{-2}.

spectra, which is expected to be close to one too. At any rate, anyone who is interested in the detail of Bendat and Piersol's statistical scheme and statistical terms should refer to Julian (1971) and Madden and Julian (1971).

Power spectra of various energies and energy conversions are displayed

Table 14.3: Winter-mean values and time-variation ranges of energies (10^5 J m^{-2}) and energy conversions (W m^{-2}) averaged between 20° and 65°N.

Variables	Winter-mean value	Time-variation range
A_Z	29.1	3~9
A_E	9.2	2~6
K_E	12.3	2~6
K_Z	13.4	1~4
$C(A_Z, A_E)$	3.82	0~8
$C(A_E, K_E)$	3.02	1~6
$C(K_E, K_Z)$	0.07	-2~2
$C(A_Z, K_Z)$	0.31	-2~1

Table 14.4: Coherency (coherence square) and a phase difference of cross spectra between energies and energy conversions in the frequency band with a period of about two weeks.

Quantity	Frequency (cpd)	Coherency	Phase (days)
$A_E \cdot A_Z$	6/90 (15 day)	0.83	7.83
	7/90 (12.86 day)	0.94	6.30
	8/90 (11.25 day)	0.93	5.00
$K_E \cdot A_E$	6/90 (15 day)	0.79	-0.78
	7/90 (12.86 day)	0.94	-0.51
	8/90 (11.25 day)	0.92	-0.30
$K_Z \cdot K_E$	6/90 (15 day)	0.76	7.79
	7/90 (12.86 day)	0.77	7.26
$K_Z \cdot A_Z$	6/90 (15 day)	0.71	0.09
	7/90 (12.86 day)	0.83	0.57
	8/90 (11.25 day)	0.78	5.00

in Fig. 14.12. Based upon the two criteria set previously, a distinct signal emerges for most power spectra inside the frequency band with a period of about two weeks. To illustrate the temporal relation between energy variables, only all possible cross spectra (solid lines) between two neighboring energy reservoirs of the Lorenz energy cycle shown in Fig. 14.13 are used as examples. These cross spectra exhibit marked signals with frequencies close to a period of two weeks and with coherency values close to 1.0. Shown in Table 14.4 are phases between two energies; phases between A_Z and A_E and between K_Z and K_E are approximately half of a cycle, whereas those between A_Z and K_Z and between A_E and K_E are insignificant. These phases confirm our eyeball examination of the time series of A_Z, A_E, K_E, and K_Z (Fig. 14.10), which suggests that eddy energies or zonal energies alone vary in phase, but that eddy and zonal energies together vary out of phase.

As can be seen in Fig. 14.10, time variations of $A_Z(K_Z)$ and $A_E(K_E)$ are out of phase, but those of $A_Z(A_E)$ and $K_Z(K_E)$ are synchronous. The phase relations between time variations of zonal and eddy energies are consistent with those determined by previous studies of energy vacillation (e.g., Miller, 1974; Hunt, 1978) and coincident with alternations between high and low indices of atmospheric circulation. According to the Lorenz energy cycle, input and output energy conversions may be considered sources and sinks, respectively, of a given energy reservoir. Comparing Figs. 14.11 and 14.12, we can see that A_Z, to some extent, varies out of phase with $C(A_Z, K_Z)$ and $C(A_E, K_E)$. $C(A_Z, A_E)$ and $C(A_E, K_E)$ vary in phase with each other, but are both slightly ahead of A_E. Additionally, the time variation of $C(A_E, K_E)$ is about one day ahead of that of K_E. The phase relations of energy variables described here indicate that it takes some time for released zonal available potential energy to develop atmospheric distur-

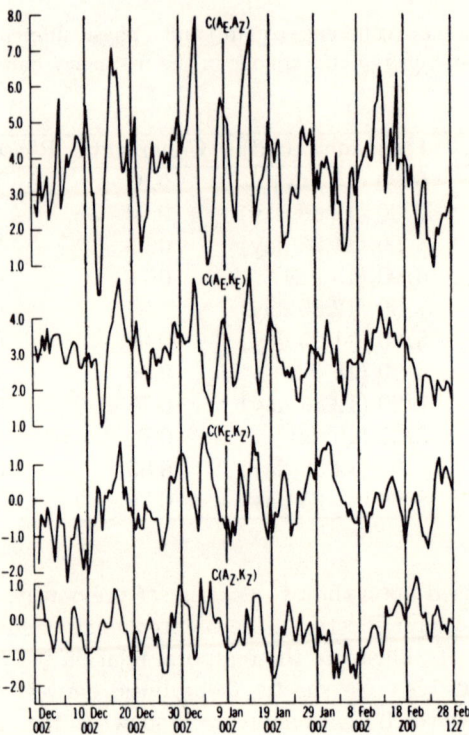

Fig. 14.11: Time series of various energy conversions in the 20-65° N belt during 1979 summer. Unit: W m^{-2}.

bances. $C(K_E, K_Z)$ varies almost in phase with K_E but is out of phase with K_Z. Moreover, $C(K_E, K_Z)$ fluctuates between positive and negative values. This fluctuation reflects the existence of a back-and-forth conversion between K_Z and K_E. It was pointed out previously that the time variations of A_Z and K_Z are in phase, but that these are out of phase with the time variation of $C(A_Z, K_Z)$.

Past studies of energy vacillation have revealed a quasi-periodic variation of two to three weeks in energy variables. Miller (1974) obtained a period of 14-16 days, while Webster and Keller (1975) obtained a somewhat longer period of 18-23 days. Later, McGuirk and Reiter (1976) and Hunt (1978) found a period of 20-26 days and 20 days, respectively. Regarding the 1978/79 winter, a low-frequency variation of energies is discernable in the energy time-series shown in Fig. 14.11. However, it is far too subjective a practice to assess the periodicity of energy variations by eyeball examinations of energy time-series. Assessing the temporal relations between energy variables is equally suitable. Thus, a time spectral analysis should be utilized as an objective method to isolate significant signals from time-series of energy variables and to determine temporal relations between

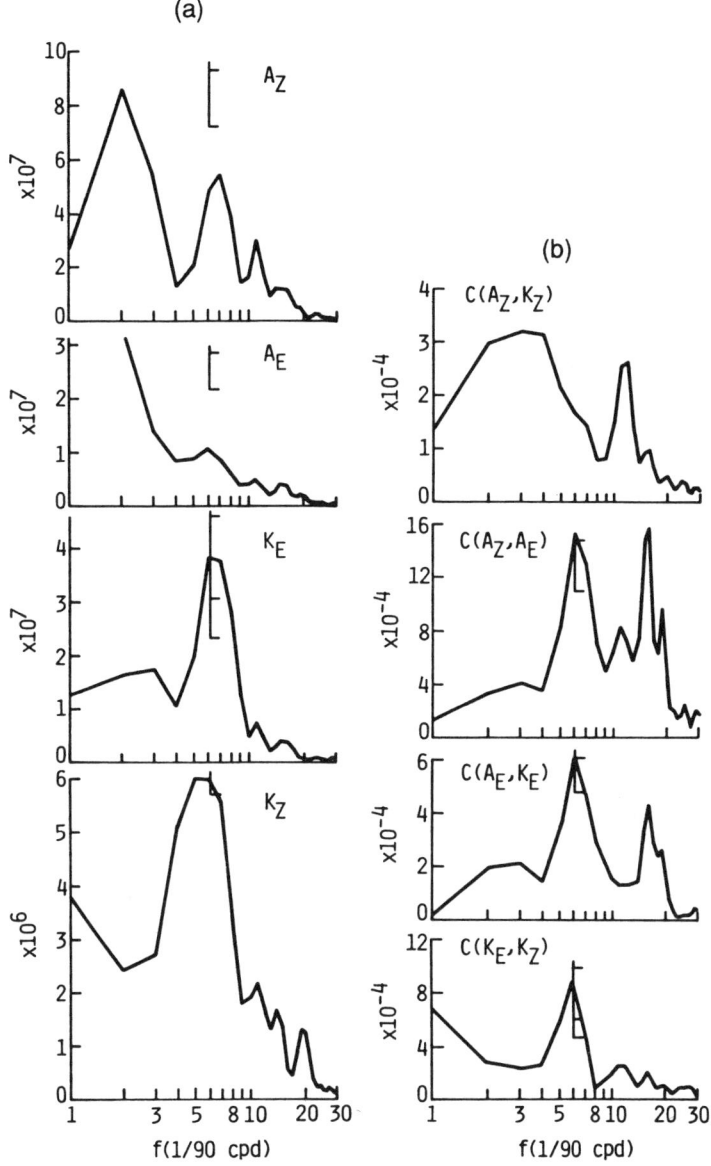

Fig. 14.12: Power spectra of (a) energies and (b) energy conversions Unit: $J^2\ m^{-4}\ day^{-1}$ for energies and $W^2\ m^{-4}\ day^{-1}$ for energy conversions.

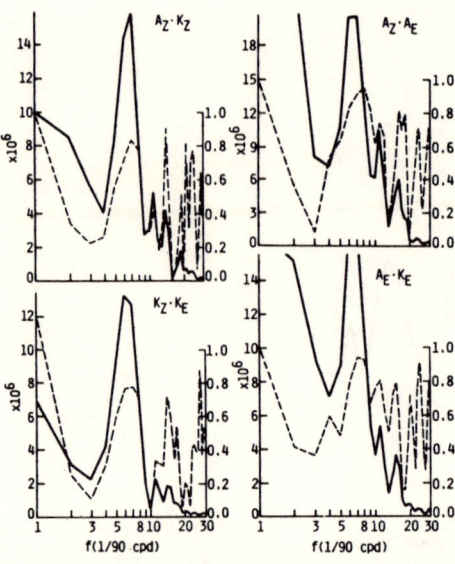

Fig. 14.13: Cross spectra (solid lines) and coherencies (dashed lines) between two energies as indicated. Values shown on the left abscissa are cross spectra, while those shown on the right abscissa are coherencies. This convention is also used in all figures of cross spectra hereafter. Units: J^2 m^{-4} day^{-1}.

them.

The time-series analysis presented above clearly shows that the vacillation period of atmospheric energetics is about two weeks, which falls within the lower bounds of the two to three week range found by previous studies. Energy vacillation appears in both available potential and kinetic energies. As indicated by a spectral analysis of energy conversions, the physical processes involved in adjustments of all energy reservoirs during energy vacillation include not only dynamic, but also thermodynamic processes. Thus, both baroclinic and barotropic processes should be contained in the vacillation of atmospheric energetics. It seems unlikely that one of the two physical processes should be favored in adjustments of atmospheric energetics from one state to another in the energy vacillation.

14.3.2 Energy Vacillation of Long- and Short-Wave Regimes

The wave theory and the spectral energetics of atmospheric circulation have shown that the long-wave regime is dominated by stationary waves and that the most unstable baroclinic waves reside in the short-wave regime. The contributions from the two different wave regimes and the effects of nonlinear interactions between them to the vacillation of atmospheric energetics will be examined in terms of the division of wave regimes: wavenumbers 1-4 (the long-wave regime) and 5-18 (the short-wave regime). The energy contents of these two wave regimes can be written as

$$A_{EL} = \sum_{n=1}^{4} A_n, \qquad A_{ES} = \sum_{n=5}^{18} A_n,$$

$$K_{EL} = \sum_{n=1}^{4} K_n, \qquad K_{ES} = \sum_{n=5}^{18} K_n.$$

Energy equations of these two wave regimes can be obtained by summing the spectral energy equations developed in Chapter 9:

$$\frac{dK_Z}{dt} = C(K_{EL}, K_Z) + C(K_{ES}, K_Z) + C(A_Z, K_Z) - D(K_Z),$$

$$\frac{dK_{EL}}{dt} = -C(K_{EL}, K_Z) + C(K_{ES}, K_{EL}) + C(A_{EL}, K_{EL}) - D(K_{EL}),$$

$$\frac{dK_{ES}}{dt} = -C(K_{ES}, K_Z) - C(K_{ES}, K_{EL}) + C(A_{ES}, K_{ES}) - D(K_{ES}),$$

$$\frac{dA_Z}{dt} = -C(A_Z, A_{EL}) - C(A_Z, A_{ES}) - C(A_Z, K_Z) + G(A_Z),$$

$$\frac{dA_{EL}}{dt} = C(A_Z, A_{EL}) - C(A_{EL}, K_{EL}) + C(A_{EL}, A_{ES}) + G(A_{ES}),$$

$$\frac{dA_{ES}}{dt} = C(A_Z, A_{ES}) - C(A_{ES}, K_{ES}) - C(A_{EL}, A_{ES}) + G(A_{ES}).$$

The time series of various energies and energy conversions in both long-wave and short-wave regimes are shown in Figs. 14.14 and 14.15. The winter-mean values and time-variation range of energy variables derived from these two figures are displayed in Table 14.5. Although $A_{EL} > A_{ES}$, other energy variables in both wave regimes are not only comparable in their winter-mean values, but also in their time-variation ranges. Evidently, contributions made by the short-wave regime to atmospheric energetics are as important as those made by the long-wave regime. In order to explore the existence of energy vacillation in both the long- and short- wave regimes, power spectra of all energies and energy conversions are displayed in Fig. 14.16. As revealed from this figure, a marked signal within the frequency band corresponding to a period of two weeks emerges from these power spectra with confidence level higher than 90%, except for A_{EL} and $C(K_{ES}, K_Z)$. A_{EL} possesses the major content of A_E, and the long-time variation of the A_E time series (Fig. 14.10) reappears in the A_{EL} time series. Surprisingly, A_{EL} power spectra do not exhibit an expected signal. However, in view of the power spectra of energies and energy conversions in the long- and short-wave regimes, it is conceivable that the short-wave regime plays a role in the energy vacillation as important as that of the long-wave regime.

218   FUNDAMENTALS OF ATMOSPHERIC ENERGETICS

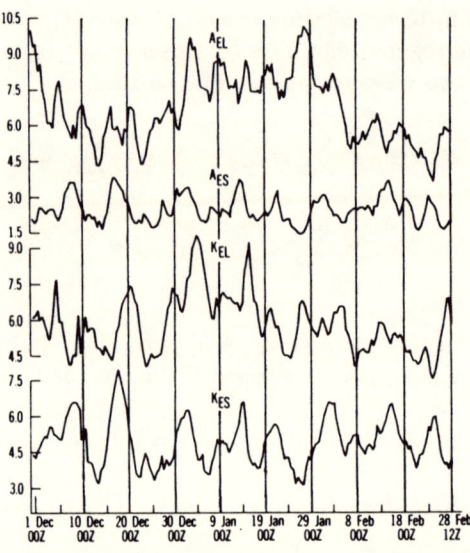

Fig. 14.14: Time-series of various energies in the long- (n=1-4) and short- (n=5-15) wave regimes. Unit: 10^5 Jm−2.

The correlations between time variations of energy variables can be illustrated in terms of a cross-spectrum analysis. Thus, one may question, for example, whether the phase relationship between energies of the total wave regime (eddy) described in Section 14.3.1 can be preserved in both long-wave and short-wave regimes. To answer this question, a cross-spectrum

Table 14.5: Winter-mean values and time-variation ranges of energies (10^5 Jm^{-2}) and energy conversions (W m^{-2}) of the long- and short-wave regimes averaged between 20° and 65°N.

Variable	Winter-mean values	Time-vartiation range
A_{EL}	6.8	$1 \sim 4$
A_{ES}	2.5	$0 \sim 2$
K_{EL}	6.3	$2 \sim 5$
K_{ES}	6.6	$3 \sim 4$
$C(A_Z, A_{EL})$	2.05	$-1 \sim 4$
$C(A_Z, A_{ES})$	1.77	$0 \sim 4$
$C(A_{EL}, K_{EL})$	1.26	$0 \sim 3$
$C(A_{ES}, K_{ES})$	1.76	$0 \sim 4$
$C(K_{EL}, K_Z)$	0.01	$-2 \sim 2$
$C(K_{ES}, K_Z)$	0.06	$-1.5 \sim 15$
$C(A_{EL}, A_{ES})$	0.27	$-1.5 \sim 1$
$C(K_{ES}, K_{EL})$	0.25	$-2 \sim 3$

analysis of neighboring energies within the context of the Lorenz energy cycle in both wave regimes should be performed. Marked cross spectra corresponding to a period of approximately two weeks with significant coherencies stand our clearly in most of the cross spectra shown in Fig. 14.17.

A visual examination of the energy time series in the two wave regimes (Fig. 14.4) indicates that A_{ES} and A_{EL} vary in phase with K_{ES} and K_{EL}, respectively. This observation is essentially in agreement with the phases of cross spectra between these energies in the frequency band of a two-week period. Another interesting feature in the time series of K_{ES} and K_{EL} is that the times series of the former lead those of the latter by a couple of days for most peak values. Interestingly, phases between K_{ES} and K_Z (Table 14.6) are about 1~2 days shorter than those between K_{EL} and K_Z. Consistently, the phases of cross spectra between the time variations of energy variables in the total-wave regime also exist in the long- and short-wave regimes. In other words, the characteristics of cross spectra between the energies of these two wave regimes substantiate our previous conclusion that both wave regimes are equally important in the vacillation of atmospheric energetics.

Table 14.6: Same as Table 14.4 except for long-wave (A_{EL}) and short-wave (A_{ES}) available potential energy and for long-wave (K_{EL}) and short-wave (K_{ES}) kinetic energy.

Quantity	Frequency (cpd)	Coherency	Phase (days)
$A_{ES} \cdot A_Z$	5/90 (18 day)	0.81	8.22
	6/90 (15 day)	0.96	7.76
	7/90 (12.86 day)	0.78	6.32
	8/90 (11.25 day)	0.81	5.03
$K_{EL} \cdot A_{EL}$	7/90 (12.86 day)	0.78	-0.51
$K_{ES} \cdot A_{ES}$	5/90 (18 day)	0.83	-1.03
	6/90 (15 day)	0.94	0.69
	7/90 (12.86 day)	0.92	0.10
	8/90 (11.25 day)	0.98	0.21
$K_Z \cdot K_{EL}$	5/90 (18 day)	0.79	9.19
	6/90 (15 day)	0.77	7.95
	7/90 (12.86 day)	0.78	7.77
$K_Z \cdot K_{ES}$	7/90 (12.86 day)	0.92	6.75
	8/90 (11.25 day)	0.98	6.20
$K_{EL} \cdot K_{ES}$	5/90 (18 day)	0.74	0.20
	6/90 (15 day)	0.81	-0.52

Finally, the function of nonlinear interactions between waves in the vacillation of atmospheric energetics should be clarified. Chapter 9 demonstrated that available potential energy cascades from long-wave to short-wave regimes and that kinetic energy cascades from synoptic-scale waves

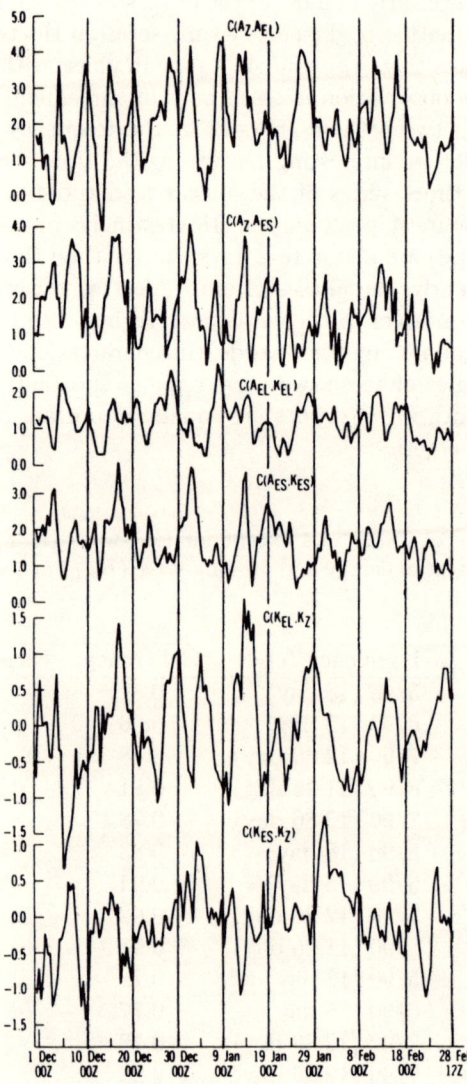

Fig. 14.15: Time series of various energy conversions in the long- and short-wave regimes. Unit: W m^2.

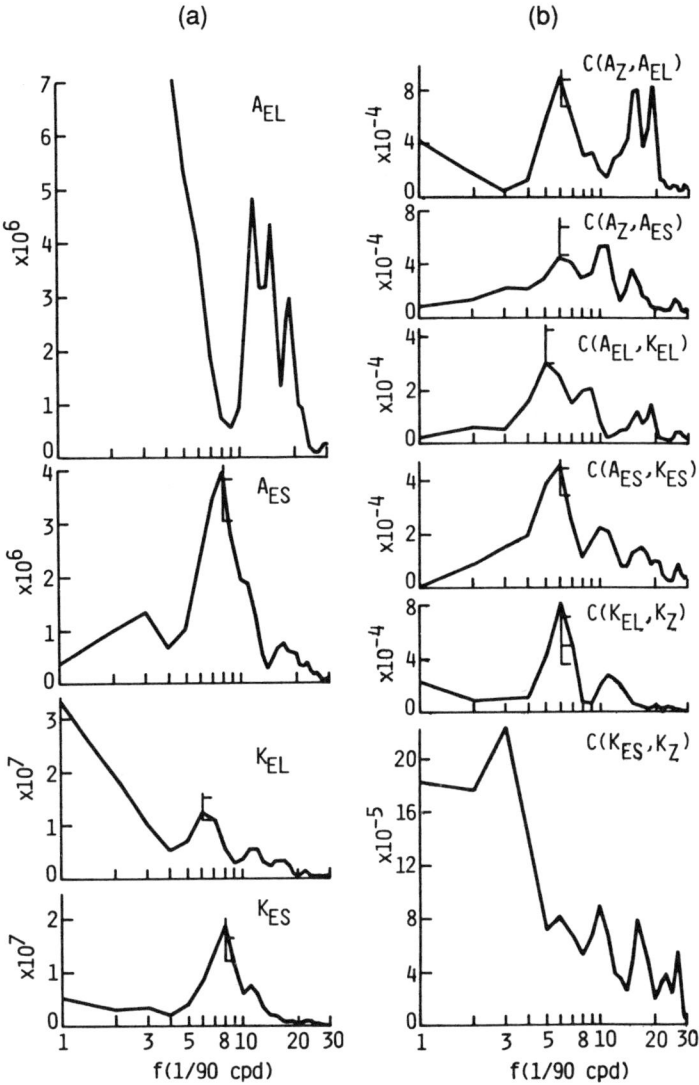

Fig. 14.16: Power spectra of (a) energies and (b) energy conversions in both the long- and short-wave regimes. Unit: $W^2\ m^{-4}\ day^{-1}$ for energies and $W^2\ m^{-4}\ day^{-1}$.

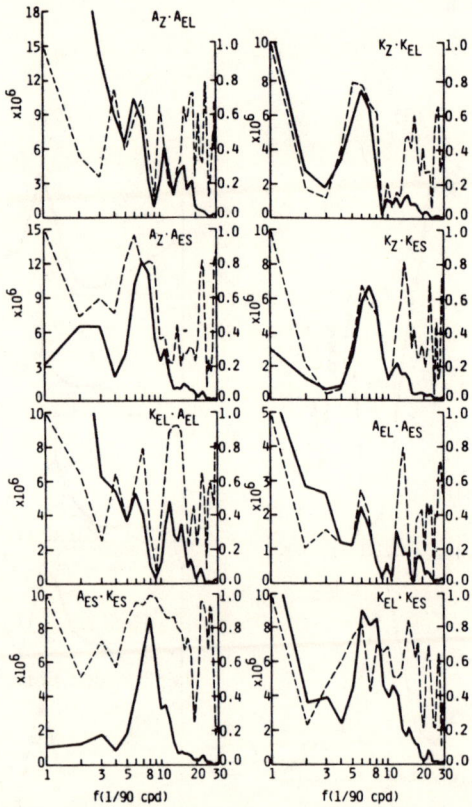

Fig. 14.17: Cross spectra (solid lines) and coherencies (dashed lines) between energies of zonal flow and long (or short) waves. Unit: $J^2\ m^{-4}\ day^{-1}$.

to longer and shorter waves. The winter-mean values of $C(A_{EL}, A_{ES})$ and of $C(K_{ES}, K_{EL})$ averaged between 20° and 65° N are consistent with the previous diagnostic analyses of spectral energetics. Although winter-mean $C(A_{EL}, A_{ES}) >$ winter-mean $C(K_{ES}, K_{EL})$, fluctuations in the magnitude of the latter are greater than those in the former (Fig. 14.16). The time series of $C(K_{ES}, K_{EL})$ usually possesses a peak value between the peak values of K_{EL} and K_{ES}. Contrarily, the time variation of $C(A_{EL}, A_{ES})$ does not correspond to an energy vacillation with a two-week period. Seemingly, the nonlinear interaction of available potential energy between long wave and short wave regimes is not vitally important to energy vacillation.

The power spectra of $C(K_{ES}, K_{EL})$ (Fig. 14.18), but not those of $C(A_{EL}, A_{ES})$, have a significant signal in the frequency band of a two-week period and have a confidence level higher than 95%. Because K_{ES} is nonlinearly transferred to K_{EL}, it is not surprising that a significant signal appears in the cross spectra between K_{EL} and $C(K_{ES}, K_{EL})$ with coheren-

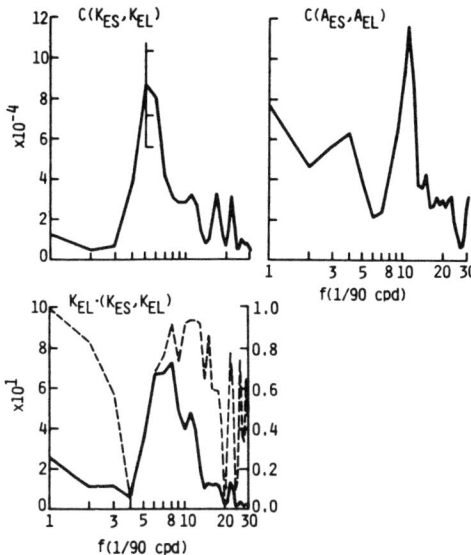

Fig. 14.18: Power spectra of $C(K_{ES}, K_{EL})$ and $C(A_{ES}, A_{EL})$ and cross spectra (solid line) and coherencies (dashed line) between K_{EL} and $C(K_{ES}, K_{EL})$. Units: W^2 m^{-2} day^{-1} for power spectra and J W m^{-2} day^{-1} for cross spectra.

cies indicating a high significance level. The phase of cross spectra between these two energy variables in the frequency band of a two-week period is from one up to a few days. In contrast, there is no significant signal in the cross spectra between A_{EL} and $C(A_{ES}, A_{EL})$. Apparently, $C(A_{ES}, A_{EL})$ does not contribute to the energy vacillation, but $C(K_{ES}, K_{EL})$ does somewhat. Although the time variation of $C(A_{EL}, A_{ES})$ is highly correlated with either A_{EL} or A_{ES}, $C(K_{ES}, K_{EL})$ does contribute somewhat to the energy vacillation.

15

ENERGETICS OF THE TROPICS: PLANETARY SCALE

Convection in the tropical cloud is locally intense, owing to the thermodynamic instability, and takes place essentially in the vertical. In the middle latitudes, by contrast, the dominant process may be termed "slantwise convection".... Thus, although air parcels may undergo vertical displacement of several kilometers in a typical disturbance..., they are at the same time subject to horizontal displacements of hundreds of thousands of kilometers. (Palmén and Newton, 1969)

Thus far, the focus has been on atmospheric energetics of the mid-latitudes. One may argue that the contributions to hemispherically averaged energetics are not only from the mid-latitudes, but are also from the tropics. As a matter of fact, magnitudes of disturbance for most meteorological variables except wind are generally an order of magnitude smaller in the tropics than in the mid-latitudes (e.g., Holton, 1992). The large-scale dynamics of atmospheric circulation in the mid-latitudes are well depicted by the quasi-geostrophic system, which fails in the tropics. According to a large number of heat-budget diagnostic studies after FGGE III-b data became available, the largest contribution to the diabatic heating of the atmosphere occurs in the tropics. It is well understood that tropical diabatic heating is due primarily to the latent heat released by cumulus convection and is the main forcing of the planetary-scale circulation in the tropics. As revealed by the previously quoted statement by Palmén and Newton (1969), large-scale dynamics in the tropics evidently differ from those of the mid-latitudes. Surely, the characteristics of large-scale motions in the tropics cannot be illustrated in terms of hemispheric energetics. It is the purpose of this chapter to differentiate between an energy analysis of the dynamics of large-scale motion in the tropics and in the mid-latitudes.

15.1 Overview of Tropical Planetary-scale Circulation

The major purpose of this chapter is to analyze the energetics of large-scale motions in the tropics in terms of the different diagnostic schemes

ENERGETICS OF THE TROPICS 225

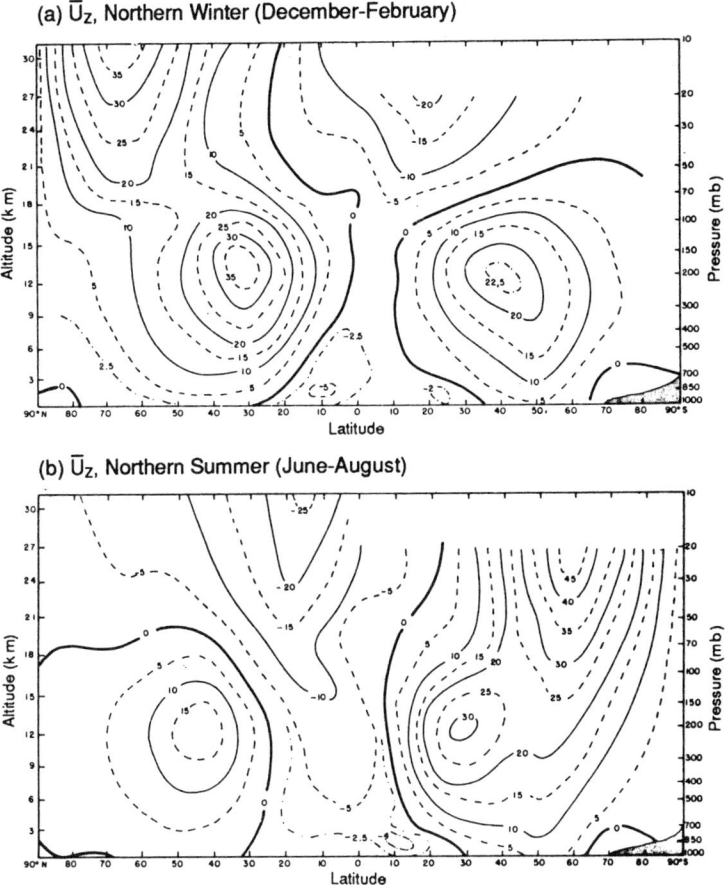

Fig. 15.1: Latitude-pressure cross section of zonal-mean wind (m s^{-1}) during (a) northern winter (December-Feburary) and (b) northern summer (June-August) (Newell et al, 1972).

proposed previously. Because the major purpose of atmospheric energetics is to facilitate understanding of the physical processes maintaining atmospheric large-scale circulation, a brief description of the tropical general circulation is necessary. The zonal-mean circulations have long been used to describe low-latitude general circulations. In contrast, the asymmetric component of the tropical general circulation was not well understood until Krishnamurti (1971a) and Krishnamurti et al. (1973a) portrayed the east-west circulation. Therefore, the *zonal-mean* and *asymmetric* components of the tropical general circulation should be described here.

The latitude-pressure cross sections of u_z for a northern winter and northern summer are shown in Fig. 15.1. Maximum westerlies exist in the upper troposphere of the mid-latitudes at around 30° (45°) of both hemi-

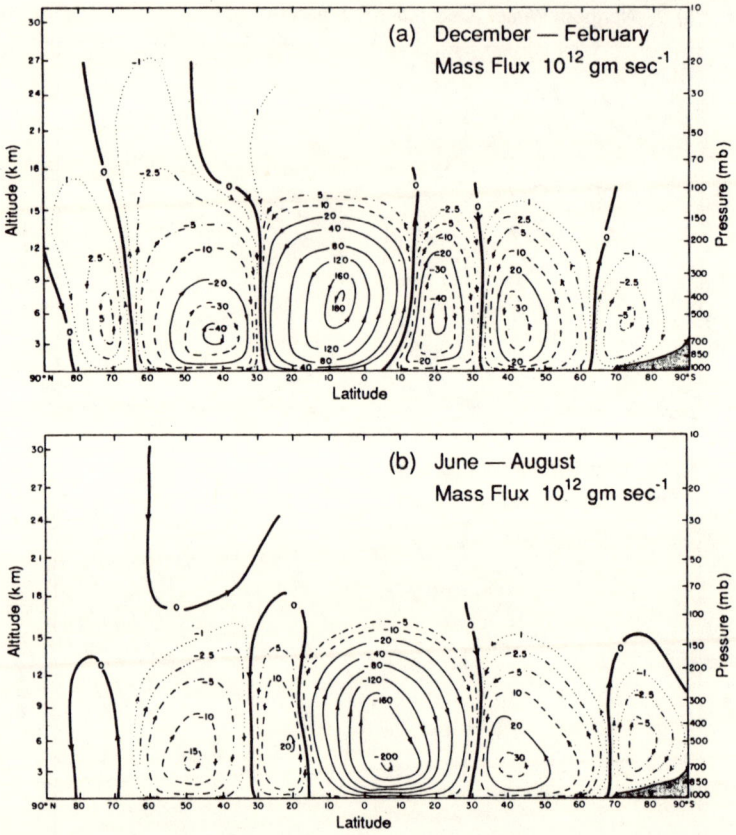

Fig. 15.2: Latitude-pressure cross section of mass flux function (gm·s^{-1}) during (a) northern winter (December-Feburary) and (b) northern summer (June-August) (Newell et al. 1972).

spheres during winter (summer). The maximum westerlies are stronger in winter than in summer, but this annual variation of u_z in the mid-latitudes is much greater in the Northern Hemisphere. In contrast, zonal-mean easterlies prevail in the tropics and are much more intense during the northern summer than during the northern winter. The summer intensification of the tropical easterlies in the upper troposphere is attributed to the tropical easterly jet south of the Tibetan plateau.

As indicated by the zonal kinetic energy equation, the time variation of zonal kinetic energy is determined by the interaction between the zonal-mean flow and eddies, and by the release of zonal available potential energy through zonal-mean meridional circulation. The former energy process is related to the asymmetric component of the tropical general circulation, whereas the latter is determined by the correlation between the Hadley circulation and the thermal structure of the tropical zonal-mean circulation. Zonal-mean meridional circulations depicted in terms of the mass flux function for northern winter and northern summer are displayed in Fig. 15.2 with a three-cell structure in each hemisphere. The Hadley circulation is always the dominant cell straddling the tropics. During the northern winter (summer), the overhead sun moves to the Southern (Northern) Hemisphere. Consequently, the ascending branch of the Hadley circulation migrates with the overhead sun, and the direction of this circulation reverses once a year. Apparently, the Hadley circulation is thermally direct. The zonal-mean temperature associated with the ascending (descending) branch of the Hadley circulation over the summer (winter) tropics is slightly warmer (colder). This positive correlation between tropical zonal-mean temperatures and the Hadley circulation vertical motions gives an important contribution to the release of zonal-mean available potential energy, $C(A_Z, K_Z)$, in the tropics.

Using the ship data, the asymmetric structure of the tropical circulation at the surface – for example , trade wind belts – intertropical convergence zone (ITCZ), oceanic subtropical highs, and monsoons, was relatively well documented before 1970. The upper-level asymmetric structure of the tropical planetary-scale circulation was analyzed in a piecewise fashion. Sadler (1967) and Aspliden et al. (1966) delineated the mid-oceanic troughs over the Pacific and Atlantic. Koteswaram (1958) examined both structure and time evolution of the tropical easterly jet located south of the Tibetan anticyclone, the thermal maintenance of which was analyzed by Flohn (1968). These three salient features and other well-known elements of the tropical circulations in the upper troposphere, e.g., the Mexican high and the subtropical westerly jet in the Southern Hemisphere, are well depicted by streamlines at 200 mb (Fig.15.3).

Krishnamurti (1971b) pointed out that a significant contrast in upper-troposphere temperatures exists between the warm Asian monsoon region and the cold Pacific oceanic trough. This temperature contrast is synoptically expected because the ascending warm air over the monsoon region converges towards the oceanic trough resulting in descending cold air. In other words, there may be a thermally direct circulation to link these important elements of the tropical asymmetric circulation. As indicated by the summer-mean velocity potential ($\overline{\chi}$) at 200 mb (Fig.15.3c), air diverges in all directions from the three tropical continents (equatorial Africa, the Asian monsoon region, and central America) towards the Atlantic Ocean and the Pacific Ocean and Somalia to form east-west circulations in the zonal direction and to establish local Hadley-type circulations in the meridional direction. The outgoing long-wave radiation (OLR), used as a proxy of cumulus convection, also exhibits three deep convection centers over the three tropical continents. In the tropics, this coincidence of divergent upper-troposphere circulation centers and OLR centers indicates that east-west and local Hadley-type circulations are thermally direct and important to the kinetic energy generation of quasi-stationary planetary-waves and zonal-mean flow.

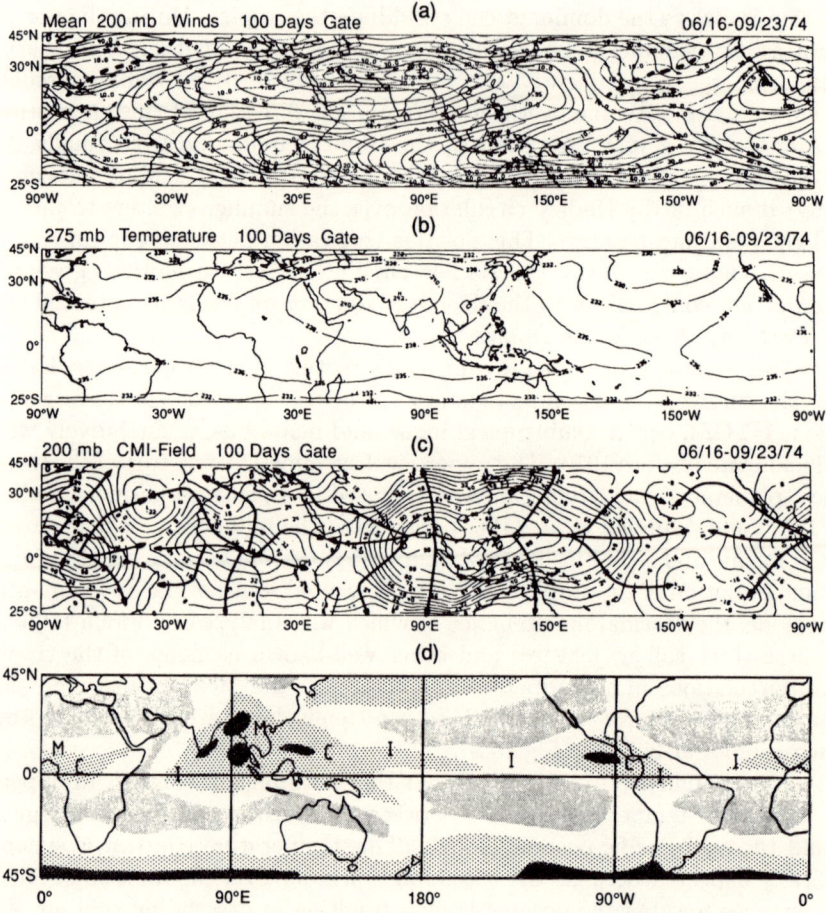

Fig. 15.3: (a) Streamlines (solid lines) and isotachs (dashed lines, knots) at 200 mb; the mid-oceanic troughs are marked with the heavy dashed lines, (b) temperature (K) at 275 mb. (c) Velocity potential (thin solid lines) at 200 mb with contour interval of 8×10^5 m^2 s^{-1} for the 100 days of GATE (Krishnamurti, 1979). (d) The northern summer-mean (June-August) IR effective temperature over the period 1974-78; darkest shading denotes temperature $< 270°$ K, stippled $> 250°$ K, clear regions $< 270°$ K and grey regions $> 250°$ K. I, C, and M denote locations of ITCZ, equatorial convective zone and monsoon regions, respectively (Webster, 1983).

The oceanic troughs over the Pacific and Atlantic consist primarily of wavenumbers 1 and 2 and exhibit a pronounced southwest-northeast tilting, which signifies momentum transport by planetary-scale waves towards mid-latitudes of the Northern Hemisphere. The energy conversion between zonal-mean flows and eddies is measured in terms of the covariance between the zonal-mean flow and north-south divergence of eddy momentum flux. The synoptic structure of oceanic troughs suggests a kinetic energy conversion from planetary-scale waves to maintain the zonal-mean flow. Additionally, eastward propagating medium-scale waves (around wavenumber 8) possess notable variance of upper-level streamfunction in the tropics (Krishnamurti et al., 1973). It is likely that the nonlinear interaction between planetary-scale and shorter waves is an energy source for the latter.

The structure of asymmetric atmospheric circulation in the tropical upper troposphere during the northern winter (Fig. 15.4a) differs from that during the northern summer. A quasi-stationary, three-wave pattern of subtropical jet streams is located in the Northern Hemisphere, in three regions of maximum winds: the southeastern United States, south of the Mediterranean Sea, and off the coast of Japan (Krishnamurti, 1961). Mid-latitude flows in the southern oceans in the upper troposphere exhibit three mid-oceanic troughs in the Atlantic Ocean, the Indian Ocean, and the Pacific Ocean (van Loon, 1972). The regions of the most active convection along the ITCZ during the northern winter exist in equatorial Africa, the maritime continent, and tropical South America; as indicated by the OLR distribution (Fig. 15.4c), these three regions coincide with the divergent centers of upper-level circulation (Fig. 15.4b). The air mass diverges from these three centers towards the three subtropical jet streams in the Northern Hemisphere and towards the three mid-oceanic troughs in the southern oceans. "Thus, the regions of convections provide a major link between flows in the two hemispheres" (Krishnamurti et al., 1973).

The possible energy relations between the subtropical jet streams in the winter Northern Hemisphere and tropical convection through planetary-scale divergent circulation are discussed in Section 13.1. Moreover, as in the northern summer, there should be a kinetic energy generation of quasi-stationary planetary-scale waves if the ascending branches of the east-west circulations are warmer than the descending branches. The three mid-oceanic troughs over the southern Atlantic, Indian, and Pacific Oceans tilt from northwest to southeast and diverge eddy momentum flux towards the mid-latitudes of the Southern Hemisphere. Tropical zonal-mean flow during the northern winter is westerly. Therefore, it is expected that the interaction between tropical zonal-mean flow and planetary-scale waves extracts energy from the former to maintain the latter.

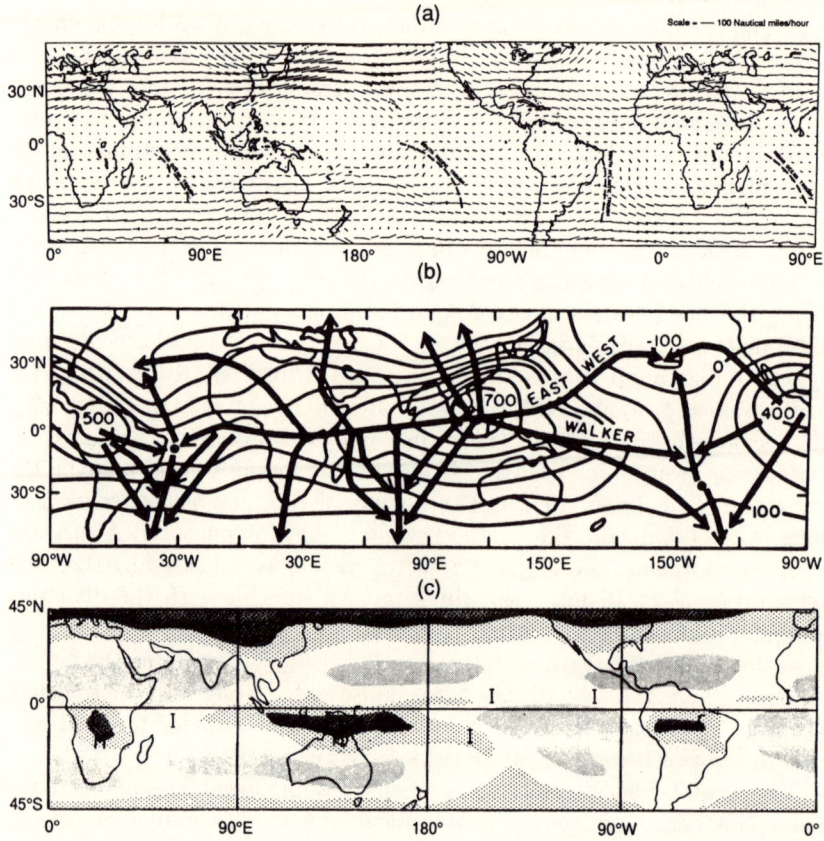

Fig. 15.4: (a) The wind vectors at 200 mb during December 1969; the mid-oceanic troughs are marked with thick dashed lines. (b) The 1969 winter-mean velocity potential at 200 mb with the contour interval of 10^6 m^2 s^{-1} (Krishnamurti et al., 1973). (c) The northern winter-mean (December-Feburary) IR effective temperature over the period 1974-78 (Webster, 1983); notation and shadings used in Fig. 15.3 are applicable here.

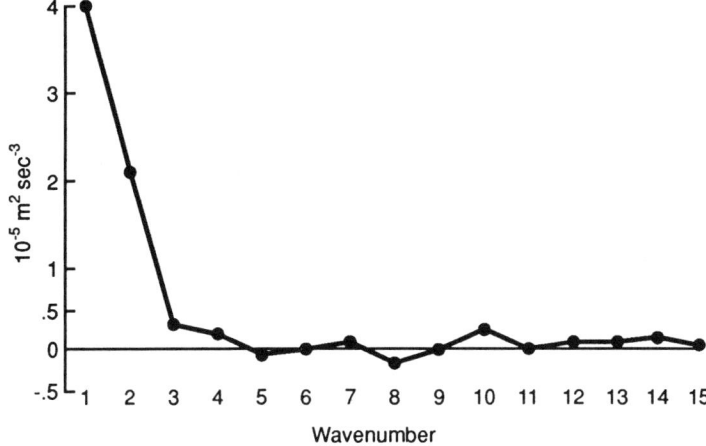

Fig. 15.5: The spectral distribution of $C(A_n, K_n)$ averaged between 10° S and 30° N at 250 mb during GATE. Unit: 10^{-6} m^2 s^{-3}.

15.2 Conventional Spectral Energetics

We discussed in Section 15.1 possible relations between important elements of the tropical general circulation in the upper troposphere and the energetics. These relations may be substantiated by a diagnostic analysis of spectral energetics. Over the past two decades, the spectral energy scheme (Saltzman, 1957) formulated in Chapter 9 has been used for this purpose. The intensity of the east-west circulation is stronger during the northern summer because of the Asian monsoon (Krishnamurti et al., 1973). Thus, efforts made in the past two decades to analyze tropical spectral energetics in the upper troposphere focused primarily on the northern summer (Kanamitsu, 1972; Depradine, 1980; Chen, 1985). Although certain analyses of tropical spectral energetics for the northern winter will be highlighted, we shall concentrate our attention on the northern summer.

Centers of the east-west circulations for the 1974 northern summer (Fig. 15.3c) are located around 10° ~ 15° N. For this reason, Deparadine (1980) analyzed the spectral energetics for the tropical belt between 10° S and 30° N. The spectral distribution of $C(A_n, K_n)$ (Fig. 15.5) shows that some minor conversions from available potential to kinetic energy occur in the high wavenumber regime with the scale of easterly waves, and some minor opposite conversion occurs around wavenumber 8. As noted previously, the upward (downward) branches of east-west circulations coincide with warm (cold) air. It is not surprising that in the tropics a pronounced release of available potential energy by thermally direct overturning takes place in the low-wavenumber (1-3) regime, thus maintaining quasi-stationary planetary-scale wave motions. Shown in Fig. 15.6a are

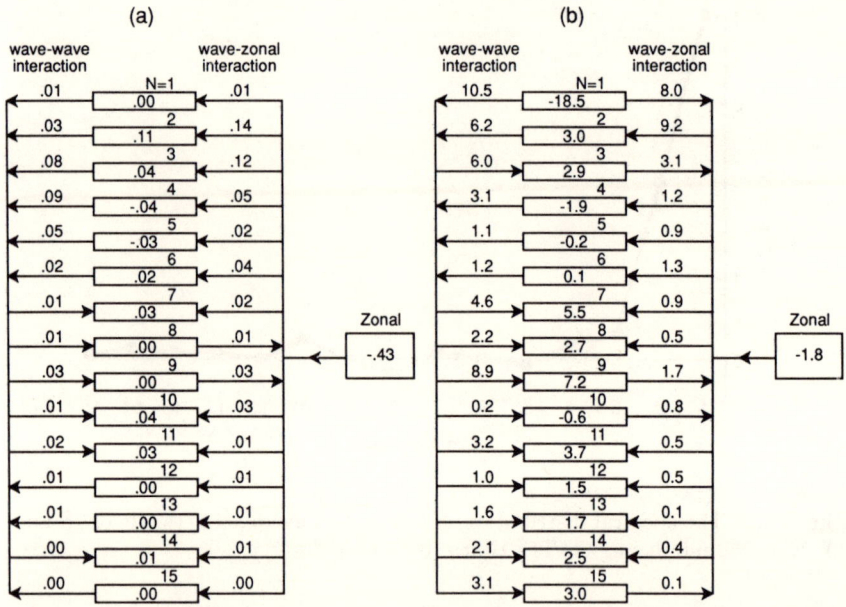

Fig. 15.6: The 250-mb conversions of (a) available potential energy and (b) kinetic energy by wave-wave and zonal-wave interactions averaged between 10° S and 30° N during GATE Unit: 10^{-6} m^2 s^{-3} (Deparadine, 1980).

the nonlinear wave-wave and zonal-wave interactions of available potential energy in the tropics. The available potential energy cascades nonlinearly from long waves to short waves. Moreover, except for wavenumbers 8 and 9, zonal available potential energy is converted to wave components.

Synoptic analysis revealed that large-scale vertical sensible heat transport in the mid-latitudes is relatively well correlated with the meridional sensible heat transport, which is most significant in the mid-latitudes. In essence, the vertical components of the large-scale motions follow the mid-latitude "slantwise" convection. Because of the correlation between these two components of sensible heat transport, the magnitudes of $C(A_Z, A_n)$ and $C(A_n, K_n)$ are expected to be comparable in the mid-latitudes. This argument has been substantiated by numerous studies – that is, Chen (1982a) – in which the direction of major energy conversions was established: $G(A_Z) \rightarrow C(A_z, A_n) \rightarrow C(A_z, K_z)$. In contrast to the mid-latitudes, temperature perturbations in the tropics are small, and meridional sensible heat transport there is insignificant, too. From the energetics viewpoint, the asymmetric thermal perturbation of the large-scale circulation in the tropics may not be maintained primarily by $C(A_Z, A_n)$. The difference in magnitude between $C(A_n, K_n)$ and $C(A_Z, A_n)$, as well as the coincidence between the centers of tropical cumulus convection and the divergent centers of the east-west circulation, supports this fact: tropical eddy available potential energy generated by east-west differential heating is released by east-west circulations to maintain quasi-stationary wave motions in the tropics with a scale of the east-west circulation.

Wave-wave and zonal-wave interactions of kinetic energy are shown in Fig. 15.6b. The southwest-northeast tilt of both mid-oceanic troughs over the Pacific Ocean and the Atlantic Ocean suggests that the kinetic energy of ultralong waves in the tropics may be converted to maintain the zonal-mean flow. The conversion of kinetic energy from wavenumbers 1 and 3 to the zonal-mean flow substantiates the synoptic argument, although most shorter waves (wavenumbers 4-15) receive kinetic energy from the zonal-mean flow. In the mid-latitudes, available potential energy is *effectively* released by synoptic-scale waves to form eddy kinetic energy and is then cascaded nonlinearly to long waves and to short waves. Regarding the nonlinear cascade of kinetic energy, synoptic-scale motions in the mid-latitudes indeed function as kinetic energy sources of other wave regimes. According to Figs. 15.5 and 15.6b, tropical atmospheric motions with wavenumbers greater than 3 do not have baroclinic energy sources. When the magnitudes of $CK(n|m, \ell)$ are compared to those of other energy conversions, the nonlinear cascade of kinetic energy in the tropics becomes a major kinetic energy source of wave motions with wavenumbers greater than 2. Evidently, the tropical quasi-stationary wave motions of wavenumbers 1 and 2, to which east-west circulations furnish kinetic energy, are major energy sources for the nonlinear cascade of kinetic energy in the tropics.

Based upon Figs. 15.5 and 15.6, the energetics of the upper-tropospheric circulation in the tropics from both long-wave regimes (L, wavenumbers 1-3)- and short-wave regimes (S, wavenumbers 4-15) are summarized in Fig. 15.7. A distinct contrast between available potential and kinetic energy budgets emerges immediately from this energy diagram, in which the former budget does not seem significant. It has long been implied that large-scale wave motions in the tropics are characteristically barotropic. In view of the minor role played by the available potential energy budget in tropical energetics, this "barotropy" argument seems plausible. However, the east-west circulation's crucial function in tropical energetics is not consistent with this barotropic nature. In fact, the insignificance of the available potential energy budget, as well as the importance of $C(A_n, K_n)$ associated with east-west circulation in the tropics, clearly demonstrates that large-scale tropical circulations are similar to the monsoons.

We have engaged in lengthy discussions of the various elements of upper-troposphere tropical circulations and of the physical processes involved in tropical spectral energetics. To relate them, a schematic diagram depicting the tropical energetics is presented in Fig. 15.8 (Krishnamurti et al., 1973). Zonal available potential energy (A_Z) is generated by north-south differential heating (H_Z) and is released by weak Hadley circulations to maintain the zonal kinetic energy (K_Z). On the other hand, ultralong-scale east-west differential heating (H_L) generates ultralong-scale available potential

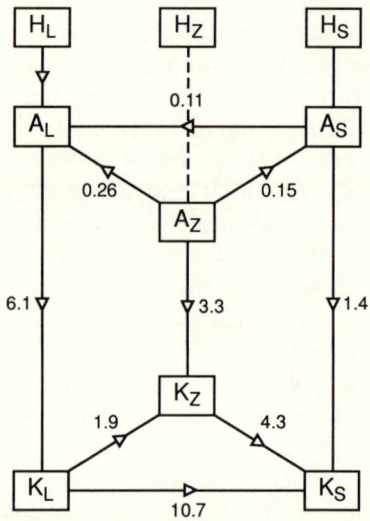

Fig. 15.7: Energy diagram of long-wave regimes (L, wavenumbers 1-3) and short-wave regimes (S, wavenumbers 4-15) in the upper troposphere; unit: 10^{-6} m^2 s^{-3}.

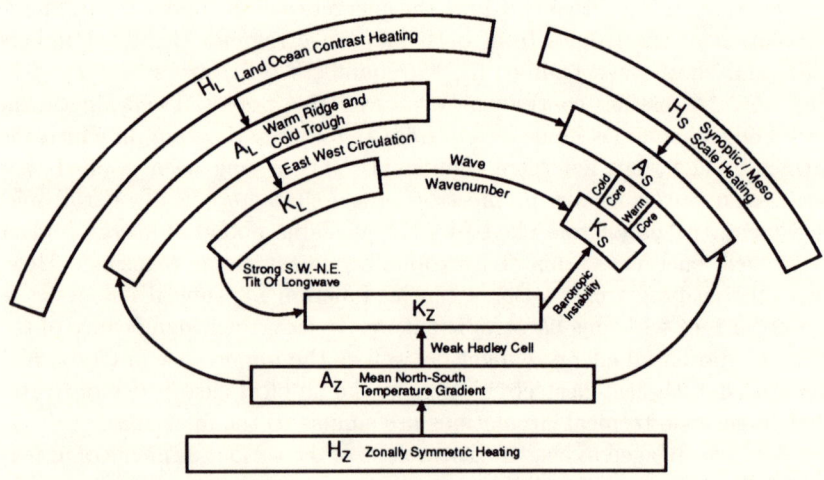

Fig. 15.8: Schematic diagram of energy transfers among long waves L, short waves S and zonal flows Z, H, A, and K denote diabatic heating, available potential energy, and kinetic energy, respectively. The transfer direction is represented by the arrow.

ENERGETICS OF THE TROPICS 235

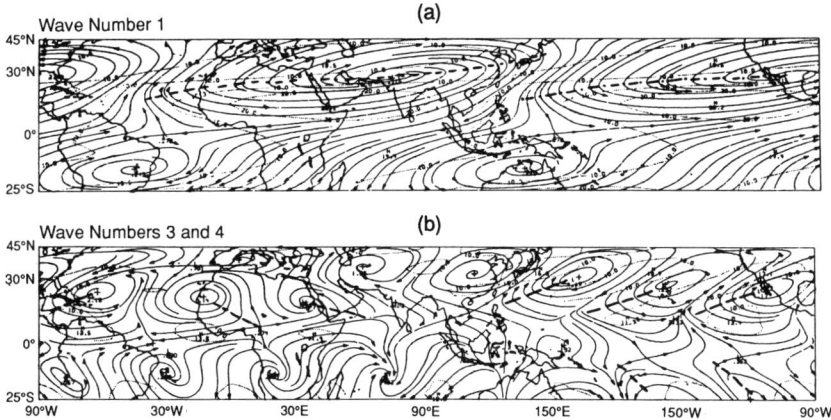

Fig. 15.9: Composite streamlines (solid lines) and isotachs (dashed lines, knots) for (a) wavenumber 1 and (b) wavenumbers 3 and 4 when the triad interaction <1,3,4> was minimum.

energy (A_L), which is released by thermally direct east-west circulations to support quasi-stationary ultralong-scale wave motions. In essence, the ultralong waves constitute an energy scale of tropical circulations. Furthermore, the zonal-mean flow receives energy from southwest-northeast tilting long waves through wave-wave interactions. Short waves may be formed as a result of barotropic instability of the zonal-mean flow. Through thermal overturning they may gain (lose) energy from (to) thermal perturbations, as indicated by A_S.

In Section 15.1, the synoptic structure of the upper-tropospheric circulation in the tropics related to several major energy conversions was delineated. The wave-wave interaction (nonlinear cascade) of kinetic energy, which consists of a triad (three-waves) of interactions, was shown in Fig. 15.6b to be an important kinetic energy source of short waves. It would definitely be of interest to explore the synoptic conditions of waves involved in such triadic interactions. The time series of daily triadic interactions examined by Krishnamurti (1979), which indicate that the (1,3,4) triad was one of the major contributors to $C(K_L, K_S)$ over the 1974 summer, are presented in Fig. 15.9. Both positive and negative centers of wavenumber-1 streamlines exhibit a pronounced southwest-northeast tilt in the Northern Hemisphere. A region of strong easterlies and strong westerlies associated with the wavenumber-1 structure appears in the regions south of this wave's streamline centers around Pakistan and the Mexican highland, respectively. For the same dates, the composite streamlines of wavenumbers 3-4 used in Fig. 15.9a have a southwest-northeast tilt over the north Pacific and a northwest-southeast tilt in the tropical easterly jet region. Superimposing the two composite streamline maps, one can see

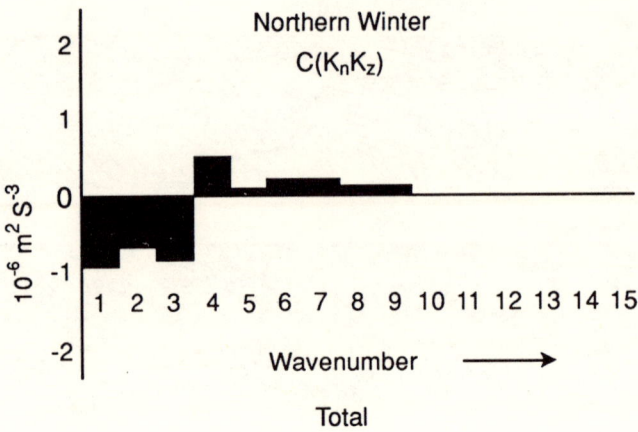

Fig. 15.10: The spectral distribution of $C(K_n, K_z)$ during the northern winter at 200 mb. Unit: 10^{-6} m^2 s^{-3} (Krishnamurti et al., 1973).

that short waves remove westerly momentum from the maximum westerly region in the north Pacific and easterly momentum from the tropical easterly jet.

Although the Australian monsoon takes place in the northern winter, the tropical spectral energetics of this season have not been analyzed in detail, except for certain preliminary computations made by Krishnamurti et al. (1973). As indicated by the intensity of the mass flux function shown in Fig. 15.2, the Hadley circulation during the northern winter is as strong as that during the northern summer. Zonal available potential energy is apparently released by the Hadley circulation to maintain the zonal flow. During the northern winter, cumulus convection centers are located over three tropical continents: the maritime continent, equatorial South America, and equatorial Africa (Fig. 15.4b). The three centers of planetary-scale divergent circulation coincide with cumulus convection centers, and air is diverged from these three centers towards the three mid-oceanic troughs in the southern oceans and towards the three subtropical jet streams in the Northern Hemisphere. Thermally direct divergent circulations are expected to release eddy available potential energy, which is generated by east-west asymmetric differential heating to support upper-level asymmetric motions. The typical value of $C(A_E, K_E)$ at 300 mb is about 10×10^{-5} m^2 s^{-1}.

Zonal-mean westerlies prevail in the tropical upper troposphere between 15° S and 15° N during the northern winter. As can be inferred from the three oceanic troughs in the southern oceans, the ultralong waves there should tilt from northwest to southeast. Westerly momentum is transported out of the tropical belt by ultralong waves, and this momentum transport results in the conversion of K_Z to K_E. The spectral distribution

ENERGETICS OF THE TROPICS

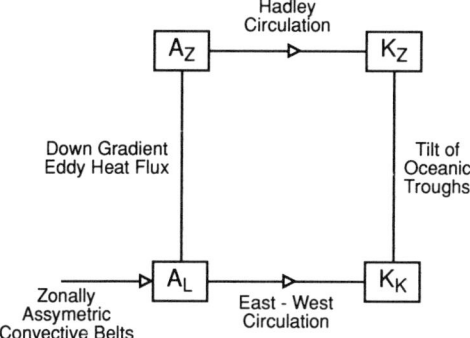

Fig. 15.11: Schematic energy exchanges of the upper troposphere for northern winter between long waves and zonal flow in the tropics. After Krishnamurti et al., 1973.

of $C(K_n, K_Z)$ (Fig. 15.10) shows that wavenumbers 1-3 receive kinetic energy from zonal-mean flow, but that shorter waves supply kinetic energy to zonal-mean flow. In fact, during the northern winter the contrast between low- and high-wavenumber regimes in terms of $C(K_n, K_Z)$ is opposite to that during the northern summer.

Based upon preliminary energy analyses of the tropics during the northern winter, a schematic energy diagram of the long-waves is suggested in Fig. 15.11. To substantiate the long-wave energetics suggested in this figure as well as the short-wave energetics, further efforts are urged.

15.3 Low-Frequency Variation of Tropical Energetics

In Chapter 14, we discussed a quasi-periodic variation of atmospheric energetics with a period of two to three weeks in the mid-latitudes. This low-frequency variation of atmospheric energetics was coined *vacillation*. Sections 15.1 and 15.2 illustrated that both the dynamics and energetics of large-scale circulations in the tropics differ from those in the mid-latitudes. A pronounced low-frequency oscillation with a period of 30-60 days in the tropics was identified by Madden and Julian (1971; 1972). This low-frequency oscillation is well portrayed by global-scale divergent circulation, which was shown to be important in the tropical energetics. For example, in Fig. 15.3b, the thermal high over the Asian monsoon region and thermal troughs over the Pacific Ocean and the Atlantic Ocean coincide with divergent and convergent contours of east-west circulations, respectively. Chen and his collaborators (Chen, 1987; Chen and Yen, 1991b,c) demonstrated that the thermal trough over the Pacific Ocean and the southern part of the tropical easterly jet exhibit a marked 30-60 day oscillation following the

Indian monsoon life cycle. The spectral energetics of the tropics (Section 15.2) reveal that ultralong waves dominate tropical energetics. Note that the important elements of the tropical circulation, which possesses a significant low-frequency oscillation, are associated with the low-wavenumber regime. Evidently, a low-frequency variation should exist in tropical energetics.

The time evolution of tropical spectral energetics was pursued by Chen (1985) using the ECMWF FGGE III-b data within the tropical belt between 10° S and 20° N for the 1979 summer (June-August). Shown in Fig. 15.12 are time series of A_Z, A_E, A_1, and A_{2-18} and time series of their kinetic energy counterparts in the tropics. Although the magnitudes of zonal energies are several factors smaller than those of eddy energies, a pronounced time variation emerges in the time series of each energy variable. With a least-squares fit curve superimposed on the time series of energies, we observe clearly that both A_E and K_E not only vary almost in phase, but also exhibit a bimodal time variation over the 1979 summer; the time span from one minimum to the other is about $1\frac{1}{2}$ months. Time series of A_Z and K_Z have only one major maximum over this summer and are more or less out of phase with those of eddy energies. Thus, the correlation between time variations of zonal and eddy energies is similar to the vacillation of atmospheric energetics in the mid-latitudes, except that the latter phenomenon has a shorter period (two to three weeks).

In the last section, it was shown that wavenumber 1 dominates tropical spectral energetics. It is not surprising to see the quasi-periodic variations of A_E and K_E, shown in Fig. 15.12. Because the short-wave regime ($n \geq 2$) does not contribute significantly to the $1\frac{1}{2}$-month variation of A_E and K_E, we can focus our attention on wavenumber 1 to search for the mechanism maintaining the low-frequency variation of tropical energetics. Additionally, $C(A_Z, K_Z)$, $C(A_1, K_1)$, $C(K_1, K_Z)$, and $CK(1/m, l)$ dominate the tropical energetics of the upper troposphere. Integrating the spectral energetics budget over the entire tropical troposphere, Chen and Marshall (1984) pointed out that the salient features of tropical energetics in the upper troposphere are maintained throughout the troposphere. The exception is the nonlinear transfer of kinetic energy in which wavenumber 2, $CK(2/m, l)$, is most important. To facilitate our discussion of the maintenance of low-frequency variation of tropical energetics, we shall focus on energy conversions of great magnitude.

In Fig. 15.13, the time series of $C(A_Z, K_Z)$ and $C(A_E, K_E)$ exhibit a clear bimodal variation in time over the 1979 summer. Recall that A_Z and K_Z time series have only one major maximum value that occurs in the middle of this summer. Because $C(A_Z, K_Z)$ represents the release of A_Z by the Hadley circulation, we may expect this energy conversion to peak when the zonal-mean circulation does. Surprisingly, it is otherwise, as shown by comparing Figs. 15.12 and 15.13. That is, the release of A_Z is minimal when the intensity of the zonal-mean circulation reaches its peaks. Interestingly, from the correlation of zonal energies with $C(A_1, K_1)$

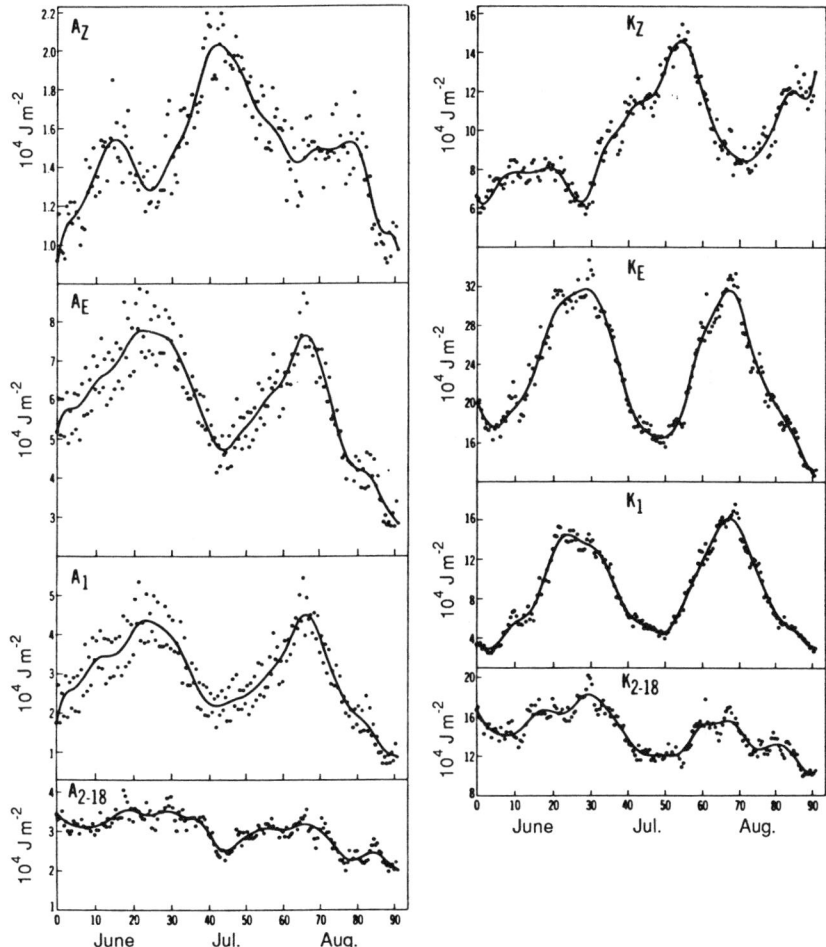

Fig. 15.12: Time series of (a) A_Z, A_E, A_1, A_{2-18} and (b) K_Z, K_E, K_1, K_{2-18} averaged between 10° S and 20° N during the 1979 northern summer.

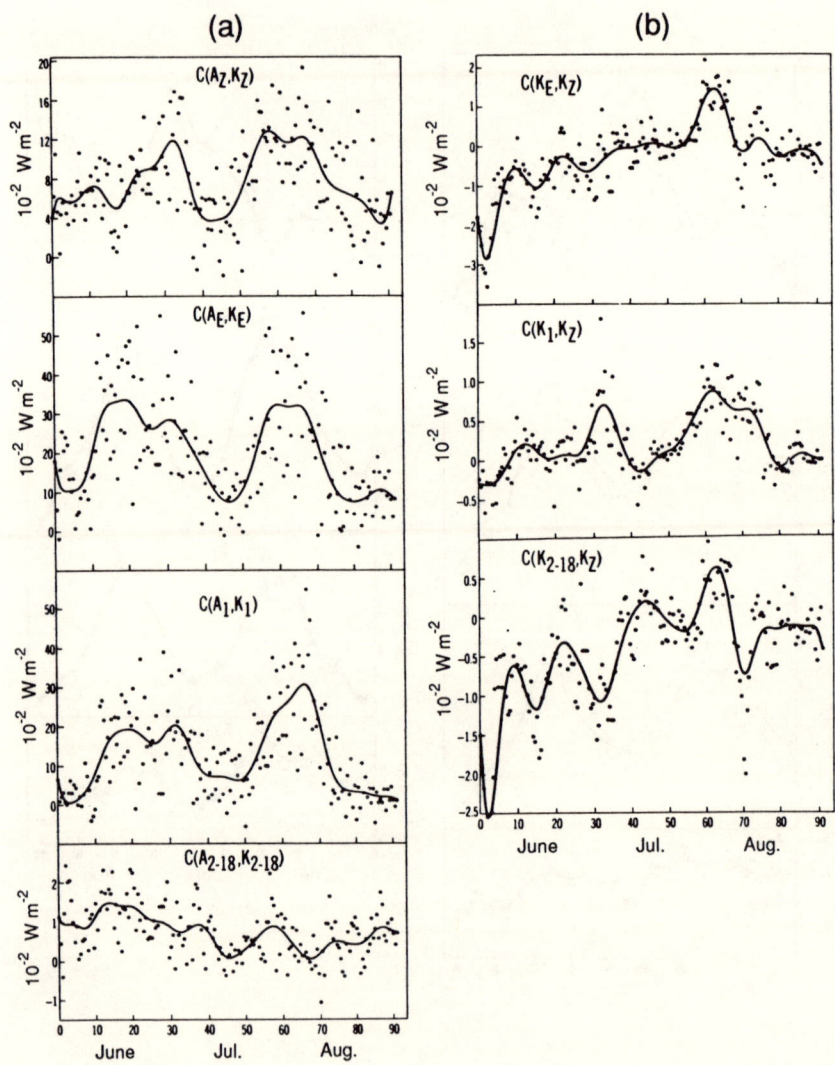

Fig. 15.13: Time series of various energy conversions averaged between 10° S and 20° N as indicated in each time series during the 1979 northern summer.

and $C(A_{2-18}, K_{2-18})$, we find that the time variation of $C(A_E, K_E)$ is contributed primarily by wavenumber 1. The release of eddy available potential energy, which is generated by east-west asymmetric heating, is the major energy source maintaining the tropical wave motions. Thus, it is inferred that wavenumber 1 is responsible for the low-frequency oscillation of tropical energetics. In spite of small magnitudes, no energy conversions except $C(K_1, K_Z)$ suggest a quasi-periodic $1\frac{1}{2}$-month variation.

The quasi-periodic $1\frac{1}{2}$-month variation of the tropical energetics during the northern summer has been established. But, how the low-frequency oscillation affects the quasiperiodic $1\frac{1}{2}$-month variation of tropical energetics is not yet quite clear. To answer this question, we shall examine the time evolution of the structure of wavenumber 1 and the transport properties of this wave associated with important energy conversions. Note that $A_1 \sim T_1^2$ and $K_1 \sim U_1^2$. The effect of a low-frequency oscillation in A_1 and K_1 is illustrated in terms of five phases of the 10-day averaged structures of $T_1(200\text{mb})$ and $U_1(200\text{mb})$ following time evolutions of A_1 and K_1 (Fig. 15.12); three phases of minimum (June 1-10, July 13-22, August 22-31) and two phases of maximum (June 21-30, August 1-10) are seen.

The summer-mean $T_1(200\text{mb})$ is shown at the bottom of Fig. 15.14a for reference. Within the computational domain (10° S-20° N), the summer-mean $T_1(200\text{mb})$ amplitude north of the equator is greater than that south of the equator and is negligible in the vicinity of the equator. The thermal ridge and the troughs of $T_1(200\text{mb})$ do not tilt north of the equator, but do tilt SE-NW south of the equator. The general features of $T_1(200\text{mb})$ during various phases of the tropical energetics are similar to those of the summer-mean $T_1(200\text{mb})$, except the amplitude of $T_1(200\text{mb})$ exhibits a low-frequency oscillation. The quasi-stationary locations of the thermal ridge over the monsoon region and the quasi-periodic variation of $T_1(200\text{mb})$ should be located in the monsoon region, and the latter varies in its intensity with a period of about $1\frac{1}{2}$ months. The July diabatic heating center computed by Johnson et al. (1985) is consistent with the $T_1(200\text{mb})$ thermal ridge, but calculation of time variation in the intensity of this heating center was not attempted over the entire northern summer.

The summer-mean U_1 (200 mb) (bottom of Fig. 15.14b) has easterly ($U_1 < 0$) anomalies located in the monsoon region and westerly ($U_1 > 0$) anomalies located outside that region. The comparision between T_1 (200 mb) and U_1 (200 mb) indicates that U_1 (200 mb) > 0 [U_1 (200 mb) < 0] is associated with the T_1 (200 mb) thermal ridge (trough). Note that the core of the tropical easterly jet is situated in the southern Indian continent. The coincidence between easterly anomalies of U_1 (200 mb) and the tropical easterly jet suggests that U_1 is an important ingredient of this jet. South of the equator and within the computational domain, easterly anomalies of U_1 (200 mb) are centered around Indonesia and westerly anomalies of U_1 (200 mb) are located over the African continent. The low-frequency variation of U_1 (200 mb) is particularly noticeable around 20° N and 20° S, where both maximum and minimum centers of U_1 (200 mb) exist. The

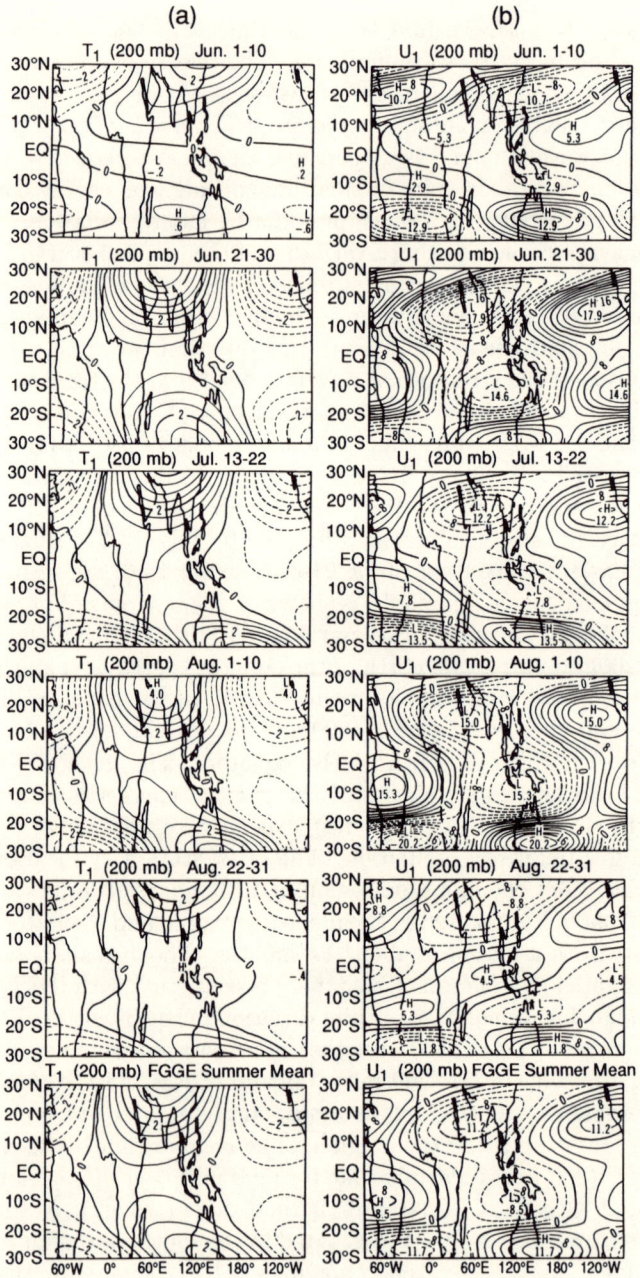

Fig. 15.14: Horizontal distribution of the 10-day averaged temperature T_1 (200 mb) and zonal wind U_1 (200 mb) for five phases indicated in each panel and for the 1979 summer-mean. The contour intervals of T_1 (200 mb) and U_1 (200 mb) are $0.5°$ C and 2 m s^{-1}, respectively.

maximum (minimum) center of U_1 (200 mb) at 20° N is almost stationary, whereas that at 10°N moves somewhat eastward (westward) around its summer-mean locations.

The differences in terms of both structure and mobility of wavenumber 1 between regions north and south of the equator suggest that the physical processes maintaining the $1\frac{1}{2}$-month variation in the tropical energetics differ in these regions. The release of A_1 to support K_1 by thermally direct overturning is determined by vertical sensible heat transport $(\omega_1 T_1)$. The conversion from K_1 to K_Z to maintain the former kinetic energy is measured by the covariance between the horizontal momentum $(U_1 V_1)$ and the meridional gradient of zonal-mean flow. It has been shown that the $1\frac{1}{2}$-month variation of A_1 and K_1 in the tropics may be maintained by these two transport properties.

Within 10° S and 20° N, $\omega_1 T_1$ (200 mb) (Fig. 15.15) becomes significant north of the equator when eddy energies peak and maximum centers of $\omega_1 T_1$ (200 mb) appear around 20° N. It was stressed previously that the large-scale tropical circulation is monsoonlike. Vertical sensible heat transport by large-scale motions in the tropics should extend locally over the whole troposphere, and vertical distributions of $\omega_1 T_1$ at 20° N during the two phases of maximum eddy energies show that this is so. The geographical distribution and temporal variation in magnitudes of $\omega_1 T_1$ coincide with those of T_1 (200 mb) and U_1 (200 mb) in the region north of the equator. This coincidence indicates that the $1\frac{1}{2}$-month variation responds to that of A_1 through the thermally direct overturning of large-scale tropical motion. As can be seen from Fig. 15.14a, the T_1 (200 mb) amplitude south of the equator is relatively insignificant. Of course, we can not expect the tropical motion of wavenumber 1 in this region to be maintained by the release of A_1. The negligible $\omega_1 T_1$ (200 mb) distribution suggests that a different physical process is responsible for the maintenance of the wavenumber 1 motion there.

It has been pointed out that $CK(1/m, l)$ was relatively small during the 1979 summer. The only possible source maintaining K_1 south of the equator is $C(K_1, K_Z)$. Recall that $C(K_1, K_Z)$ (Fig. 15.13) has a small numerical value although time series of $C(K_1, K_Z)$ during the 1979 summer show a bimodal variation in time and coincide with those of K_1. We may suspect that $C(K_1, K_Z)$ is not a source of K_1 south of the equator. With this doubt in mind, we need to explore spatial and temporal distributions of $U_1 V_1$.

The horizontal distribution of the 10-day averaged $U_1 V_1$ (200 mb) in various phases of eddy energies is displayed in Fig. 15.16. The momentum transport by wavenumber 1 is significant when eddy energies peak, particularly south of the equator. The vertical cross-sections of 10-day averaged $U_1 V_1$ at 10° S that correspond to horizontal $U_1 V_1$ (200 mb) indicate that the maximum momentum transport by wavenumber 1 occurs in the upper troposphere. Between the equator and 10°N over the Arabian Sea, the core of the tropical easterly jet is relatively steady over the entire summer.

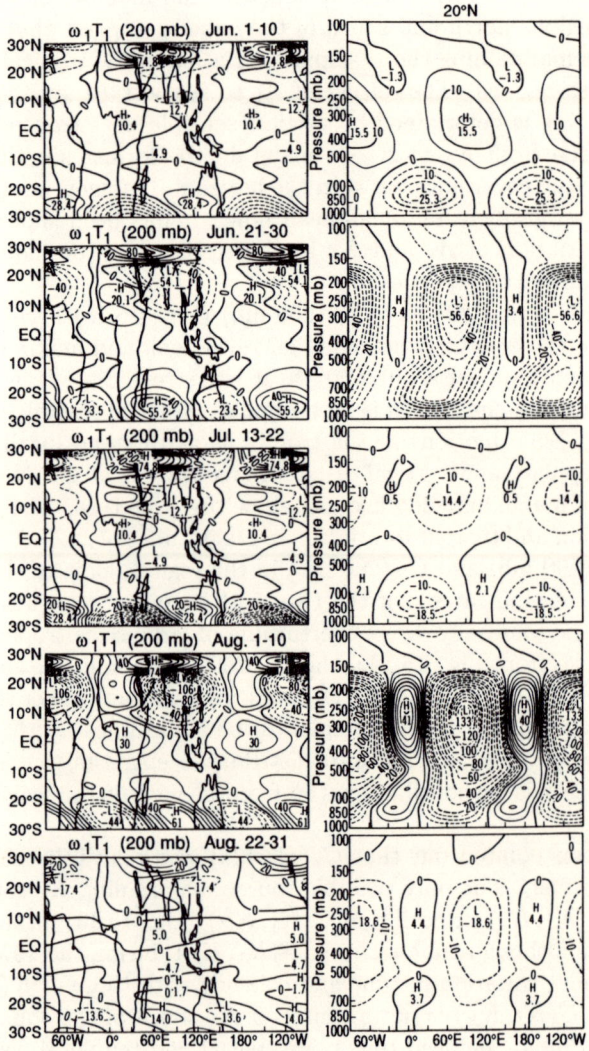

Fig. 15.15: Horizontal distribution of the 10-day averaged vertical sensible heat transport of wave 1 at 200 mb, $\omega_1 T_1$, at 20° N (second column) for five phases indicated in each panel. Unit: 10^{-7} degree mb s^{-1}.

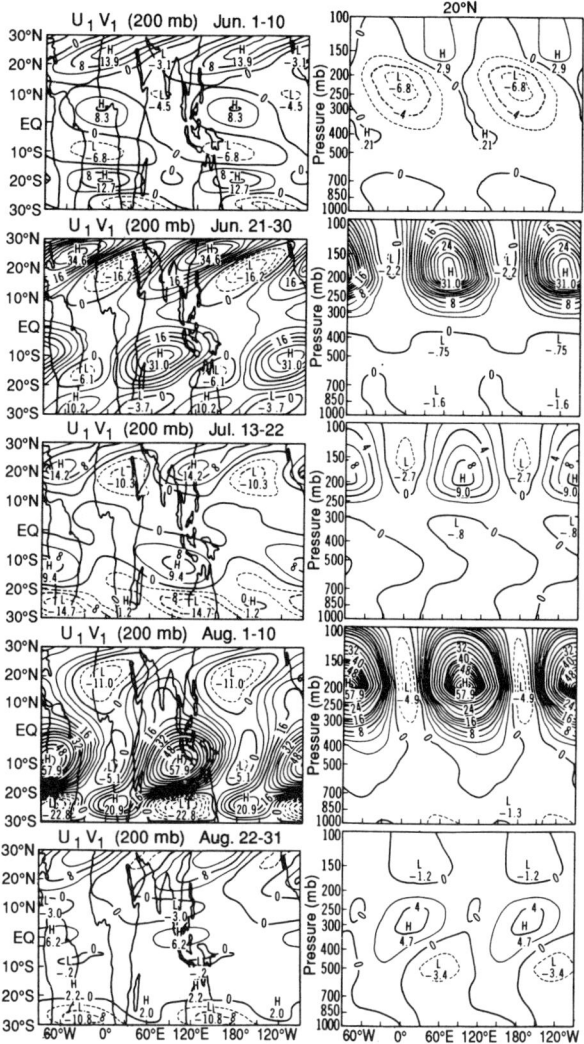

Fig. 15.16: Horizontal distribution of the 10-day averaged momentum transport of wave 1 at 200 mb, U_1V_1 (200 mb), (first column) and longitude-pressure cross section of U_1V_1 at 10° S (second column) for five phases indicated on each panel. Unit: m² s⁻¹.

Moreover, the north-south wind shear of the zonal-mean flow changes its sign between the equator and 10° N. As can be inferred from Fig. 15.16, the zonal-mean U_1V_1 is usually northward. Therefore, $C(K_1, K_Z)$ changes its sign with the north-south wind shear of the zonal-mean flow. In other words, $C(K_1, K_Z)$ is positive (negative) south (north) of the equator. The cancellation of positive and negative values in the latitudinal integration results in a small numerical value for $C(K_1, K_Z)$. Nevertheless, the time variation of U_1V_1 south of the equator, and the variance between U_1V_1 and the north-south gradient of zonal-mean flow are maintained by $C(K_1, K_Z)$.

Presumably, the $1\frac{1}{2}$-month variation of T_1 and U_1 fields in the tropics during the northern summer is induced by the eastward-propagating intraseasonal oscillation, as inferred by Chen (1987). The $1\frac{1}{2}$-month variation of tropical energetics reflects an important response of the large-scale circulation in the tropics to this intraseasonal oscillation.

15.4 Spectral Energetics of Baroclinic and Barotropic Flows

In Section 15.2, the spectral energetics of the tropical upper-troposphere circulation indicated that the primary energy source driving the quasi-stationary planetary-wave motions comes from the release of available potential energy by the east-west circulation. Evaluation of this energy conversion can be made in terms of two factors: the covariance between temperature and vertical motion (ω) and the work by the pressure gradient force. The ω field is usually generated indirectly and may be subject to data or methodological bias. In addition, the magnitudes of temperature and of pressure perturbations are an order smaller in the tropics than in the mid-latitudes. Consequently, substantial errors may be present in the computation of the conversion from available potential energy to kinetic energy. In the tropics, the observations of winds are more reliable than those of temperatures and heights. The conversion between baroclinic and barotropic kinetic energy in (5.15), which requires only the wind field, offers an alternative method of examining the energetics of large-scale motions in the tropics. It was demonstrated in Section 15.2 that the low-wavenumber (1-3) regime is, in terms of energy, a source for other wave energy regimes. To understand the role played by this wavenumber regime in the interaction between baroclinic and barotropic flows in the tropics, we should pursue the energetics of these two flow components in the spectral domain.

Following (9.1), we can expand the given meteorological variables ( ) in terms of a Fourier series:

$$( ) = C_0 + \sum_{n=1}^{N} [C_n \cos n\lambda + S_n \sin n\lambda], \qquad (15.1)$$

where C_0, C_n and S_n are Fourier coefficients and where N is the cutoff number for these coefficients. We may write K_S, K_M, and the conversion

ENERGETICS OF THE TROPICS

between them in terms of C_0, C_n, and S_n. The spectra of K_S and K_M are

$$K_M = K_0^M + \sum_{n=1}^{N} K_n^M \quad \text{and} \quad K_S = K_0^S + \sum_{n=1}^{N} K_n^S, \quad (15.2)$$

where

$$K_0^S = \frac{1}{2g(\sin\phi_2 - \sin\phi_1)} \int_0^{p_0} \int_{\phi_1}^{\phi_2} \left(UC_0^{S^2} + VC_0^{S^2}\right) \cos\phi \, d\phi \, dp,$$

$$K_n^S = \frac{1}{4g(\sin\phi_2 - \sin\phi_1)} \int_0^{p_0} \int_{\phi_1}^{\phi_2} \left(UC_n^{S^2} + VC_n^{S^2}\right) \cos\phi \, d\phi \, dp,$$

$$K_0^M = \frac{1}{2g(\sin\phi_2 - \sin\phi_1)} \int_0^{p_0} \int_{\phi_1}^{\phi_2} \left(UC_0^{M^2} + VC_0^{M^2}\right) \cos\phi \, d\phi \, dp,$$

$$K_n^M = \frac{1}{4g(\sin\phi_2 - \sin\phi_1)} \int_0^{p_0} \int_{\phi_1}^{\phi_2} \left(UC_n^{M^2} + VC_n^{M^2}\right) \cos\phi \, d\phi \, dp,$$

$$(15.3)$$

and where ϕ_1 and ϕ_2 are the boundary latitudes of the computational domain.

To obtain a spectral form, the conversion from K_s to K_M, (5.19) and (5.20), may be rewritten in the scalar form

$$C(K_S, K_M) = -\frac{1}{g} \int_0^{p_0} \int_S [\zeta_S (v_M u_S - u_M v_S)$$
$$C_{ND}(K_S, K_M)$$
$$+ D_S (u_M u_S + v_M v_S)] \, dp \, ds. \quad (15.4)$$
$$C_D(K_S, K_M).$$

The integrand of (15.4) contains products of three variables. A detailed derivation for the general spectral expression of triple products with C_0, C_n, and S_n can be found in Wiin-Nielsen and Drake (1965). The spectral form of $C(K_S, K_M)$ may be written in the symbolic form

$$C(K_S, K_M) = C^0(K_S, K_M) + \sum_{n=1}^{N} C^n(K_S, K_M)$$
$$= C_{ND}^0(K_S, K_M) + \sum_{n=1}^{N} C_{ND}^n(K_S, K_M)$$
$$+ C_D^0(K_S, K_M) + \sum_{n=1}^{N} C_D^N(K_S, K_M). \quad (15.5)$$

The detailed formulation of the symbolic expressions in (15.5) is

$$C_{ND}^0(K_S, K_M) = \frac{1}{2g(\sin\phi_2 - \sin\phi_1)} \int_0^{p_0} \int_{\phi_1}^{\phi_2} T_0(\phi) \cos\phi \, d\phi \, dp, \quad (15.6)$$

where

$$\begin{aligned}
T_0(\phi) = \ & 2\left(UC_0^M \bullet ZC_0^S \bullet VC_0^S - VC_0^M \bullet ZC_0^S \bullet UC_0^S\right) \\
& + UC_0^M \sum_{n=1}^N \left(VC_n^S \bullet ZC_n^S + VS_n^S \bullet ZS_n^S\right) \\
& - VC_0^M \sum_{n=1}^N \left(UC_n^S \bullet ZC_n^S + US_n^S \bullet ZS_n^S\right).
\end{aligned}$$

$$\begin{aligned}
C_{ND}^n(K_S, K_M) = \ & \frac{1}{2g(\sin\phi_2 - \sin\phi_1)} \Bigg[\\
& \int_0^{p_0} \int_{\phi_1}^{\phi_2} \left(-T_n^1 + T_n^2 - T_n^3 + T_n^4\right) \cos\phi \, d\phi \, dp \\
& + \int_0^{p_0} \int_{\phi_1}^{\phi_2} \left(T_n^5 - T_n^6\right) \cos\phi \, d\phi \, dp \Bigg], \quad (15.7)
\end{aligned}$$

$$\begin{aligned}
T_n^1 &= UC_0^S \left(VC_n^M \bullet ZC_n^S + VS_n^M \bullet ZS_n^M\right), \\
T_n^2 &= ZS_0^S \left(UC_n^M \bullet VC_n^S + US_n^M \bullet VS_n^S\right), \\
T_n^3 &= ZS_0^S \left(VC_n^M \bullet UC_n^S + VS_n^M \bullet US_n^S\right), \\
T_n^4 &= VC_0^S \left(VC_n^M \bullet ZC_n^S + US_n^M \bullet ZS_n^m\right),
\end{aligned}$$

$$\begin{aligned}
T_n^5 = \ & \sum_{m=1}^n \left\{ UC_n^M \left[VC_m^S \left(ZC_{n+m}^S + ZC_{n-m}^S\right) + VS_m^S \left(ZS_{n+m}^S - ZS_{n-m}^S\right)\right] \right. \\
& \left. + US_n^M \left[VC_m^S \left(ZS_{n+m}^S + ZS_{n-m}^S\right) - VS_m^S \left(ZC_{n+m}^S - ZC_{n-m}^S\right)\right] \right\} \\
+ & \sum_{m=n+1}^N \left\{ UC_n^M \left[VC_m^S \left(ZC_{m+n}^S + ZC_{m-n}^S\right) + VS_m^S \left(ZS_{m+n}^S + ZS_{m-n}^S\right)\right] \right. \\
& \left. + US_n^M \left[VC_n^S \left(ZS_{m+n}^S - ZS_{m-n}^S\right) - VS_m^S \left(ZC_{m+n}^S - ZC_{m-n}^S\right)\right] \right\}
\end{aligned}$$

and

$$\begin{aligned}
T_n^6 = \ & \sum_{m=1}^n \left\{ VC_n^M \left[UC_m^S \left(ZC_{n+m}^S + ZC_{n-m}^S\right) + US_m^S \left(ZS_{n+m}^S - ZS_{n-m}^S\right)\right] \right. \\
& \left. + VS_n^M \left[UC_m^S \left(ZS_{n+m}^S + ZS_{n-m}^S\right) - US_m^S \left(ZC_{n+m}^S - ZC_{n-m}^S\right)\right] \right\}
\end{aligned}$$

$$+ \sum_{m=n+1}^{N} \{VC_n^M \left[UC_m^S \left(ZC_{m+n}^S + ZC_{m-n}^S\right) + US_m^S \left(ZS_{m+n}^S + ZS_{m-n}^S\right)\right]$$
$$+ VS_n^M \left[UC_m^S \left(ZS_{m+n}^S - ZS_{m-n}^S\right)\right]\} .$$

In (15.4), a comparison of the mathematical expressions $C_{ND}(K_S, K_M)$ and $C_D(K_S, K_M)$ indicates that the latter can be obtained by replacing ζ_S, v_M, and $-u_M$ in the former with D_S, u_M, and v_M, respectively. Accordingly, by substituting the Fourier coefficients of ζ_S, v_M, and $-u_M$ in (15.7) with D_S, u_M, and v_M, we are able to formulate the mathematical expression of $C_D^n(K_S, K_M)$.

Diagnostic computations of (15.2) and (15.5) were performed by Chen (1983) with the ECMWF FGGE III-b wind data of the 1979 summer (June-August) for the tropical belt between 15° S and 15° N. The truncation of the Fourier expansion was $N = 18$. The summer-mean values of K_S and K_M in the tropical belt are respectively, 2.05×10^5 J m^{-2} and 1.29×10^5 J m^{-2}, which are about one to two orders of magnitude smaller than those in mid-latitudes (Chapter 5). K_S/K_M is about 0.38 all year around north of 15° N, whereas this ratio for the summer tropics is 159%. The sharp contrast in K_S/K_M ratios between these two regions results from a dramatic change in the circulation regimes.

According to Fig. 15.3a, the most pronounced features of the upper-troposphere circulation in the tropics between 15° N and 15° S are the tropical easterly jet and the mid-oceanic troughs with which westerlies are associated. In the tropical lower troposphere, the Indian monsoon westerlies extend from the western Indian Ocean to the South China Sea, and the trades are located underneath the westerlies of the mid-oceanic troughs. In fact, this planetary-scale monsoon circulation regime in the tropics yields to the mid-latitude circulation regime in which upper-troposphere westerlies commonly extend to the lower troposphere. Therefore, the great universal value of the K_S/K_M ratio in the tropics does not imply strong baroclinicity, but reflects the monsoonal character of tropical large-scale circulation.

The energy content of the K_n spectrum does not depend significantly upon wavenumber in the low-wavenumber regime, but does decrease dramatically when $n \geq 7$. Because atmospheric circulation in the tropics has a monsoonal nature, the energy content of the tropical K_n spectrum differs from its mid-latitude counterpart. The energy content of the tropical K_n spectrum (Fig. 15.17) decreases monotonically as the wavenumber increases. Ultralong waves corresponding to the east-west circulation constitute the energetics scale in the tropics. It is not surprising that wavenumber 1 dominates the tropical K_n spectrum. In the mid-latitudes, the K_M^n spectrum has a greater energy content in the low-wavenumber regime than does the K_M^n spectrum elsewhere, except in the wavenumber regime $n \geq 13$. Unlike in the mid-latitudes, the K_S^n spectrum in the tropics

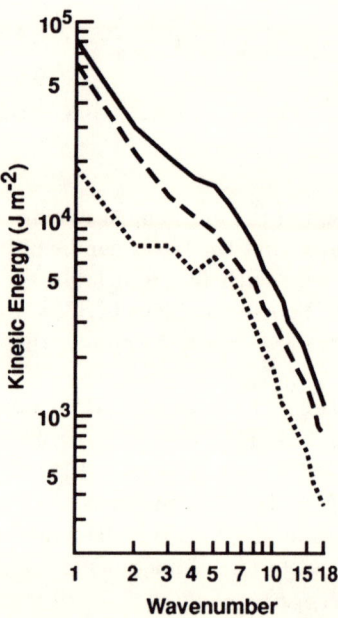

Fig. 15.17: Spectra of K (solid line), K_S (dashed line) and K_M (dotted line) in the tropics (15° S-15° N) during the 1979 northern summer (June-August).

always has a greater energy content than the tropical K_M^n spectrum with the same wavenumber, particularly in the low- wavenumber regime.

It was shown in Section 15.2 that the main energy source maintaining quasi-stationary ultralong waves in the tropics is the release of available potential energy by the east-west circulation. Large K_S in the tropics reflects the monsoonal characteristics of large-scale motions. Thus, we expect energy to convert from K_S to K_M in the tropics. The numerical value of $C(K_S, K_M) = 0.44$ W m^{-2} averaged between 15° S and 15° N confirms our argument. Additionally, Chapter 5 demonstrated that $C_{ND}(K_S, K_M)$ is the only energy conversion between K_S and K_M in a quasi-geostrophic system. Regarding $C(K_S, K_M)$, the difference between the quasi-geostrophic and the primitive equation models is $C_D(K_S, K_M)$. The numerical values of $C_{ND}(K_S, K_M)$ and of $C_D(K_S, K_M)$ in the tropical belt are 0.48 W m^{-2} and -0.04 W m^{-2}, respectively. The small value of $C_D(K_S, K_M)$, i.e. $C_D(K_S, K_M)/C_{NC}(K_S, K_M) \ll 1$, indicates that the quasi-geostrophic nature of large-scale motions persists in the tropics.

The role played by the east-west circulation in the tropics was made clear by the spectral energetics analysis discussed in Section 15.2. To clarify our inference that K_S and K_M energetics are linked to the east-west circulation, the $C^n(K_S, K_M)$ spectrum is displayed in Fig. 15.18. The $C^n(K_S, K_M)$ spectrum is separated by $C^3(K_S, K_M)$ into a bimodal spectral distribution with two wavenumber regimes: $n = 0 - 2$, and $n \geq 4$.

ENERGETICS OF THE TROPICS

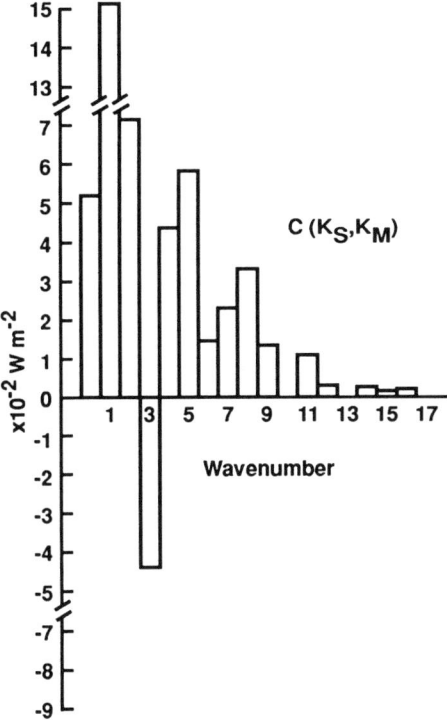

Fig. 15.18: Spectral distribution of $C(K_S, K_M)$ in the tropics (15° S-15° N) during the 1979 northern summer (June-August).

The dominant contribution from the ultralong-wave regime to $C(K_S, K_M)$, especially from wavenumber 1, is consistent with the energetics of the east-west circulation. The $C^n(K_S, K_M)$ of the larger wavenumber regime ($n = 4 - 18$) is small, but not negligible. The positive value of $C(K_S, K_M)$ in this wavenumber regime supports the argument inferred from $C(A_n, K_n)$ (Fig. 15.5) that warm-core synoptic systems dominate the tropics during summer. Finally, the negative value of $C^3(K_S, K_M)$ is a peculiar feature of the FGGE summer. The spectral energetics analysis of the tropics that is discussed in Section 15.2 indicates that an energy source of this wave may be the nonlinear cascade from other waves.

15.5 Spectral Energetics of Tropical Divergent and Rotational Flows

The divergent and rotational flow kinetic energy budgets formulated in Chapter 8 show that available potential energy is converted first into divergent kinetic energy (K_3) and then into rotational kinetic energy (K_2). Compared to rotational kinetic energy values, divergent kinetic energy values always remain small. This energetics characteristic of divergent and rotational flows implies that $C(A, K_3)$ and $C(K_3, K_2)$ are approximately equal at all times and that K_3 is a necessary energy reservoir through which available potential energy is converted to maintain the rotational flow. In Section 15.2 we have argued that the tropical available potential energy generated by the east-west differential heating is released by the east-west circulation to support quasi-stationary planetary-scale wave motions. The east-west circulation is a part of the planetary-scale divergent circulation depicted by the rotational flow. Although the diagnostic computation of $C(A_n, K_n)$ has been used to infer the maintenance of quasi-stationary planetary-scale monsoon circulation by the east-west circulation, it would certainly be natural to explore this matter by evaluating the direct interaction between divergent and rotational flows through $C(K_3, K_2)$. To this end, Chen (1980) analyzed the energetics of tropical divergent and rotational flows in the upper troposphere between 15° S and 20° N with the 1967 summer 100 mb wind data compiled by Krishnamurti (1971b). As done in Section 15.4, the energetics of both tropical divergent and rotational flows are presented in the spectral domain in such a way that the energetics roles played by different scales of the tropical motions can be understood.

Let us follow the Fourier expansion approach adopted in Section 15.4 to formulate spectral representations of K_3, K_2, and $C(K_3, K_2)$. The K_3 and K_2 spectra are

$$K_2 = K2_0 + \sum_{n=1}^{N} K2_n$$

$$\text{and} \quad K_3 = K3_0 + \sum_{n=1}^{N} K3_n, \tag{15.8}$$

respectively, where

$$K3_0 = \frac{1}{2((\sin\phi_2 - \sin\phi_1)} \int_{\phi_1}^{\phi_2} V3C_0^2 \cos\phi \, d\phi, \tag{15.9}$$

$$K3_n = \frac{1}{4(\sin\phi_2 - \sin\phi_1)} \int_{\phi_1}^{\phi_2} \left(U3C_n^2 + U3S_n^2 + V3S_n^2 + V3S_n^2\right) \cos\phi \, d\phi,$$

in which ϕ_1 and ϕ_2 have the same meaning as earlier. Note that $(v_2)_Z = (\partial\psi/\partial x)_Z$ and $(u_3)_Z = (\partial\chi/\partial x)_Z$ vanish. The $n = 0$ Fourier coefficients of v_2 and u_3 are zero – that is, $V2C_0 = 0$ and $U3C_0 = 0$ in (15.9). For convenience in a spectral formulation, $C(K_3, K_2)$ (8.20) is expressed in the scalar form

$$C(K_3, K_2) = \frac{1}{s}\int_s [f(u_2v_3 - u_3v_2) + \zeta_2(u_2v_3 - u_2v_3)$$
$$\underbrace{}_{C(K_3,K_2)1} \quad \underbrace{}_{C(K_3,K_2)2}$$
$$-\frac{1}{2}D_3(u_2u_2 + v_2v_2)]\, ds, \qquad (15.10)$$
$$\underbrace{}_{C(K_3,K_2)3}$$

where the term involving ω is neglected and $D_3 = \nabla \bullet V_3$. Using Fourier coefficients $u_2, v_2, u_3, v_3, \zeta_2$, and D_3, we may write (15.10) in the spectral form

$$C(K_3, K_2) = C^0(K_3, K_2) + \sum_{n=1}^{N} C^n(K_3, K_2)$$

$$= \underbrace{\left[C^0(K_3, K_2)1 + \sum_{n=1}^{N} C^n(K_3, K_2)1\right]}_{C(K_3,K_2)1}$$

$$+ \underbrace{\left[C^0(K_3 + K_2)2 + \sum_{n=1}^{N} C^n(K_3, K_2)2\right]}_{C(K_3,K_2)2}$$

$$+ \underbrace{\left[C^0(K_3, K_2)3 + \sum_{n=1}^{N} C^n(K_3, K_2)3\right]}_{C(K_3,K_2)3}, \qquad (15.11)$$

where

$$C^0(K_3, K_2)L = \frac{1}{2(\sin\phi_2 - \sin\phi_1)} \int_{\phi_1}^{\phi_2} CLI_0 \cos\phi\, d\phi \qquad (15.12)$$

and

$$C^n(K_3, K_2)L = \frac{1}{2(\sin\phi_2 - \sin\phi_1)} \int_{\phi_1}^{\phi_2} CLI_n \cos\phi\, d\phi, \qquad (15.13)$$

in which $L = 1, 2$, and 3. $C(K_3, K_2)2$ and $C(K_3, K_2)3$ contain triple products of Fourier coefficients. To express CLI^0 and CLI^n, we shall employ the same approach used for the spectral formulation of $T_0, T^{(1)}, \ldots T^{(6)}$ in Section 15.4:

$$C1I_0 = f \bullet U2C_0 \bullet V3C_0,$$

$$C1I_n = f[(U2C_n \bullet V3C_n + U2S_n \bullet V3S_n)$$
$$- (U3C_n \bullet V2C_n + U3S_n \bullet V2S_n)], \qquad (15.14)$$

$$C2I_0 = 2U2C_0 \bullet Z2C_0 \bullet V3C_0 \qquad (15.15)$$
$$+ U2C_0 \bullet \sum_{n=1}^{N} (Z2C_n \bullet U3C_n + Z2S_n \bullet V3S_n),$$

$$C2I_n = C2I_n^{(1)} + C2I_n^{(2)} + C2I_n^{(3)} + \frac{1}{2}\left[C2I_n^{(4)} + C2I_n^{(5)}\right]. \qquad (15.16)$$

$$C3I_0 = 2UC_0 \bullet D3C_0 \bullet U2C_0 \sum_{n=1}^{N} (D3C_n \bullet U2C_n + D3S_n \bullet U2S_n),$$

$$C3I_n = C3I_n^{(1)} + C2I_n^{(2)} - C3I_n^{(3)} + \frac{1}{2}\left[C3I_n^{(4)} + C3I_n^{(5)}\right]. \qquad (15.17)$$

The formation of $C2I_n^{(s)}$, where $s = 1,\ldots,5$, with Fourier coefficients in (15.16) as

$$C2I_n^{(1)} = V3C_0 (U2C_n \bullet Z2C_n + U2S_n \bullet Z2S_n),$$
$$C2I_n^{(2)} = Z2C_0 (U2C_n \bullet V3C_n + U2S_n \bullet V3S_n),$$
$$C2I_n^{(3)} = Z2C_0 (V2C_n \bullet U3C_n + U3S_n \bullet V2S_n),$$
$$C2I_n^{(4)} = \sum_{m=1}^{n} \{U2C_n [Z2C_m (V3C_{n+m} + V3C_{n-m})$$
$$+ Z2S_m (V3S_{n+m} - V3S_{n-m})]$$
$$+ U2S_n [Z2C_m (V3S_{n+m} + V3S_{n-m})$$
$$- Z2S_m (V3C_{n+m} - V3C_{n-m})]\}$$
$$+ \sum_{m=n+1}^{N} \{U2C_n [Z2C_m (V3C_{m+n} + V3C_{m-n})$$
$$+ Z2S_m (V3S_{n+m} + V3S_{m-n})]$$
$$+ U2S_n [Z2C_m (V3S_{n+m} - V3S_{m-n})$$
$$- Z2S_m (V3C_{n+m} - V3C_{m-n})]\}, \qquad (15.18)$$

$$C2I_n^{(5)} = \sum_{m=1}^{n} \{V2C_n [Z2C_m (U3C_{n+m} + U3C_{n-m})$$
$$+ Z2S_m (U3S_{n+m} - U3S_{n-m})]$$
$$+ V2S_n [Z2C_m (U3S_{n+m} + U3S_{n-m})$$
$$- Z2S_m (U3C_{n+m} - U3C_{n-m})]\}$$
$$+ \sum_{m=n+1}^{N} \{V2C_n [Z2C_m (U3C_{n+m} + U3C_{m-n})$$
$$+ Z2S_m (U3S_{n+m} - U3C_{n-m})]$$
$$+ V2S_n [U2C_m (U3S_{n+m} - U3S_{m-n})$$

ENERGETICS OF THE TROPICS

$$-Z2S_m \left(U3C_{m+n} - U3C_{m-n}\right)]\}. \tag{15.19}$$

The spectral expression of $C3I_n^{(s)}$ can be formulated simply by replacing the Fourier coefficients ζ_2, v_3, and u_3 in $C2I_n^{(2)}$ with $-\frac{1}{2}D_3$, u_2, and $-v_2$, respectively.

Table 15.1: Kinetic energies K, K_2, K_3 and ratios K_2/K and K_3/K in the tropics (15° S-20° N) at 200 mb during the 1967 summer. Unit: m² s⁻² (J kg⁻¹)

K	K_2	K_3	K_2/K	K_3/K
88.2	75.0	8.5	85.0%	9.6%

It was pointed out in Chapter 8 that, in a global-mean sense, the major part of K is explained by K_2 and that K_3 is about two orders of magnitude smaller than K_2. The numerical values of K_2 and K_3 averaged between 15° S and 20° N for the 1967 summer at 200 mb (Table 15.1) reveal that $K_3/K \sim 10\%$ in the tropics, where undoutedly K_3 is still much smaller than K_2. But the magnitude between K_2 and K_3 shows clearly that the tropical divergent circulation is more important than its counterpart in mid-latitudes. A minor difference between K and $K_2 + K_3$ is seen in Table 15.1. This difference is attributed to the nonvanishing $V_2 \bullet V_3$ term. Displayed in Fig. 15.19 are K_2 and K_3 spectra. The main contributors to K_2 are the zonal-mean components and the wavenumber 1. The energy content of K_2 decreases dramatically as the wavenumber increases in the wavenumber regime $n \geq 2$. The great amount of K_2 in wavenumber 1 is contributed mainly by the broad belt of easterlies that extends from the Date Line to the western Atlantic Ocean and in which the tropical easterly jet is embedded. The K_3 spectrum (Fig. 15.19a) differs from the K_2 spectrum in the sense that the former does not have a great magnitude either in the zonal-mean flow or in the wavenumber contributions, but it does have considerable spread.

Table 15.2: Energy conversion $C(K_3, K_2)$ and its components $C(K_3, K_2)1$, $C(K_3, K_2)2$ and $C(K_3, K_2)3$ in the tropics (15° S-20° N) at 200 mb during the 1967 summer. Unit: 10^{-5} m² s⁻³ (W kg⁻¹)

$C(K_3, K_2)$	$C(K_3, K_2)1$	$C(K_3, K_2)2$	$C(K_3, K_2)3$
15.2	18.3	-8.3	5.2

The numerical value of $C(K_3, K_2)$ ($= 15.2 \times 10^{-5}$ m²s⁻³) shown in Table 15.2 is greater than that of $C(A_E, K_E)$ ($= 7.5 \times 10^{-5}$ m² s⁻³). There may be two reasons for the difference between these energy conversions. $C(A_E, K_E)$ is computed in terms of the covariance between vertical motion and temperature. The former variable cannot be directly measured

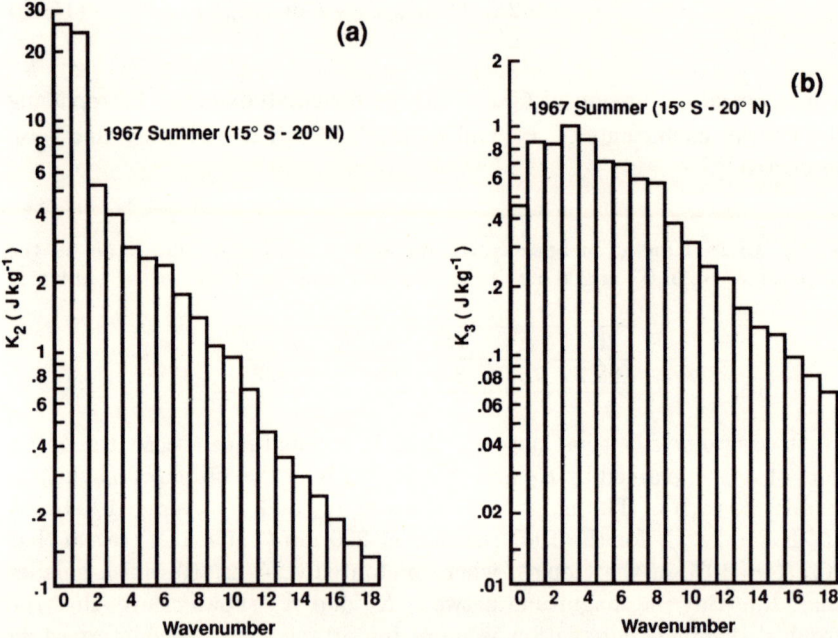

Fig. 15.19: Spectra of (a) K_2 and (b) K_3 averaged between 15° S and 20° N at 200 mb during the 1967 northern summer (June-August). Units: J kg^{-1}.

and must be indirectly estimated so that $C(A_E, K_E)$ may be underestimated. Additionally, the tropical easterly jet undergoes a pronounced interannual variation. The intensity of this jet in 1967 was stronger than that in the 1974 summer (Chen and van Loon, 1987). It is likely that some of this difference comes from the interannual variation of the tropical circulation during the summer. Both Krishnamurti (1971) and Kanamitsu et al. (1972) showed that the 200 mb kinetic energy in the tropics during the 1967 summer was generated by wavenumbers 1 and 2. The spectra of $C(K_3, K_2)$ shown in Fig. 15.20 confirms their findings that these two ultralong waves are the major energy source of the tropical circulation in summer. Moreover, the computation of $C(K_3, K_2)$, which requires only horizontal winds, provides an alternative for evaluating the generation of kinetic energy of large-scale motions in the tropics.

According to (15.10), $C(K_3, K_2)$ consists of several terms. $C(K_3, K_2)1$ was shown in Chapter 8 to be the only term with which conversion between K_3 and K_2 can take place in a quasi-geostrophic system. As illustrated by Table 15.2, the non-negligible contributions from $C(K_3, K_2)2$ and $C(K_3, K_2)3$ to $C(K_3, K_2)$, compared to the contribution from $C(K_3, K_2)1$, indicate that the ageostrophic effect becomes relatively important in the tropics. It has been demonstrated in Section 15.2 that available poten-

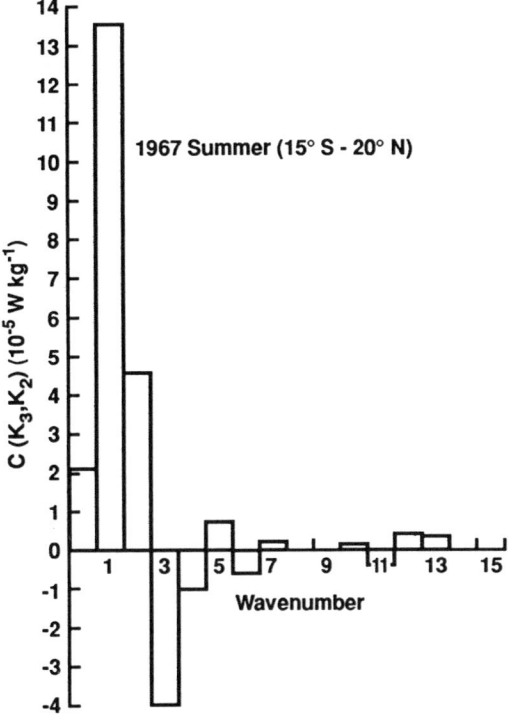

Fig. 15.20: Spectral distribution of $C(K_3, K_2)$ in the 1967 northern summer (June-August). Units: 10^{-4} W kg^{-1}.

tial energy released by thermally direct east-west circulation occurs primarily in wavenumbers 1 and 2, particularly in the former. We have argued that the K_3 value is small at all times and that $C(K_3, K_2)$ and $C(A_E, K_E)$ are closely correlated. It is therefore not surprising to find from the $C^n(K_3, K_2)$ spectrum that wavenumbers 1 and 2 dominate the contributions to $C(K_3, K_2)$. Additionally, the $C(K_3, K_2)$ contributed by wavenumbers 3 and 4 is negative, particularly that contributed by the former. After analyzing the spectra of the 200 mb kinetic energy budget, Kanamitsu et al. (1972) suggested that nonlinear wave-wave interactions are the most likely energy source for wavenumber 3.

15.6 Spectral Analysis of the Tropical Enstrophy

In this chapter the energetics of the large-scale tropical circulation have been diagnostically analyzed in terms of three different spectral energy schemes. These spectral analyses are surely illustrative for understanding the maintenance of large-scale tropical circulation. The vorticity budget is another diagnostic approach often used for the same purpose, particularly to examine the dissipation mechanism of synoptic-scale motions in the tropics. The role played by planetary-scale motions in the vorticity budget may be realized through a spectral analysis. This spectral vorticity budget requires examination of the maintenance of both phase speed and amplitude of a given wave component by various physical processes given in a spectral form of the vorticity equation. Comparing the budget analyses of spectral energy and of spectral vorticity reveals that the former deals only with the maintenance of energy, but the latter deals with maintenance of two variables – phase speed and amplitude. Thus, the spectral vorticity budget is more cumbersome to treat.

Half of the area-mean vorticity squared is defined as *enstrophy*. Thus, the relation between vorticity and enstrophy is similar to that between velocity and kinetic energy. To circumvent the disadvantage of the spectral vorticity budget, a spectral enstrophy budget of large-scale motions in the tropics is a natural alternative. In turbulence theory, the nonlinear transfer of enstrophy between different spectral components has been used to search for the existence of an inertial subrange. However, we shall analyze diagnostically the spectral enstrophy budget in detail rather than just examine the nonlinear cascade of enstrophy. Enstrophy is defined as

$$E = \frac{1}{S} \int_S \frac{1}{2} \zeta^2 \, ds.$$

For the practical computation, we get

$$E = \frac{1}{2\pi a^2 (\sin \phi_2 - \sin \phi_1)} \int_{\phi_1}^{\phi_2} \int_0^{2\pi} \frac{1}{2} \zeta^2 a^2 \cos \phi \, d\phi \, d\lambda. \qquad (15.20)$$

Vorticity, ζ, and enstrophy, e, of a longitudinal strip with unit latitudinal width may be expressed as

$$\zeta = \zeta_z + \zeta_E$$

and

$$e = \frac{1}{2\pi} \int_0^{2\pi} \frac{1}{2} \zeta^2 \, d\lambda = e_z + e_E, \qquad (15.21)$$

where

$$e_Z = \frac{1}{2} \zeta_z^2 \quad \text{and} \quad e_E = \frac{1}{2} \zeta_E^2.$$

$(\ )_z$ and $(\ )_E$ are a zonal average and the departure from it, respectively, of the quantity $(\ )$. Substituting into (15.20), we can rewrite this equation

$$E = E_z + E_E, \qquad (15.22)$$

ENERGETICS OF THE TROPICS

where

$$E_z = \frac{1}{(\sin\phi_2 - \sin\phi_1)} \int_{\phi_1}^{\phi_2} e_z \cos\phi \, d\phi,$$

$$E_E = \frac{1}{(\sin\phi_2 - \sin\phi_1)} \int_{\phi_1}^{\phi_2} e_E \cos\phi \, d\phi. \tag{15.23}$$

The time variation of enstrophy is

$$\frac{dE}{dt} = \frac{dE_z}{dt} + \frac{dE_E}{dt}, \tag{15.24}$$

where

$$\frac{dE_z}{dt} = \frac{1}{(\sin\phi_2 - \sin\phi_1)} \int_{\phi_1}^{\phi_2} \frac{\partial e_z}{\partial t} \cos\phi \, d\phi, \tag{15.25}$$

$$\frac{dE_E}{dt} = \frac{1}{(\sin\phi_2 - \sin\phi_1)} \int_{\phi_1}^{\phi_2} \frac{\partial e_E}{\partial t} \cos\phi \, d\phi. \tag{15.26}$$

Let us first form the physical enstrophy equation, which we can use to verify the spectral equation. The complete vorticity equation and the continuity equation are

$$\frac{\partial \zeta}{\partial t} + \frac{u}{a\cos\phi}\frac{\partial \zeta}{\partial \lambda} + \frac{v}{a}\frac{\partial s}{\partial p} + \omega\frac{\partial \zeta}{\partial p} + \beta v$$
$$= -(f+\zeta)\left(\frac{1}{a\cos\phi}\frac{\partial u}{\partial \lambda} + \frac{1}{a}\frac{\partial v}{\partial \phi} - \frac{\tan\phi}{a}v\right)$$
$$+ \left(\frac{\partial u}{\partial p}\frac{1}{a}\frac{\partial \omega}{\partial \phi} - \frac{\partial v}{\partial p}\frac{1}{a\cos\phi}\frac{\partial \omega}{\partial \lambda}\right) + d \tag{15.27}$$

and

$$\frac{1}{a\cos\phi}\frac{\partial u}{\partial \lambda} + \frac{1}{a\cos\phi}\frac{\partial}{\partial \phi}(v\cos\phi) + \frac{\partial \omega}{\partial p} = 0, \tag{15.28}$$

respectively. d is dissipation of vorticity.

First, we decompose (15.27) and (15.28) into zonal and eddy components. Then, we follow (15.21) to formulate the zonal enstrophy equation as

$$\frac{\partial e_z}{\partial t} = - \underbrace{\frac{1}{a\cos\phi}\frac{\partial}{\partial \phi}\left[\left(v_z\frac{\zeta_z^2}{2} + \zeta_z(v_E\zeta_E)_z\right)\cos\phi\right]}_{F_{EH}(0)}$$
$$- \underbrace{\frac{\partial}{\partial p}\left[\omega_z\frac{\zeta_z^2}{2} + \zeta_z(\omega_E\zeta_E)_z\right]}_{F_{EW}(0)} + \underbrace{\frac{(v_E\zeta_E)_z}{a}\frac{\partial \zeta_z}{\partial \phi} + (\omega_E\zeta_E)_z\frac{\partial \zeta_z}{\partial p}}_{C_E(n|0)} - \underbrace{\beta v_z\zeta_z}_{\beta_E(0)}$$

$$- \underbrace{(f+\zeta_z)\zeta_z D_z - \zeta_z\,(\zeta_E D_E)_z - \frac{\zeta_z}{a}\frac{\partial\omega_z}{\partial\phi}\frac{\partial u_z}{\partial p}}_{G_E(0)}$$

$$- \zeta_z\underbrace{\left[\frac{1}{a\cos\phi}\left(\frac{\partial\omega_E}{\partial\lambda}\frac{\partial v_E}{\partial p}\right)_z - \frac{1}{a}\left(\frac{\partial\omega_E}{\partial\phi}\frac{\partial u_E}{\partial p}\right)_z\right]}_{T_E(0)} + d_z \qquad (15.29)$$

and the eddy enstrophy equation as

$$\frac{\partial e_E}{\partial t} = -\underbrace{\frac{1}{a\cos\phi}\frac{\partial}{\partial\phi}\left[\left(v_z\frac{(\zeta^2)}{2} + \frac{1}{2}(v_E\zeta_E^2)_z\right)\cos\phi\right]}_{F_{EH}(n)}$$

$$- \underbrace{\frac{\partial}{\partial p}\left[\omega_z\frac{(\zeta_E^2)_z}{2} + \frac{1}{2}(\omega_E\zeta_E^2)_z\right]}_{F_{EW}(n)} - \underbrace{\frac{(v_E\zeta_E)_z}{a}\frac{\partial\zeta_z}{\partial\phi} - (\omega_E,\zeta_E)_z\frac{\partial\zeta_z}{\partial p}}_{C_E(n|0)}$$

$$- \underbrace{\beta(v_E\zeta_E)_z}_{\beta_E(n)} - \underbrace{(f+\zeta_z)(D_E\zeta_E)_z - D_z\,(\zeta_E^2)_z - (D_E\zeta_E^2)_z}_{G_E(n)}$$

$$- \underbrace{\left[\left(\frac{\zeta_E}{a\cos\phi}\frac{\partial\omega}{\partial\lambda}\right)_z\frac{\partial v_z}{\partial p} - \left(\zeta_E\frac{\partial\omega}{\partial\phi}\right)_z\frac{\partial u_z}{\partial p}\right.}_{T_E(n)}$$

$$+ \left.\left(\frac{\zeta_E}{a\cos\phi}\frac{\partial\omega}{\partial\lambda}\frac{\partial v_E}{\partial p}\right)_z - \left(\frac{\zeta_E}{a}\frac{\partial\omega_E}{\partial\phi}\frac{\partial u_E}{\partial p}\right)_z\right] + d_E. \qquad (15.30)$$

Let us now turn to the formulation of the spectral enstrophy equation. Following (15.21), we may describe enstrophy as

$$e = \frac{1}{2\pi}\int_0^{2\pi}\frac{1}{2}\zeta^2\,d\lambda = \sum_{n=-N}^{N}\frac{1}{2}Z(n)Z(-n)$$

$$= \sum_{n=0}^{N}\left(1 - \frac{1}{2}\delta_{0,n}\right)Z(n)Z(-n) = \sum_{n=0}^{N}e(n), \qquad (15.31)$$

where $\delta_{0,n} = 1$ if $n = 0$ or where $\delta_{0,n} = 0$ if $n \neq 0$, and where $Z(n)$ is the Fourier coefficient of ζ. The spectral form of the vorticity and continuity equations are

$$\frac{dZ(n)}{dt} = \sum_{m=-M}^{M}\left(\frac{im}{a\cos\phi}U(n-m)Z(m) - \frac{1}{a}V(n-m)\frac{dZ(m)}{d\phi}\right.$$

$$\left.- W(n-m)\frac{\partial Z(m)}{\partial p}\right) - \beta V(n) - fD(n) - \sum_{m=-M}^{M}Z(n-m)D(m)$$

ENERGETICS OF THE TROPICS

$$+ \sum_{m=-M}^{M} \left(\frac{\partial}{\partial p} U(n+m) \frac{\partial}{a \partial \phi} W(m) - \frac{\partial}{\partial p} V(n+m) \frac{im}{a \cos \phi} W(m) \right)$$

(15.32)

and

$$\frac{in}{a \cos \phi} U(n) + \frac{1}{a \cos \phi} \frac{d}{d\phi} [V(n) \cos \phi] + \frac{\partial W(m)}{\partial p} = 0. \quad (15.33)$$

Applying (15.31) and (15.33) to the time variation of enstrophy obtained from the result of substituting (15.30) into (15.20), we obtain the spectral enstrophy equations of the zonal flow and of wavenumber n:

$$\frac{dE(0)}{dt} = -C_E(n|0) + \beta_E(0) + G_E(0) + T_E(0) + F_{EH}(0)$$
$$+ F_{EW}(0) + D_E(0) \quad (15.34)$$

and

$$\frac{dE(n)}{dt} = C_E(n|m, n-m) + C_E(n|0) + \beta_E(n) + G_E(n) + T_E(n)$$
$$+ F_{EH}(n) + F_{EW}(n) + D_E(n), \quad (15.35)$$

respectively. The detailed mathematical expressions of symbolic forms used in these two equations are as follows:

$$\beta_E(0) = \frac{-1}{(\sin \phi_2 - \sin \phi_1)} \int_{\phi_1}^{\phi_2} \beta(\phi) V(0) Z(0) \cos \phi \, d\phi,$$

$$G_E(0) = \frac{-1}{(\sin \phi_2 - \sin \phi_1)} \int_{\phi_1}^{\phi_2} \left\{ [f(\phi) + Z(0)] D(0) Z(0) \right.$$
$$\left. + Z(0) \sum_{m=1}^{N} [D(m) Z(-m) + D(-m) Z(m)] \right\} \cos \phi \, d\phi,$$

$$T_E(0) = \frac{-1}{(\sin \phi_2 - \sin \phi_1)} \int_{\phi_1}^{\phi_2} \left\{ \frac{1}{a} \frac{dW(0)}{d\phi} Z(0) \frac{dU(0)}{dp} \right.$$
$$+ \sum_{m=1}^{N} \left[\frac{im}{a \cos \phi} Z(0) \left(W(-m) \frac{dV(m)}{dp} Wm \frac{dV(-m)}{dp} \right) \right.$$
$$\left. \left. + \frac{1}{a} Z(0) \left(\frac{dW(-m)}{d\phi} \frac{dV(m)}{dp} + \frac{dW(m)}{d\phi} \frac{dW(-m)}{dp} \right) \right] \right\} \cos \phi \, d\phi,$$

$$F_{EH}(0) = \frac{-1}{(\sin \phi_2 - \sin \phi_1)} \int_{\phi_1}^{\phi_2} \left\{ \frac{1}{a \cos \phi} \frac{\partial}{\partial \phi} \left[\left(\frac{V(0)}{2} Z(0) Z(0) \right. \right. \right.$$

$$+Z(0)\sum_{m=1}^{N}[V(m)Z(-m)+V(-m)Z(m)]\bigg)\cos\phi\bigg]\bigg\}\cos\phi\,d\phi,$$

$$F_{EW}(0) = \frac{-1}{(\sin\phi_2-\sin\phi_1)}\int_{\phi_1}^{\phi_2}\frac{\partial}{\partial p}\bigg(\frac{W(0)}{2}Z(0)Z(0)$$

$$+Z(0)\sum_{m=1}^{N}[W(m)Z(-m)+W(-m)Z(m)]\bigg)\cos\phi\,d\phi,$$

$$D_E(0) = \frac{1}{(\sin\phi_2-\sin\phi_1)}\int_{\phi_1}^{\phi_2}d_e(0)\cos\phi\,d\phi,$$

$$C_E(n|m,n-m) = \frac{1}{(\sin\phi_2-\sin\phi_1)}\int_{\phi_1}^{\phi_2}\bigg\{\sum_{m=1}^{M}\frac{iZ(m)}{a\cos\phi}$$

$$[(n+m)U(n-m)Z(-n)+(m-n)U(-n-m)Z(n)]$$

$$+\frac{1}{2a}\bigg[V(n-m)\bigg(Z(-n)\frac{dZ(m)}{d\phi}-Z(m)\frac{dZ(n)}{d\phi}\bigg)$$

$$+V(-n-m)\bigg(Z(n)\frac{dZ(m)}{d\phi}-Z(m)\frac{dZ(n)}{d\phi}\bigg)\bigg]$$

$$+\frac{1}{2}\bigg[W(n-m)\bigg(Z(-n)\frac{dZ(m)}{dp}-Z(m)\frac{dZ(-n)}{dp}\bigg)$$

$$+W(-n-m)\bigg(Z(n)\frac{dZ(m)}{dp}-Z(m)\frac{dZ(n)}{dp}\bigg)\bigg]\bigg\}\cos\phi\,d\phi,$$

$$C_E(n|0) = \frac{1}{(\sin\phi_2-\sin\phi_1)}\int_{\phi_1}^{\phi_2}\bigg\{[V(n)Z(-n)+V(-n)Z(n)]\frac{1}{a}\frac{DZ(0)}{d\phi}$$

$$+[W(n)Z(-n)+W(-n)Z(n)]\frac{dZ(0)}{dp}\bigg\}\cos\phi\,d\phi,$$

$$\beta_E(n) = \frac{-1}{(\sin\phi_2-\sin\phi_1)}\int_{\phi_1}^{\phi_2}\beta(\phi)\,[V(n)Z(-n)+V(-n)Z(n)]\cos\phi\,d\phi,$$

$$G_E(n) = \frac{-1}{(\sin\phi_2-\sin\phi_1)}\int_{\phi_1}^{\phi_2}\{[f(\phi)+Z(0)]\,[D(n)Z(-n)$$

$$+D(-n)Z(n)]+D(0)Z(n)Z(-n)$$

$$+\sum_{m=1}^{N}[D(n-m)Z(-n)+D(-n-m)Z(n)]\,Z(m)\bigg\}\cos\phi\,d\phi,$$

$$T_E(n) = \frac{1}{(\sin\phi_2-\sin\phi_1)}\int_{\phi_1}^{\phi_2}\bigg\{\frac{in}{a\cos\phi}[W(n)Z(-n)+Z(-n)Z(n)]\frac{dV(0)}{dp}$$

$$+\bigg(\frac{dW(n)}{ad\phi}Z(-n)+\frac{dW(-n)}{ad\phi}Z(n)\bigg)\frac{dU(0)}{dp}$$

ENERGETICS OF THE TROPICS

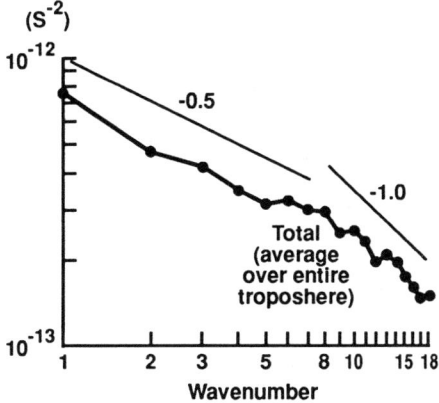

Fig. 15.21: Enstrophy spectra averaged over the entire tropical troposphere between 10° S and 20° N in the 1979 northern summer.

$$F_{EH}(n) = \left(\frac{-1}{(\sin \phi_2 - \sin \phi_1)} \int_{\phi_1}^{\phi_2} \left\{ -\sum_{m=1}^{N} \left[\frac{1}{a \cos \phi} [(n-m)W(n-m)Z(-n) \right.\right.\right.$$
$$\left.\left.\left. -(n+m)W(-n-m)Z(n)] \frac{dV(m)}{dp} - \frac{1}{a} \left(\frac{dW(n-m)}{d\phi} Z(-n) \right.\right.\right.\right.$$
$$\left.\left.\left.\left. + \frac{dW(-n-m)}{d\phi} Z(n) \right) \frac{dU(m)}{dp} \right] \right\} \cos \phi \; d\phi,$$
$$+ \frac{-1}{(\sin \phi_2 - \sin \phi_1)} \int_{\phi_1}^{\phi_2} \frac{1}{a \cos \phi} \frac{\partial}{\partial \phi} \{V(0)Z(n)Z(-n)$$
$$+ \sum_{m=1}^{N} \frac{V(m)}{2} [Z(n-m)Z(-n)$$
$$+ Z(-n-m)Z(n)] \cos \phi\}) \cos \phi \; d\phi,$$

$$F_{EW}(n) = \frac{-1}{(\sin \phi_2 - \sin \phi_1)} \int_{\phi_1}^{\phi_2} \frac{\partial}{\partial p} \{W(0)Z(n)Z(-n)$$
$$+ \sum_{m=1}^{N} \frac{W(m)}{2} [Z(n-m)Z(-n) + Z(-n-m)Z(n)] \} \cos \phi \; d\phi,$$

$$D_E(n) = \frac{-1}{(\sin \phi_2 - \sin \phi_2)} \int_{\phi_1}^{\phi_2} d_e(n) \cos \phi \; d\phi.$$

Contributions of different scales of tropical motions between 15° S and 20° N to enstrophy can be seen in the enstrophy spectrum of the 1979 northern summer (Fig. 15.21). The enstrophy spectrum decreases monotonically as wavenumber increases. As with the tropical kinetic energy

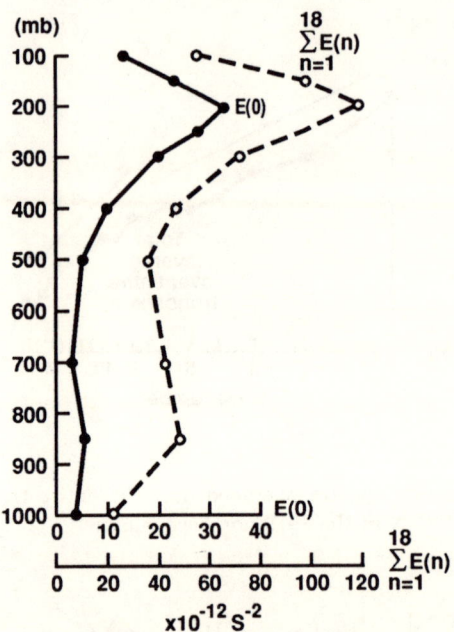

Fig. 15.22: Vertical distributions of $E(0)$ and $\sum_{n=1}^{18} E(n)$ averaged between 10° S and 20° N in the 1979 northern summer.

spectrum (Fig. 15.17 or Fig. 15.19a), wavenumber 1 has the greatest enstrophy. Although we are not searching for the inertial subrange of tropical wave disturbances, it is of interest to determine the slope of the enstrophy spectra in different wavenumber regimes. Slopes of -0.5 and -1.0 exist, respectively, in the small ($n = 1 - 7$) and the large ($n > 7$) wavenumber regimes. It may be inferred from vertical temperature structures and from zonal velocity perturbations in the tropics that A_E and K_E are bimodally distributed (Section 15.3). The vertical structure of enstrophy (Fig. 15.22) behaves in the same manner as A_E or K_E. Despite the bimodal distribution of enstrophy in the vertical, the enstrophy spectra of Fig. 15.22 are typical of those over the entire layer of the tropical troposphere (Chen, 1985).

The vertically averaged spectral enstrophy budget of (15.35) is displayed in Fig. 15.23. The beta effect, $\beta_E(n)$, and the nonlinear transfer, $C_E(n|m,l)$, are evidently dominant processes. As revealed from the $C_E(n|m,l)$ spectrum, tropical enstrophy is nonlinearly transferred out of the long-wave regime ($n = 1 - 7$) to the short-wave regime ($n = 10 - 18$), with minor roles played by wavenumbers 8 and 9. In view of this nonlinear transfer property of wave enstrophy, it can be inferred that as tropical enstrophy is generated in the long-wave regime, the $C_E(n|m,l)$ process be-

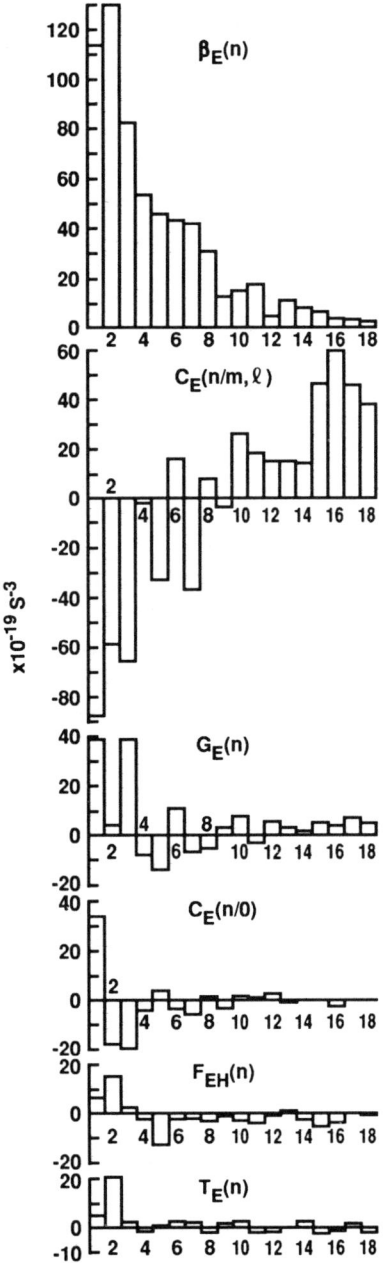

Fig. 15.23: Spectral distribution of various terms in the spectral enstrophy budget equations (15.35) and (15.36) averaged between 10° S and 20° N in the 1979 northern summer (June-August).

comes an enstrophy source of the $\beta_E(n)$ process. The vertical average of $F_{EW}(n)$ is zero and the numerical values of other quantities (15.35) in, $C_E(n|0)$, $F_{EH}(n)$, $T_E(n)$, and $G_E(n)$ are insignificant, except in ultralong waves. It was demonstrated in Section 15.2 that the barotropic interaction between ultralong waves and the zonal flow is important in tropical energetics. Nonetheless, this enstrophy interaction does not seem to be very important in the spectral enstrophy budget.

Recall that the maximum wave enstrophy in the tropics occurs in the upper troposphere. The beta effect was recognized as an enstrophy source in the vertically averaged spectral enstrophy budget, but this effect may not be the only source maintaining the maximum wave enstrophy. To explore the maintenance of maximum wave enstrophy in the tropical upper troposphere, the vertical distribution of the spectral enstrophy budget should be examined.

It has been stressed that the large-scale tropical circulation is monsoonal and the vertical motions associated with east-west circulations are important for the release of the tropical available potential energy generated by the east-west differential heating. Thus, the in-situ generation of enstrophy by vortex stretching should be important in the tropics. Fig. 15.23 shows that the vertically averaged $G_E(n)$ spectrum contributes little to the tropical enstrophy budget, except for wavenumbers 1 and 3. Actually, this vertically averaged $G_E(n)$ spectrum does not shed much light on its function in the tropical enstrophy budget. Remember that the Coriolis parameter f is vertically constant and that the horizontal wind divergence changes its sign in the middle of the troposphere. $G_E(n)$ may therefore change its sign vertically one time. This argument is substantiated by the vertical distribution of $\sum_{n=1}^{18} G_E(n)$ (Fig. 15.24), which is characterized by a large positive (negative) value in the lower (upper) troposphere. Consequently, vertically averaged $G_E(n)$ results in small numerical values due to cancellation between the upper and lower tropospheres.

The beta effect $\sum_{n=1}^{18} \beta_E(n)$ (Fig. 15.24) has significant values over the entire tropical troposphere, with maxima in both the upper and lower troposphere. The dissipation of wave enstrophy, $\sum_{n=1}^{18} D_E(n)$ computed by the residual method, exhibits a *bimodal* vertical distribution, as does the beta effect. In the lower troposphere, both $\sum_{n=1}^{18} \beta_E(n)$ and $\sum_{n=1}^{18} G_E(n)$ work coherently to counterbalance dissipation. In contrast, $\sum_{n=1}^{18} G_E(n)$ changes its sign in the upper troposphere so that it cannot function constructively with the beta effect to counterbalance dissipation there. Undoubtedly, the maximum wave enstrophy in the tropical upper troposphere cannot be maintained in situ only by the counterbalance between the contribution from the two enstrophy sources and the dissipation of enstrophy. There must be another enstrophy source serving this purpose.

The vertical transport of wave enstrophy flux, $\sum_{n=1}^{18} F_{EW}(n)$, has negative (positive) values in the lower (upper) troposphere. This vertical distribution of $\sum_{n=1}^{18} F_{EW}(n)$ indicates that wave enstrophy is transported from

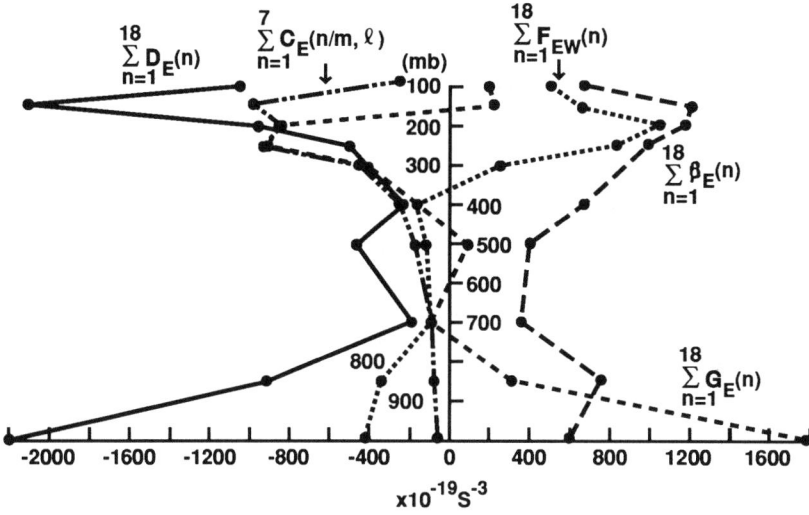

Fig. 15.24: Vertical distributions of $\sum_{n=1}^{18} \beta_E(u)$, $\sum_{n=1}^{18} G_E(u)$, $\sum_{n=1}^{18} F_{EW}(n)$, $\sum_{n=1}^{18} D_E(n)$, and $\sum_{n=1}^{18} C_E(n/m,\ell)$.

the lower to the upper troposphere. Evidently, part of the wave enstrophy generated by the beta effect and the vortex stretching in the lower troposphere is transported upward by large-scale vertical motions to compensate for their part of the enstrophy lost in the upper troposphere due to both vortex stretching and dissipation. Additionally, nonlinear wave enstrophy transfer in the long-wave regime ($n = 1 - 7$), $\sum_{n=1}^{7} C_E(n|m,l)$, is significant only in the upper troposphere. Again, part of the wave enstrophy generated in the upper troposphere by the beta effect and by the vertical transport in the long-wave regime is consumed by this nonlinear transfer.

The zonal enstrophy budget is maintained by seven physical processes, but only the vertical distributions of significant quantities are shown in Fig. 15.25. Both the beta effect, caused by the north-south transport of planetary vorticity resulting from the Hadley circulation, and the dissipation of zonal enstrophy $D_E(0)$, estimated by the residual method, possess a bimodal vertical distribution with maxima in the upper and lower troposphere. The generation, $G_E(0)$, and the horizontal zonal flux, $F_{EH}(0)$, of zonal enstrophy are significant only in the upper troposphere. Comparing the vertical distributions of these four quantities, we conclude that zonal enstrophy produced by the beta effect is lost through dissipation in the lower troposphere. On the other hand, part of the zonal enstrophy generated by both the beta effect and the vortex stretching in the upper troposphere is transported horizontally out of the tropics.

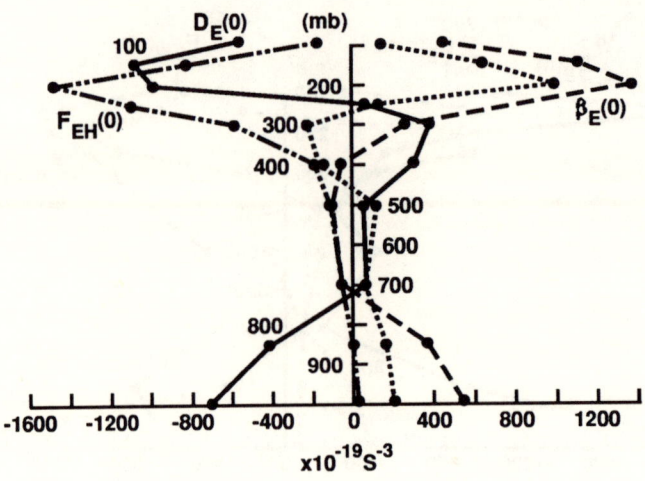

Fig. 15.25: Vertical distributions of $\beta_E(0)$, $G_E(0)$, $F_{EW}(0)$, and $D_E(0)$.

15.7 Kinetic Energy Budget of the Tropical Easterly Jet

Koteswaram (1958) pointed out in an analysis of average July rainfall that heavy precipitation occurs north of the entrance of the tropical easterly jet and that the opposite occurs in the exit of the jet. Large-scale ascent generally occurs over regions of heavy precipitation. It thus seems that a thermally direct (indirect) cross-jet circulation occurs in the entrance (exit) region of the tropical easterly jet. Analyzing the summer-mean meridional wind, Chen (1982b) showed that a southward flow occurs over the Bay of Bengal, on the upstream side of the tropical easterly jet, and that northward flow occurs over equatorial Africa, on the downstream side of the jet. Koteswaram's argument is consistent with the summer-mean meridional wind-distribution data. In fact, the relation between the cross-jet circulations and the tropical easterly jet is the same as that between these circulations and subtropical jets in the mid-latitudes. The mechanism maintaining subtropical jet streams in the mid-latitudes (Section 13.1) may be applied to the tropical easterly jet. On the other hand, the U_1 (200 mb) distribution (Section 15.3) indicated that the tropical easterly jet is associated with wavenumber 1 in the tropics. According to tropical spectral energetics, the release of available potential energy by thermally direct overturning is the primary energy source maintaining this wave. However, it is not clear how the east-west circulation contributes to the regional maintenance of the tropical easterly jet. To clarify this issue, we shall examine the kinetic energy budget of this jet in light of the relation between the east-west circulation and the divergent component of kinetic energy flux in association with the tropical easterly jet.

ENERGETICS OF THE TROPICS

The kinetic energy equation is

$$\frac{\partial k}{\partial t} = -\nabla \cdot (Vk) - \frac{\partial}{\partial p}(\omega k) - V \cdot \nabla \phi + D(k). \qquad (15.36)$$

The vertical transport term is generally small and negligible. The horizontal kinetic energy flux Vk can be split into divergent and rotational components:

$$Vk = (Vk)_D + (Vk)_R. \qquad (15.37)$$

In the equilibrium state, the kinetic energy budget of the tropical easterly jet may reach the following balance

$$-\nabla \cdot (Vk)_D - V \cdot \nabla \phi \simeq 0. \qquad (15.38)$$

The dissipation term, $D(k)$, can be obtained by the difference between the kinetic energy generation $-V \cdot \nabla \phi$ and the divergence of the divergent kinetic energy flux $-\nabla \cdot (Vk)_D$. As a matter of fact, $D(k)$ is usually much smaller in magnitude than $-V \cdot \nabla \phi$ and $-\nabla \cdot (Vk)_D$.

The available potential energy released by the divergent circulation can be transformed to support the tropical easterly jet through work done by cross-contour flow – that is, by kinetic energy generation. Although kinetic energy is an indicator of the motion intensity, it is a scalar. Thus, $(Vk)_D$ is essentially determined by the divergent wind V_D. According to (15.38), the generated kinetic energy can be locally balanced by the divergent component of kinetic energy flux. The local maintenance of the tropical easterly jet by the east-west circulation may be illustrated through the kinetic energy budget analysis of (15.38). For this purpose, the kinetic energy budget of the tropical easterly jet during the 1970 summer by Chen and van Loon (1987) is used here.

As indicated by the 10 m s^{-1} isotach in the tropics (Fig. 15.26), the tropical easterly jet extends from the western Pacific, across the Indian Ocean, to the eastern Atlantic. This jet is sandwiched between the Tibetan high to the north and the western part of Australian anticyclone to the south and may reach about 25 m s^{-1} in its core. An interesting relation between the tropical easterly jet and the east-west circulation emerges by comparing total wind vectors (Fig. 15.26a) and divergent wind vectors (Fig. 15.26b): the strongest divergent center of the east-west circulation is located on the upstream side of the jet and the downstream side west of India is associated with the convergence region of the west branch of the east-west circulation.

As pointed out previously, the magnitude of geopotential perturbations in the tropics is an order of magnitude smaller than in mid-latitudes. An accurate computation of the kinetic energy generation $-V \cdot \nabla \phi$ may not be easily accomplished. However, the divergent component of the kinetic energy flux $(Vk)_D$ may not be as difficult to compute accurately as $-V \cdot \nabla \phi$ if the divergent wind is well represented in the total wind field. The

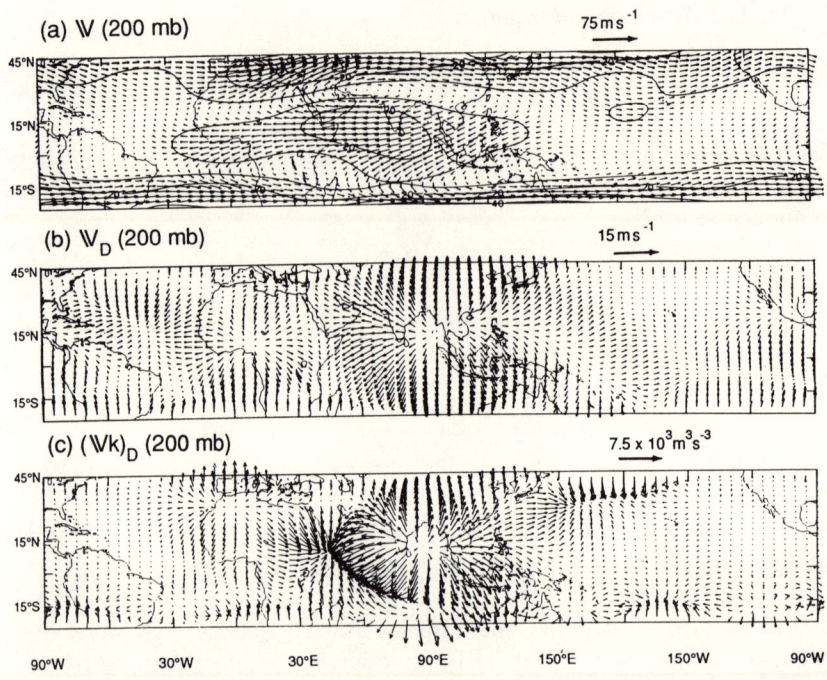

Fig. 15.26: (a) Mean wind vectors V (200 mb) and isotachs (contour interval is 10 m s^{-1}) (b) mean divergent wind vectors V_D (200 mb) and (c) mean divergent kinetic energy flux $(Vk)_D$ (200 mb) of the 1970 summer.

divergent component of kinetic energy flux at 200 mb, $(Vk)_D$ (200 mb) (Fig. 15.26c), indicates that kinetic energy is transported from the Bay of Bengal, where the divergent center of the east-west circulation is located, toward the east coast of Africa, which is on the downstream side of the tropical easterly jet associated with the convergence centers of the east-west circulation. In view of the $(Vk)_D$ distribution, it is clear that the kinetic energy of the tropical jet stream is generated (destroyed) in the upstream (downstream) region, according to (15.38). In other words, the relation between the structure of the east-west circulation and the tropical easterly jet develops a concise picture for the maintenance of this jet. Finally, an interesting comment should be made. The east-west circulations undergo interannual variations. Chen and van Loon (1987) also demonstrated that the interannual variation of the tropical easterly jet results from that of the east-west circulation, as suggested by the interannual variation of divergent kinetic energy flux associated with this jet.

15.8 Exchange of Kinetic Energy between Low and Middle Latitudes

It was illustrated in Chapter 5 that the momentum of the westerlies in the mid-latitudes is maintained by the upgradient transport of momentum from the tropics to mid-latitudes. Accordingly, we may expect that atmospheric kinetic energy is also transported from low to middle latitudes in association with the atmospheric kinetic energy between these two regions. One may formulate the kinetic energy equation in such a way that it includes a term measuring the meridional transport of kinetic energy. This requirement may be accomplished by integrating the kinetic energy equation along latitudinal circles.

$$\frac{\partial [k]_z}{\partial t} = -\frac{\partial}{a \partial \varphi}[vk]_z - \frac{\partial}{\partial p}[\omega k]_z - [V \cdot \nabla \phi]_z + [D(k)]_z. \qquad (15.39)$$

On the right-hand side of (15.39), the first term is the convergence of the meridional kinetic energy flux, which provides the means for the kinetic energy exchange between low and middle latitudes. The second term is the convergence of the vertical kinetic energy flux, which redistributes kinetic energy vertically. $[D(k)]_z$, the dissipation of kinetic energy, exists wherever there is motion. Because the long-term mean atmospheric circulation is so steady, there must be generation of kinetic energy, the third term, to provide kinetic energy for redistribution and to counterbalance dissipation.

If atmospheric kinetic energy is exported from the tropics to mid-latitudes as suggested previously, atmospheric kinetic energy should be generated in low latitudes and destroyed in mid-latitudes so that a steady-state atmospheric general circulation can be maintained. This hypothesis may be substantiated by a comparison between latitude-pressure cross-sections of the convergence of the meridional kinetic energy flux and the generation of kinetic energy. The generation of kinetic energy is attributed to the work done by cross-isobaric (ageostrophic) flow. As discussed earlier in this chapter, the height perturbation in the tropics is generally an order of magnitude smaller than in mid-latitudes. It is thus difficult, if not impossible, to compute $-V \cdot \nabla \phi$ accurately in low latitudes. A full-physics general circulation model may offer us an alternative to circumvent this difficulty although the simulated model atmosphere is not real.

Using (15.39), Chen and Lee (1982) analyzed the zonally-averaged kinetic energy budget of the winter atmospheric general circulations simulated by two climate models. The latitude-pressure cross sections of $[k]_z$, $-\partial [vk]_z / a \, \partial \varphi$, and $-[V \cdot \nabla \phi]_z$ of both the models show maxima in the upper troposphere of mid-latitudes (Fig. 15.27). The convergence of meridional kinetic energy flux is negative south of the maximum $[k]_z$ and positive north of the maximum $[k_z]$; the kinetic energy of the model atmosphere is clearly transported from low to middle latitudes.

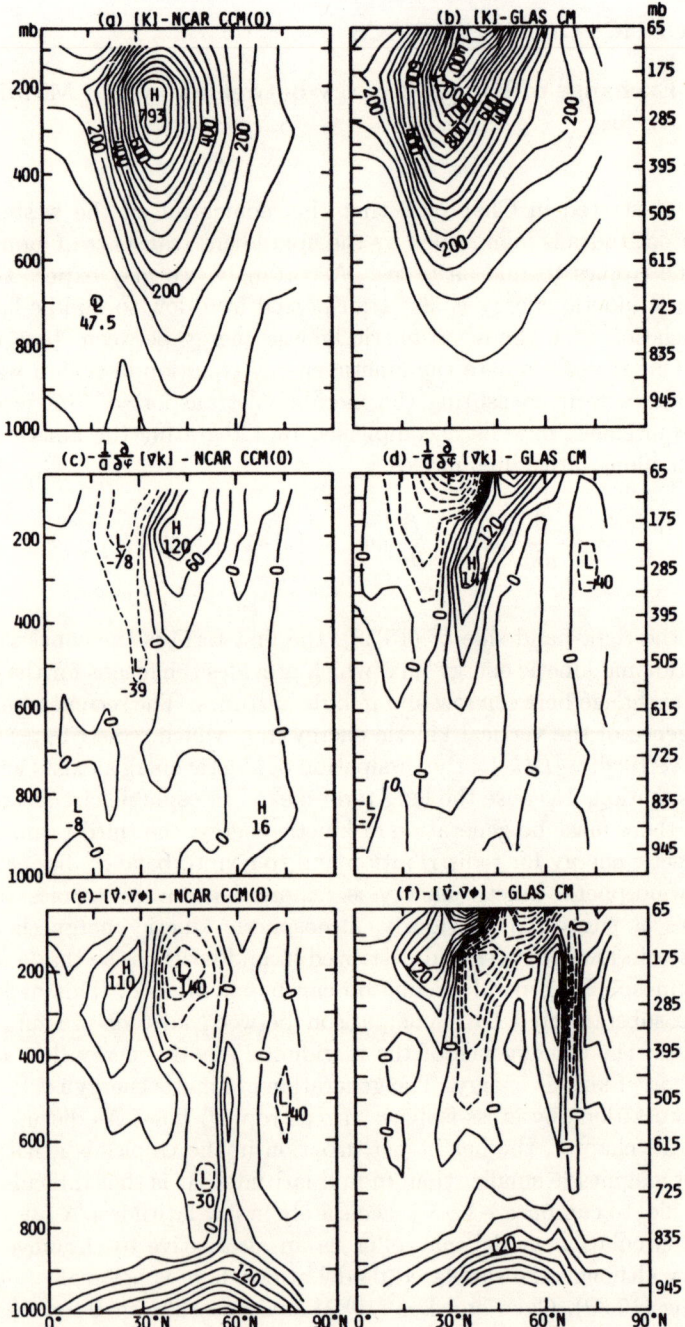

Fig. 15.27: Latitudinal-pressure distribution of kinetic energy of (a) NCAR model and (b) GLAS model, convergence of kinetic energy flux of (c) NCAR model and (d) GLAS model, and generation of kinetic energy of (e) NCAR model and (f) GLAS model; the contour intervals of energy are 5×10^2 J m^{-2} mb^{-1}, but change to 10^3 J m^{-2} mb^{-1} after k reaches 4×10^3 J m^{-2} mb^{-1}, and that of convergence of kinetic energy flux and generation of kinetic energy is 3×10^{-3} W m^{-2} mb^{-1}

ENERGETICS OF THE TROPICS 273

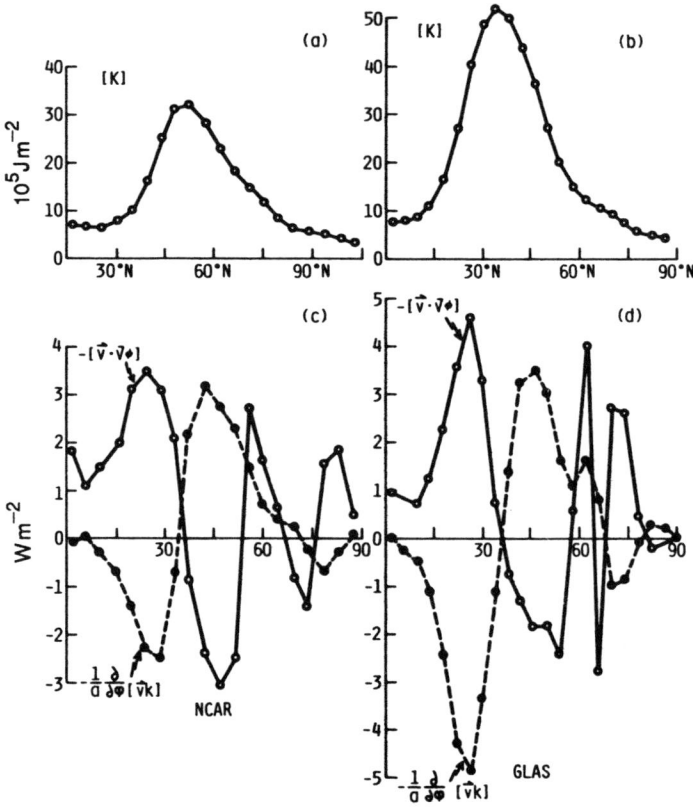

Fig. 15.28: Latitudinal distributions of energy variables shown in Fig. 14.26, and integrated from 700 mb to the model top for the NCAR model and from 725 mb to the model top for the GLAS model.

The contrast between $[k]_z$ and $-\partial[vk]_z/a\,\partial\varphi$ reveals that kinetic energy is transported upgradient on the equatorward side of maximum $[k]_z$. Indeed, the meridional kinetic energy flux is consistent with the eddy momentum transport, whose maximum occurs in the upper troposphere as well. As inferred from the convergence of the meridional kinetic energy flux, the depletion of kinetic energy in low latitudes by the flux divergence should be furnished by a kinetic energy source, and the accumulation of kinetic energy by the flux convergence in mid-latitudes should be consumed by a kinetic energy sink. It is shown in Figs. 15.27e and 15.27f that kinetic energy is generated in the middle and upper troposphere south of 30° N and destroyed north of 30° N. The distributions of $-\partial[vk]_z/a\,\partial\varphi$ and $-[V\cdot\nabla\phi]_z$ are similar in pattern and magnitude, but opposite in sign. The kinetic energy of the model atmosphere is apparently generated in low latitudes, transported northward, and destroyed in mid-latitudes. Note that the generation of kinetic energy is also very significant below 800 mb where the cross isobaric flow is generally large. The kinetic energy generated there is perhaps not removed by meridional transport, but may be dissipated by the frictional effect inside the planetary boundary layer.

In order to obtain a quantitative measurement of various energy variables in (15.40), a latitudinal distribution of the vertically integrated $[k]_z$, $-\partial[vk]_z/a\,\partial\varphi$, and $-[V\cdot\nabla\phi]_z$ from 700 mb to the top of the model atmosphere is displayed in Fig. 15.28. The counterbalance between convergence of the meridional kinetic energy flux and generation of kinetic energy in low and middle latitudes is well shown in this figure.

16

ENERGETICS OF THE TROPICS: SYNOPTIC SCALE

The characteristics of atmospheric disturbances are generally determined by the embedded large-scale environment and by energy sources. The pronounced vertical wind shear, which is an indication of baroclinicity, associated with the subtropical jet streams is maintained by the north-south temperature gradient in mid-latitudes through the thermal wind relation. The synoptic disturbances initiated by the baroclinic instability of the basic large-scale flow are characterized by slanted convection. The available potential energy of a large-scale environment is the primary energy source of synoptic disturbances in mid-latitudes through interactions between these disturbances and the basic mean state. The energy source provided through diabatic heating becomes secondary to synoptic disturbances. Hence, the synoptic disturbances in mid-latitiudes mainly exist in the regions of significant baroclinicity.

In the tropics, the meridional temperature gradient of the large-scale mean flow is relatively insignificant compared to that of mid-latitiudes. The synoptic disturbances of the tropics may not be able to derive a sufficient amount of available potential energy from the large-scale mean flow. Since the tropics receive the major portion of solar energy on the earth, the condensation heating released by cumulus convection constitutes the main energy source of synoptic disturbances in the tropics. It is therefore expected that synoptic disturbances of the tropics behave differently from their counterparts of mid-latitudes. A study of the tropical hydrological cycle reveals that water vapor converges toward the Intertropical Convergence Zone (ITCZ) and the three tropical continents: Equatorial Africa, Central America, and the monsoon-western Pacific region. Satellite observations reveal that the major cumulus convection and rainfall take place in these regions. Because of the possible generation of available potential energy by condensation heating over these tropical regions, synoptic disturbances should occur there, as shown in Fig. 16.1.

The four different types of synoptic disturbances shown in Fig. 16.1 share some common features: eastward propagation at a speed of 5-8 degrees day^{-1}, a life cycle of 3-5 days, and a cold-core thermal structure in the lower layer. The large-scale environments along the ITCZ and the three tropical continents differ from each other. Thus, it is expected that some characteristics of synoptic disturbances over these regions must be differ-

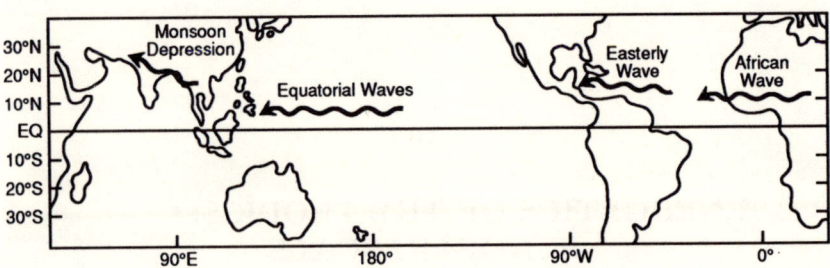

Fig. 16.1: Geographic locations of synoptic-scale disturbances in the tropics. After Nitta, 1982.

ent. To examine these differences, we shall explore the three-dimensional structure and energetics of tropical synoptic disturbances. The easterly waves over the Caribbean Sea were the first type of synoptic disturbances in the tropics to be analyzed and documented (Riehl, 1954). Perhaps it is because of the lack of quality data over the ocean that no attempt was made to analyze the energetics of the easterly waves in the Caribbean Sea. Under these circumstances, we shall focus on only the other three types of synoptic disturbances in the tropics.

16.1 Equatorial Waves over the Western Pacific

As mentioned previously, equatorial waves are active over the equatorial western Pacific during the wet season. Reed and Recker (1971) analyzed observational data of July-September 1967 over this region and found that equatorial waves propagate westward at a speed of 7° longitude day^{-1}. These waves possess a life cycle of five days and wavelength of about 3500 km. In order to examine the structure of equatorial waves, Reed and Recker applied a 2-15 day bandpass filter to isolate the signal of equatorial waves from various meteorological variables collected over a triangle of three stations (Kwajalein, Eniwetok, and Ponape) in the region (7°-10° N, 155°-170° E). Using a composite of 18 equatorial waves propagating through this triangle, Reed and Recker portrayed the structure of these waves with height cross sections of the following filtered variables: (a) meridional wind, (b) temperature, (c) horizontal divergence and (d) vertical motion (ω). Based upon the meridional wind analyses (Fig. 16.2a), eight categories are assigned to different parts of the waves. Categories 2, 4, 6, and 8 are centered, respectively, on the maximum northerly wind, the trough axis, the maximum southerly wind, and the ridge axis of the waves. Categories 1, 3, 5, and 7 are assigned to intermediate positions.

The perturbation meridional wind (Fig. 16.2a) exhibits centers in the lower troposphere near 800 mb and in the upper troposphere around 150

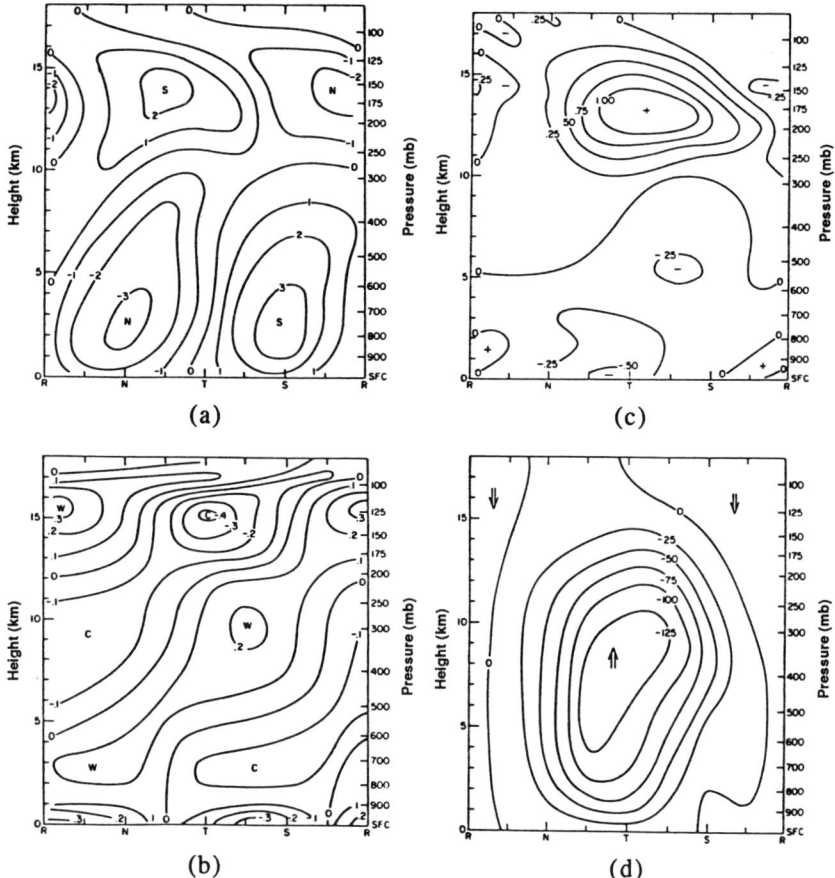

Fig. 16.2: Composite vertical cross sections of (a) meridional wind (m s^{-1}), (b) temperature (C°), (c) horizontal divergence (10^{-5} s^{-1}) and (d) vertical motion (10^{-5} mb s^{-1}) of the Pacific equatorial waves over the triangle of Kwajelein, Eniwetok, and Ponape. The letters R, N, T, and S refer to the ridge, north wind, trough, and south wind regions, respectively (Reed and Recker, 1971).

mb with weak perturbations around 300 mb. The amplitude of the perturbation meridional wind is about 2∼3 m s^{-1} in the upper troposphere and 3∼4 m s^{-1} in the lower troposphere. Perturbations tilt somewhat eastward with height and the upper and lower perturbations are almost out of phase. The vertical structure of equatorial waves differs from that of mid-latitiude synoptic-scale waves, which tilt westward during the developing stage. As inferred from the vertical structure of the perturbation meridional wind, equatorial waves have cyclonic shear in the trough and anticyclonic shear in the upper troposphere.

As expected from scale analysis, the perturbation temperature of the equatorial waves (Fig. 16.2b) is small, no larger than 0.5°C. The maximum amplitudes appear at four levels. At the surface, the maximum perturbation temperature is found around the ridge and west of the heaviest precipitation area, which exists around and somewhat west of the trough. Behind the trough is negative perturbation temperature, which may be attributed to cooling of unsaturated downdraft and cloud cover. The next maximum perturbation temperature is located at 700 mb. The negative perturbation at this level west of the trough may be caused by evaporative cooling. At 300 mb, the perturbation is out of phase with negative perturbations below a center of positive perturbation east of the trough. The highest level of perturbation centers occurs at about 125 mb. The perturbation temperature at this altitude is in phase with that in the lower troposphere. As will be illustrated later, the vertical structure of the perturbation temperature in the vicinity of the trough is affected by cumulus convection. Moreover, a comparison between perturbation meridional wind (Fig. 16.2a) and temperature (Fig. 16.2b) reveals that the thermal wind relationship holds between them over the whole depth of the equatorial waves.

The vertical structure of perturbation divergence of equatorial waves (Fig 16.2c) is not as complicated as the perturbation temperature. The perturbation convergence slopes upward to the east with a surface convergence center west of the trough and another mid-tropospheric convergence center somewhat east of the trough. To satisfy the mass continuity, a strong divergence region near 175 mb overlies the lower-level convergence region. Inspecting Figs. 16.2b and 16.2c, one can see that perturbation convergence and divergence centers are sandwiched between centers of perturbation temperatures. Vertical motions must exist between convergence and divergence centers. The release of perturbation available potential energy is measured by the covariance between temperature and vertical motion. Thus, we expect that the available potential energy of the equatorial waves may be released in the mid-troposphere. The divergence center in the upper troposphere is close to the altitude of cirrus outflow from cumulonimbus towers. These synoptic-scale divergence centers should result from the collective outflow from cumulonimbus clusters. In the ridge a weak, irregular pattern of convergence and divergence appears.

The upward motion corresponding to the perturbation divergence of equatorial waves that is described above dominates the wave over its entire depth with a maximum value larger than 100 mb day^{-1} at 400 mb above the trough. Taking density into account, the maximum vertical motion reaches 2.5 cm s^{-1} at 300 mb. Based upon scale analysis, the diabatic heating of synoptic-scale waves in the tropics is mostly balanced by adiabatic cooling. As inferred from the upward motion of the perturbation, the cumulonimbus cluster over the trough should release its latent heat with a maximum value at 400 mb. The coincidence between perturbation temperature and the inferred latent heat released by the cumulonimbus suggests that perturbation available potential energy of equatorial waves is produced in the mid-troposphere.

It was implied from the vertical structure of equatorial waves described previously that the latent heat released by cumulus convection associated with these waves may generate available potential energy in the mid-troposphere, and that the direct overturning by the warm-air rising in the mid-troposphere releases the generated available potential energy to maintain the motion of the equatorial waves. To substantiate this inference, let us follow (7.28) to formulate the energy equations of transient wave perturbations. The conventional Lorenz approach of (7.28) and (7.29) is modified in such a way that we deal with the time-departure and time-mean components instead of the eddy and zonal-mean components. Thus, we obtain

$$\frac{\partial k'}{\partial t} = -\overline{u'v'}\frac{\partial U}{\partial y} - \overline{u'\omega'}\frac{\partial U}{\partial p} - \frac{\partial}{\partial y}\overline{(v'\phi')} - \frac{\partial}{\partial p}\overline{(\omega'\phi')} - \overline{\alpha'\omega'}, \quad (16.1)$$

$$\frac{\partial A'}{\partial t} = -\frac{1}{\sigma}\overline{v'\alpha'}\frac{\partial \bar{\alpha}}{\partial y} + \overline{\alpha'\omega'} + \frac{1}{\sigma c_p}\left(\frac{R}{p}\right)\overline{\alpha'H'}, \quad (16.2)$$

where $k' = (1/2)(u'^2 + v'^2)$ and $A' = (1/2)\sigma\alpha'^2$. Note that $\overline{(\ )} = 1/T \int_0^T (\ ) \, dt$ and $(\ )' = (\ ) - \overline{(\ )}$. U is a basic mean zonal flow. Various terms in (16.1) and (16.2) can be diagnostically estimated with observational data, except terms involving ω and H.

Because ω and H cannot be directly observed, several indirect methods were introduced to evaluate them. The synoptic-scale perturbations in the tropics are not quasi-geostrophic, and the latent heat released by cumulus convection within the tropical synoptic waves is significant over a part of these waves' life cycle. It thus appears that the kinematic method with the isobaric continuity equation

$$\nabla \cdot V + \frac{\partial \omega}{\partial p} = 0 \quad (16.3)$$

is a proper one for evaluating ω. Moreover, we may also assume that $\omega = 0$ at $p = 1000$ mb and that the top pressure level can be treated as the lower and upper boundary conditions. However, some adjustment should be made on the original $\nabla \cdot V$ field (Nitta, 1970) or the computed ω field with the original $\nabla \cdot V$ (O'Brian, 1970) in such a way that the upper boundary condition can be satisfied. The isobaric thermodynamic equation has long been used to evaluate diagnostically the atmospheric diabatic heating

$$H = \frac{c_p}{(p_0/p)^k}\left(\frac{\partial \theta}{\partial t} + \nabla \cdot (V\theta) + \frac{\partial}{\partial p}(\omega\theta)\right). \quad (16.4)$$

The major deficiencies of this approach are an underestimate of $\nabla \cdot V$ in observational data and an inaccurate estimate of ω in the kinematic method.

Nitta (1972) evaluated ω and H for April-July 1956 with the upper-air data over the Marshall Islands (150° E-180°, 0°-15° N). The pressure-time cross-sections of these two variables averaged over the region during the

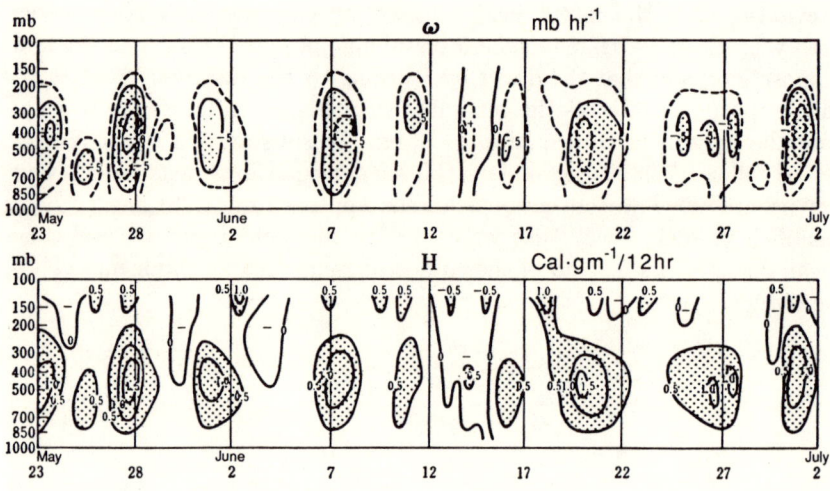

Fig. 16.3: The p-t diagrams of the vertical velocity and diabatic heating during the time period from 23 May to 2 July 1956 (Nitta, 1972).

time period of 23 May - 2 July 1956 are displayed in Fig. 16.3. The most striking features of the two cross-sections are (a) maximum values of both ω and H occuring synchronously near 400 mb and (b) a quasi-periodic five-day variation of ω and H clearly emerging. It was asserted by many studies that the life cycle of equatorial waves is about five days. The quasi-periodic variations of ω and H shown in Fig. 16.3 are certainly caused by equatorial waves. To be precise, the power spectral analyses of these two variables were performed by Nitta. Two distinct peaks of ω and H power spectra are centered at periods of 5 and 12.5 days.

Let us now turn to the energy budget analysis of equatorial waves. Shown in Fig. 16.4 are vertical distributions of the generation of perturbation available potential energy, $G(A')$, and the conversion between perturbation available potential energy and kinetic energy, $C(A', K')$, computed in terms of co-spectra between α' and H', and between α' and ω', respectively. Both $G(A')$ and $C(A', K')$ exhibit positive values between 400 and 200 mb with a maximum located at 300 mb and two distinct peaks at periods of 5 and 12.5 days, as shown in power spectra of ω and H. It was revealed from Fig. 16.2 that the perturbation temperature of equatorial waves is positive between 400 and 200 mb. Positive values of $G(A')$ and $C(A', K')$ should be expected. In contrast, the negative values of $G(A')$ and $C(A', K')$ in the upper and lower layers may be attributed to the cold-core structure of equatorial waves in these two layers.

The horizontal heat advection by synoptic-scale motions in the tropics is generally small. According to the thermodynamic equation, diabatic heating associated with synoptic waves in the tropics is almost balanced by

ENERGETICS OF THE TROPICS

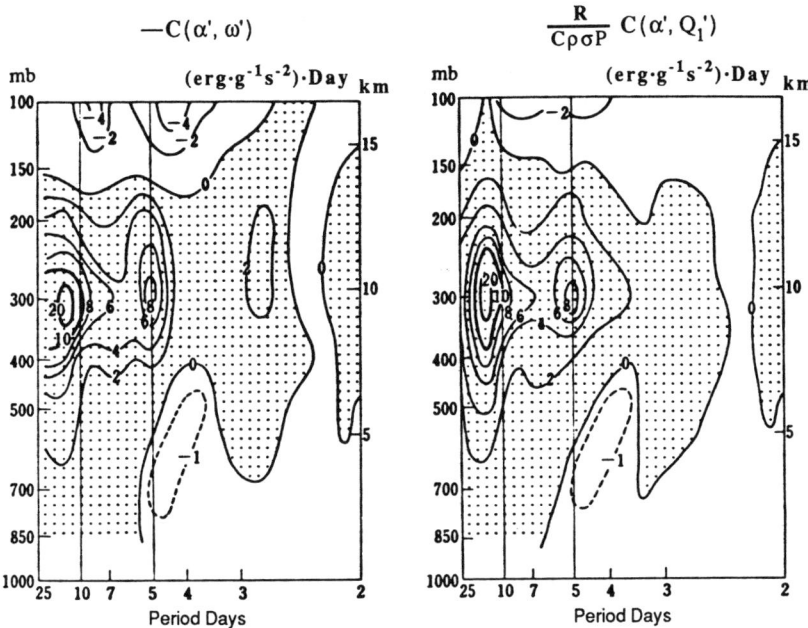

Fig. 16.4: The vertical distribution of (a) release of available potential energy and (b) generation of available potential energy in the frequency domain during the time period from 23 May to 2 July 1956 (Nitta, 1972).

adiabatic cooling. For this reason, it is conceivable that the approximation

$$-\overline{\alpha'\omega'} \simeq (1/\sigma c_p)(R/p)\overline{\alpha' H'}$$

is valid. It is not surprising that a resemblance exists between vertical distributions of $G(A')$ and $C(A', K')$ spectra shown in Fig. 16.4. Apparently, the perturbation available potential energy of equatorial waves generated by the released latent heat from cumulus convection is transformed by a direct overturning to furnish the kinetic energy of equatorial waves.

The maximum meridional wind of equatorial waves (Fig. 16.2a) occurs at 125 and 800 mb, but the maximum $C(A', K')$ appears at 300 mb, where the perturbation meridional wind of these waves is weak. One may question why the generated available potential energy of equatorial waves does not support in-situ their perturbation motions. To answer this question, vertical distributions of $C(A', K')$ and $-\overline{\omega'\phi'}$ contributed by the frequency domain with a period shorter than 20 days are shown in Fig. 16.5. A single positive maximum value of $C(A', K')$ is found at 300 mb, while above and below this level appear, respectively, positive and negative maximum values of $-\overline{\omega'\phi'}$. As indicated by (16.1), the vertical divergence of potential

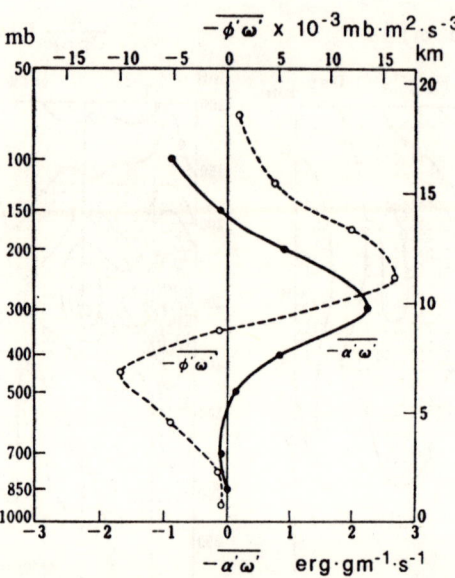

Fig. 16.5: The vertical distribution of the release of available potential energy integrated over the entire frequency domain shown in Fig. 16.4 during the time period from 23 May to 2 July 1956 (Nitta, 1972).

energy flux is one of the energetic processes maintaining perturbation kinetic energy. It is inferred from the contrast between vertical distributions of $C(A', K')$ and $-\overline{\omega'\phi'}$ that the available potential energy of equatorial waves generated by the latent heat released from cumulus convection is transported upward and downward to perturbation motions of equatorial waves in the upper and lower layers.

So far, we have discussed only three energetics terms in (16.1) and (16.2). Others are related to horizontal transports of perturbations and the zonal-mean states of basic flow. Note that the synoptic-scale waves in mid-latitudes are characterized by slanted convection in which horizontal transports of sensible heat and momentum are significant. Moreover, the north-south gradients of zonal-mean flow and temperature are large. Consequently, $C(A_Z, A')$ and $C(K_Z, K')$ are important. But contrarily, the synoptic-scale waves in the tropics are monsoonlike. The vertical transports of physical quantities by these waves are more significant than the horizontal transports. Additionally, the north-south gradients of zonal-mean flow and temperature over the Marshall Islands region are insignificant. We should not expect that energetics terms in (16.1) and (16.2) other than $G(A')$, $C(A', K')$, and $-\partial \left(\overline{\omega'\phi'} \right)/\partial p$ are important. The vertical distribution of various energy variables averaged over the Marshall Islands region and contributed by perturbations with periods smaller than 20 days are

ENERGETICS OF THE TROPICS

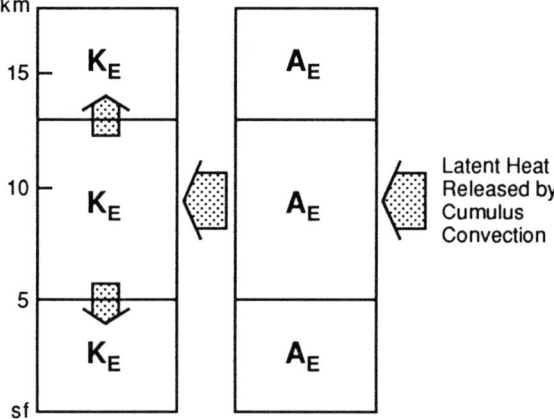

Fig. 16.6: Schematic energy diagram of the Pacific equatorial wave. After Nitta, 1982.

displayed in Table 15.1. As revealed from this table, major energy conversions are $G(A') \to C(A', K') \to -\partial(\overline{\omega'\phi'})/\partial p$; other energy quantities are one to two orders of magnitude smaller than the three energetics variables. Based upon the energy analysis of equatorial waves, a schematics energy diagram illustrating the maintenance of equatorial waves is shown in Fig. 16.6. The importance of the latent heat released by cumulus convection to equatorial waves is stressed by the large arrow in the figure.

16.2 African Waves

The African waves originate over central and western Africa and propagate westward to the eastern Atlantic. An example of these wave synoptics is shown in Fig. 16.7. These waves have wavelengths of 2500 km and periods of 3.5 days, and move westward at a speed of 8 m s^{-1} (6~ 7° longitude day $^{-1}$). The data collected during Phase III of the GARP (Global Atmospheric Research Programme) Atlantic Tropical Experiment (GATE) has been analyzed by Reed et al. (1977) and Thompson et al. (1979) to portray the structure of African waves.

During the northern summer, the strong surface heating over the Sahara to the north and the cooling of the sea water over the Gulf of Guinea to the south establish a pronounced north-south temperature gradient over central and western Africa. This temperature gradient maintains the mid-tropospheric easterly jet stream because of the easterly thermal wind, even though the surface layer is occupied by the shallow monsoon westerlies (Fig. 16.8a). Overlying them are the upper-tropospheric easterly jet stream and

Table 16.1: Energy transformations due to the disturbances with periods from 1 day to 20 days. Unit is 10^{-1} erg·gr^{-1}·s^{-1}.

MB	$-\overline{u'\omega'}\frac{\partial U}{\partial p}$	$-\overline{u'\omega'}\frac{\partial U}{\partial p}$	$-\frac{\partial}{\partial p}(\overline{\phi'\omega'})$	$-\overline{\alpha'\omega'}$	$\frac{R}{c_p\sigma_p}\overline{\alpha'Q'}$	$-\frac{1}{\sigma}\overline{v'\alpha'}\frac{\partial\bar{\alpha}}{\partial y}$
50	-0.4					-0.2
100	0.0	-0.5	6.6	-8.8	-5.3	-0.2
150	0.1	0.1	15.0	-1.1	-0.8	-0.0
200	-0.1	0.3	5.2	8.9	6.5	-0.2
300	-1.0	-0.3	-16.0	22.3	23.3	-0.1
400	-0.4	-0.1	-9.3	8.4	8.7	0.0
500	-0.2	0.5	3.1	1.3	1.9	0.0
700	0.0	0.3	2.5	-0.9	-0.7	-0.0
850	0.7	-0.2	0.2	0.0	-0.0	0.0
1000	0.1					-0.0

the southward extension of the mid-latitude westerlies at the upper troposphere. The absolute vorticity (Fig. 16.8b) of the mean-zonal wind changes the sign of its north-south gradient over the stippled area near the mid-troposphere that is embedded with African waves. This constitutes a necessary condition of barotropic instability in the vicinity of this jet and provides a preferable environment for the genesis of African waves as suggested by Burpee (1972).

The structure of African waves is delineated in Fig. 16.9 with the perturbation meridional winds, zonal winds, temperature and vertical motion.

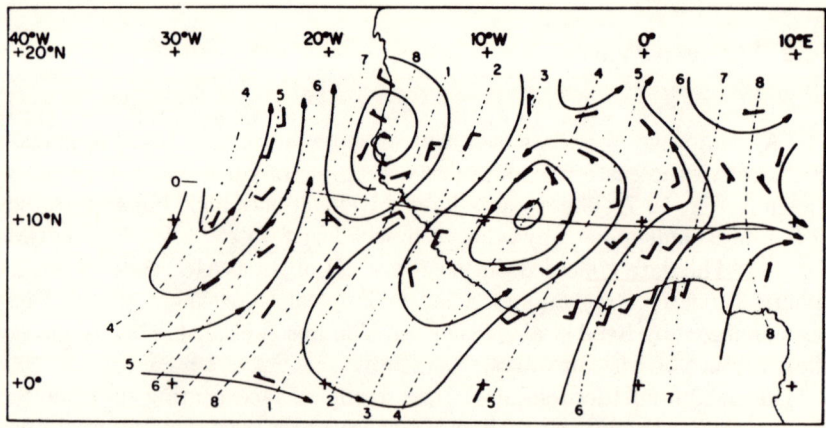

Fig. 16.7: Wind distribution of an African wave at 1200 GMT 7 September 1974: streamline (solid), track of disturbances (thin solid line), and phase line (dashed). After Reed et al. (1977).

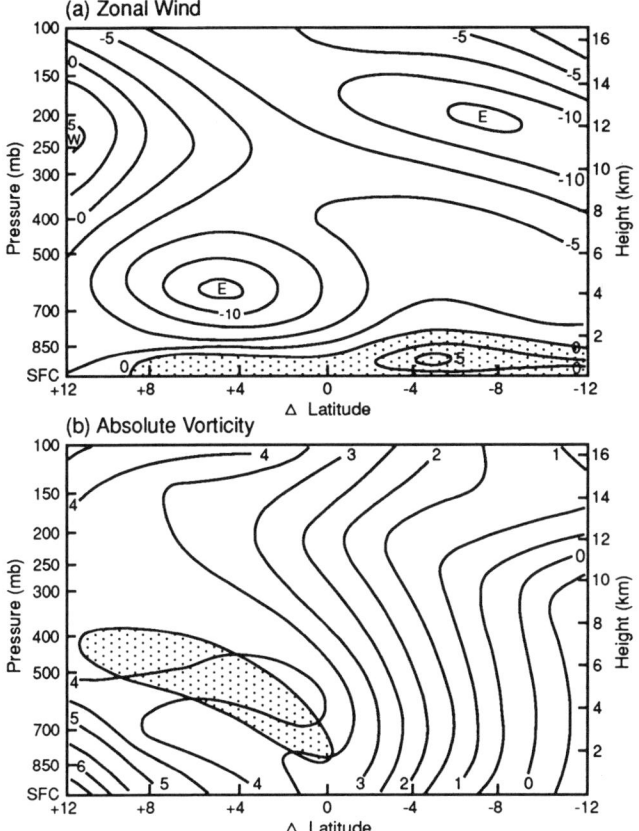

Fig. 16.8: (a) Zonal wind (m s^{-1}) and (b) absolute vorticity (10^{-5} s^{-1}) for the period 23 August to 19 September, 1974. The track of African waves is located at zero latitude, 11° N over land and 12° N over the ocean. After Reed et al. (1977).

Contrary to equatorial waves in the western Pacific, the axes of perturbation meridional wind of African waves slope westward with maximum amplitudes at 700 and 175 mb. The perturbations are almost out of phase in the upper and lower troposphere. The lower-level perturbations are transported by the mid-tropospheric easterly jet stream, which is absent in the western Pacific. The perturbation zonal wind has its maximum amplitudes appearing in the upper and lower troposphere; the lower ones are located at the level of the mid-tropospheric easterly jet stream. It is revealed from Figs. 16.9a and 16.9b that a close correlation exists between westerlies and southerlies and between easterlies and northerlies. This correlation indicates that the easterly momentum is transported southward by African waves. Since the north-south gradient of the mean zonal wind

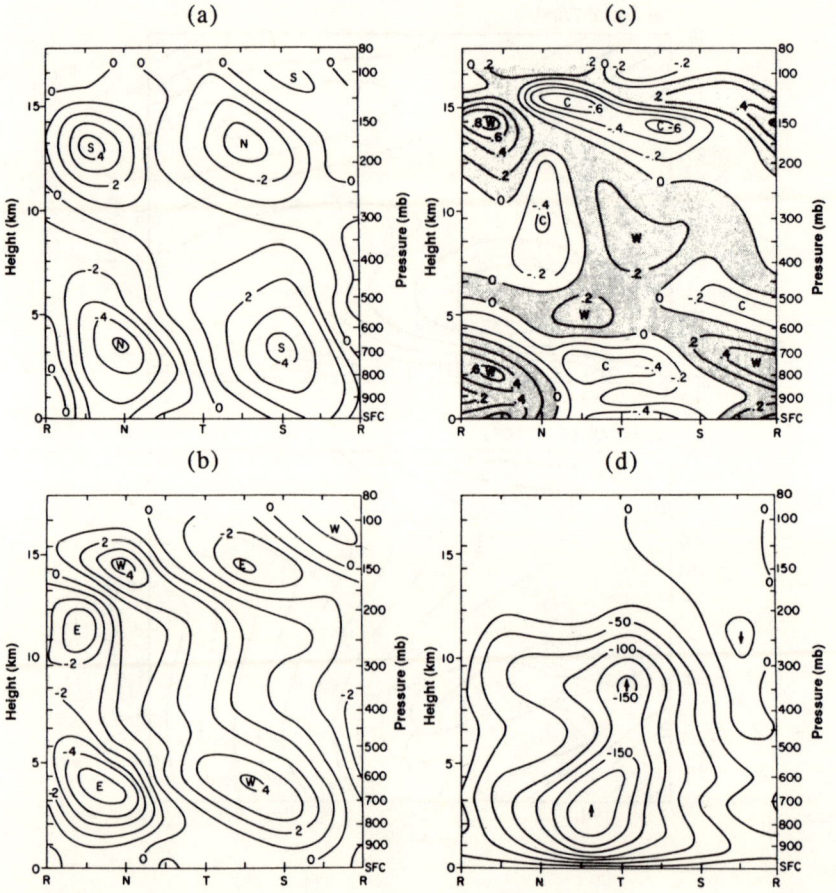

Fig. 16.9: Composite vertical cross sections of African waves: (a) meridional wind (m s^{-1}), (b) zonal wind (m s^{-1}), temperature (°C), and (d) vertical motion (10^{-5} mb s^{-1}). After Thompson et al. (1979).

is significant over the mid-tropospheric easterly jet, the southward momentum transport by African waves is important to the wave energetics.

Similar to the equatorial waves in the western Pacific, the perturbation temperature of African waves (Fig. 16.9c) is small in magnitude, less than 1°C, and complex in its vertical structure. The positive and negative temperature anomalies alternate longitudinally and vertically. Below 700 mb, the positive and negative perturbations occur in the ridge and trough, respectively. This pattern is reversed between 400 and 300 mb, and is reversed back again near 150 mb. The cold core in the lowest layer in the trough may be caused by evaporative cooling and the warm core above, between 400 and 300 mb, results from condensation heating of convective clouds. The negative perturbation in the upper layer is attributed

ENERGETICS OF THE TROPICS

to adiabatic cooling by overshooting of the cumulonimbus. A couplet of positive and negative perturbations between 600 and 500 mb may be related to the reversal of the mean-zonal temperature gradient at 650 mb through thermal advection. Although the vertical motion of African waves (Fig. 16.9d) is upward below 200 mb, two maxima are located between 800 and 700 mb somewhat ahead of the trough, and near 400 mb in the trough. In addition to the complexity in its thermal structure, the vertical motion pattern of African waves is more complicated than that of equatorial waves, which possesses only the upper maximum. The release of available potential energy is measured in terms of the covariance between perturbation temperature and vertical motion. Certainly, the release of available potential energy within African waves should be distributed in a more complicated way than that within equatorial waves.

In Section 16.1, the energetics of equatorial waves were analyzed with the energy equations of transient perturbations formulated by virtue of the time-departure and time-mean components of meteorological variables. We shall adopt the Lorenz approach [(7.27) and (7.28)] in this section from the eddy kinetic (K_E) and eddy available potential (A_E) energy equations in an open domain:

$$\frac{dK_E}{dt} = -(\alpha_E \omega_E)_Z - \left[(U_E V_E)_Z \frac{\partial U_Z}{\partial y} + (V_E V_E)_Z \frac{\partial V_Z}{\partial y} \right]$$
$$\phantom{\frac{dK_E}{dt} =} \quad C(A_E, K_E) C(K_E, K_Z)$$
$$\phantom{\frac{dK_E}{dt} =} - \left[(U_E \omega_E)_Z \frac{\partial U_Z}{\partial p} \right] + B(k_E) + B(\phi_E) - D(k_E), \tag{16.5}$$

$$\frac{dA_E}{dt} = -\left[\frac{1}{\sigma} (V_E T_E)_Z \frac{\partial T_Z}{\partial y} + \frac{1}{\sigma} (\omega_E T_E)_Z \frac{\partial T_Z^*}{\partial y} \right] + (\alpha_E \omega_E)_Z$$
$$\phantom{\frac{dA_E}{dt} =} \quad C(A_Z, A_E) -C(A_E, K_E)$$
$$\phantom{\frac{dA_E}{dt} =} +B(A_E) + G(A_E) \tag{16.6}$$

Only $C(A_E, K_E)$, $C(K_E, K_Z)$, and $C(A_Z, A_E)$ have been diagnostically computed and vertically integrated with the data of GATE phase III (Norquist et al., 1977) and displayed in Fig. 16.10. The energetics of African waves over the land and ocean are also separately analyzed.

The energetics of African waves combined over subregions of land and ocean are shown in the top energy diagram of Fig. 16.10. The comparable magnitudes of both $C(K_E, K_Z)$ and $C(A_E, K_E)$ suggest that the eddy kinetic energy of African waves is maintained by both barotropic and baroclinic processes. In addition, $C(A_Z, A_E)$ and $C(A_E, K_E)$ are also comparable in their magnitudes. If the time change in A_E is small and $B(A_E)$ is insignificant, the eddy available potential energy of African waves generated by diabatic heating, $G(A_E)$, should be negligible. Obviously, A_E of African waves is supplied by $C(A_Z, A_E)$, not by $G(A_E)$. It was shown in

288   FUNDAMENTALS OF ATMOSPHERIC ENERGETICS

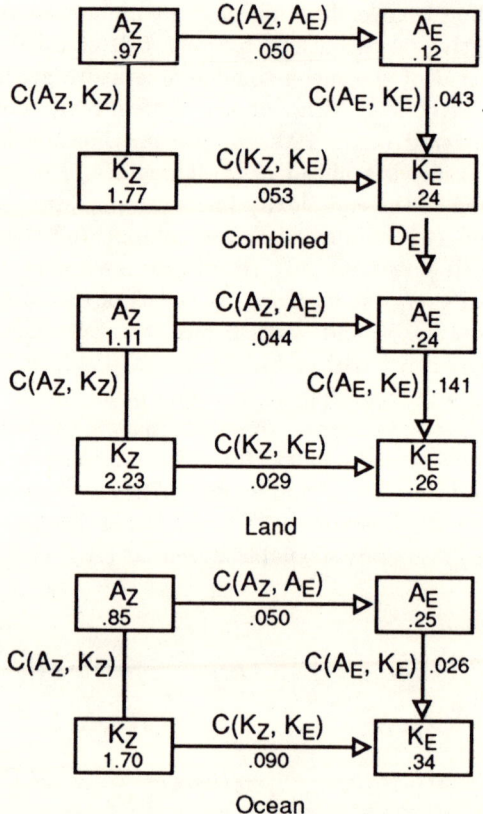

Fig. 16.10: Energy diagram of African waves for land, ocean and combined regions. Units: energy and energy conversions are 10^5 J m^{-2} and W m^{-2}, respectively. After Norquist et al., 1977.

Section 16.1 that the equatorial waves over the western Pacific gain their available potential energy through the generation by condensation heating. The motion of these waves is maintained by direct thermal overturning to release the generated available potential energy. In other words, the energetics of the Pacific equatorial waves are monsoonlike. In contrast, the existence of the pronounced north-south temperature gradient and the presence of the mid-tropospheric jet stream over western Africa make the energetics of African waves behave very differently from those of the Pacific equatorial waves.

In view of the structure and energy budget, it was realized that the characteristics of African waves are determined by the large-scale environment of western Africa. In order to shed more light along this line, the height-latitude cross sections of the three energy conversions discussed above are displayed in Fig. 16.11. The southward transport of easterly momentum

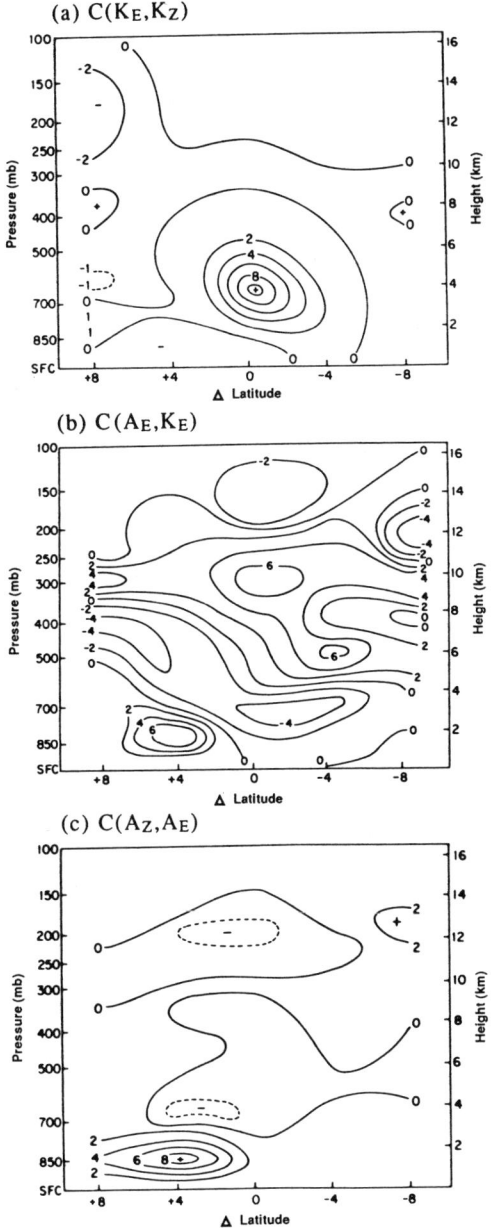

Fig. 16.11: Vertical cross sections of (a) $C(K_Z, K_E)$, (b) $C(A_E, K_E)$, and (c) $C(A_Z, A_E)$. Unit: W kg^{-1} (10^{-5} m^2 s^{-3}). After Norquist et al., 1977.

by African waves and the north-south gradient of the mean-zonal wind are significant south of the mid-tropospheric jet stream. It is not surprising that significant barotropic conversion $C(K_E, K_Z)$ (Fig. 16.11a) occurs only there. Because of the absence of the mid-tropospheric jet stream in the western Pacific, the barotropic interaction between equatorial waves and the basic mean-zonal flow is unimportant to the maintenance of these waves. The release of available potential energy is measured by the correlation between the perturbation temperature and vertical motion. Although the vertical motion of African waves is upward almost everywhere, the complicated thermal structure of these waves leads to a complicated distribution of $C(A_E, K_E)$ (Fig. 16.11b). In the trough, a negative center below 700 mb and a positive center between 600 and 200 mb correspond, respectively, to the low-level cold core and the upper-level warm core of African waves. Additionally, another positive $C(A_E, K_E)$ center is located in the lower layer ahead of the trough. The Pacific equatorial waves also possess a cold core in the lower layer without a center of upward motion as in the African waves. Thus, there is no negative $C(A_E, K_E)$ center in the lower layer of the Pacific equatorial waves.

The synoptic-scale disturbances in mid-latitudes are characterized by the slanted convection. The vertical motion of the mid-latitude synoptic waves generally responds to horizontal north-south sensible heat transport. Thus, $C(A_E, K_E)$ and $C(A_Z, A_E)$ of these waves correlate well temporally and spatially. In contrast, monsoonlike African waves are characterized by their dominant vertical motion. Although the horizontal sensible heat transport by African waves may not be significant, the north-south gradient of the mean-zonal termperature is pronounced north of the mid-tropospheric easterly jet and results in significant values of $C(A_Z, A_E)$ there. The north-south gradient of mean-zonal temperature is insignificant over the equatorial western Pacific, and $C(A_Z, A_E)$ is never a vital part of the energetics of the Pacific equatorial waves.

The African waves propagate from their genesis region over central and western Africa to the eastern Atlantic. The large-scale environments differ between the land and the ocean. This difference results in striking contrasts in the energetics of African waves over these two regions. The conversion between A_Z and A_E, $C(A_Z, A_E)$, is about the same in both regions. However, the baroclinic conversion, $C(A_E, K_E)$, is much larger over the land than over the ocean, and the opposite situation occurs with the barotropic conversion $C(K_E, K_Z)$. As revealed from the large difference between $C(A_E, K_E)$ and $C(A_Z, A_E)$ over the land, a significant generation of available potential energy of African waves by condensation heating is required. Over the ocean, the latter conversion is larger than the previous conversion. The diabatic heating effect does not supply energy to African waves over the ocean, but here $G(A)$ is negative.

ENERGETICS OF THE TROPICS   291

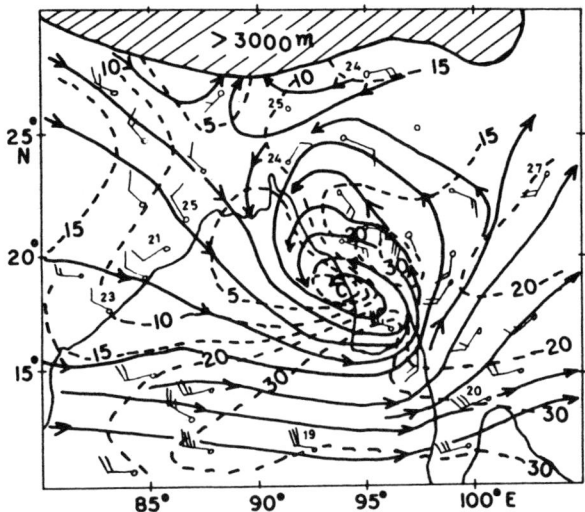

Fig. 16.12: The 850 mb wind analysis at 1200 WTC 4 July 1979. Dashed lines are isotachs with unit in knots. After Saha and Saha, 1988

16.3 Monsoon Depression

During the monsoon season, a number of monsoon disturbances (monsoon lows and depressions) move westward across India from the Bay of Bengal. The tracks of many monsoon disturbances originate in this region. The daily cloud pictures do not usually show westward-propagating cloud clusters that move from the western Pacific. It was thus believed that the monsoon disturbances form in-situ over the Head Bay of Bengal. A bay monsoon depression one day after its initiation (3 July 1979) during the Monsoon Experiment (MONEX) is shown in Fig. 16.12 as an example. However, based upon the motion of downstream amplification (illustrated with Hovmöller diagrams), Krishnamurti et al. (1977) demonstrated that the formation of the majority of monsoon depressions in the Bay of Bengal over a 43-year period (1933-76) is attributable to a slow westward propagation of wave groups initiated by typhoons or tropical storms from the western Pacific. Using maps of the 24-hr tendency of sea-level pressure during July and August over a 10-year period (1969-78), Saha et al. (1981) also established evidence that most of the bay disturbances were from nuclei of disturbances that propagate from the east. Examples of the formation of the bay disturbances by the residual lows are given in Fig. 16.13, where tracks of five tropical storms directly related to the monsoon disturbances over the Bay of Bengal during 1969-78 are shown.

An immediate question is why are the residual lows from the east cyclogenetically intensified by moving over the Head Bay of Bengal. It would

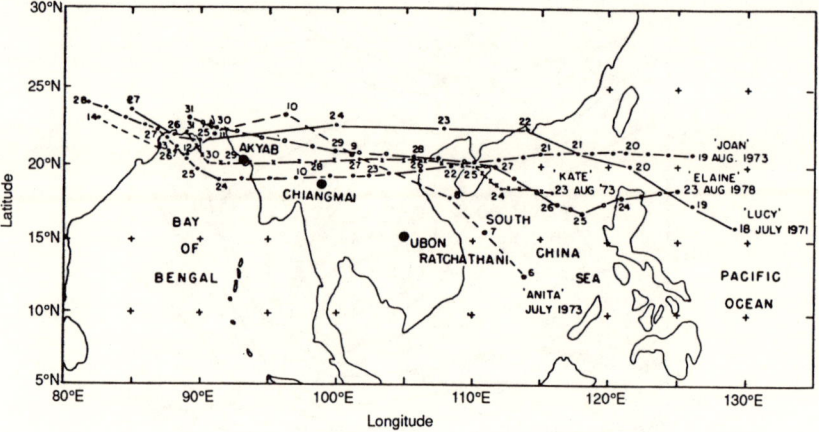

Fig. 16.13: Paths of five tropical storms directly related to monsoon disturbances in the Bay of Bengal. After Saha et al., 1981.

be natural to explore the large-scale environment in which the monsoon depressions form. As revealed from the mean-zonal wind along 85° E (Fig. 16.14), the primary circulation in the bay region is characterized in the vertical by the lower-level monsoon westerlies and the upper-level easterlies, and meridionally by the lower-level easterlies south of the equator and the monsoon westerlies north of the equator; and it is flanked by weak easterlies south of the Himalayas. Both the vertical and horizontal wind shears are strongly developed in this region. Evidently, the large-scale environment containing the monsoon depression differs significantly from those containing the Pacific equatorial and African waves described in Sections 16.1 and 16.2, respectively. It is likely that the combined barotropic-baroclinic instability is the basic genesis mechanism of the monsoon depressions. As a matter of fact, the monsoon depressions produce heavy rainfall ahead of the depression center. The joint CISK-barotropic-baroclinic instability would be more likely as a genesis mechanism of monsoon depressions (Shukla, 1978). Whatever the mechanism of initiation and intensification may be, our understanding of their synoptic structure and energetics would be helpful to explore the maintenance of monsoon depressions.

As shown in Fig. 16.12, a monsoon depression (3-8 July 1979) developed during the intensive field observation of MONEX. Various aspects of the MONEX monsoon depression were analyzed, but unfortunately its three-dimensional structure was not well documented. Godbole (1977) constructed the composite three-dimensional structure of five monsoon depressions during July-September 1973 over India, where observations were available. However, because of the lack of quality observational data, the energy analysis of a monsoon depression does not produce consistent results. Krishnamurti et al. (1976) used a multi-level primitive equation

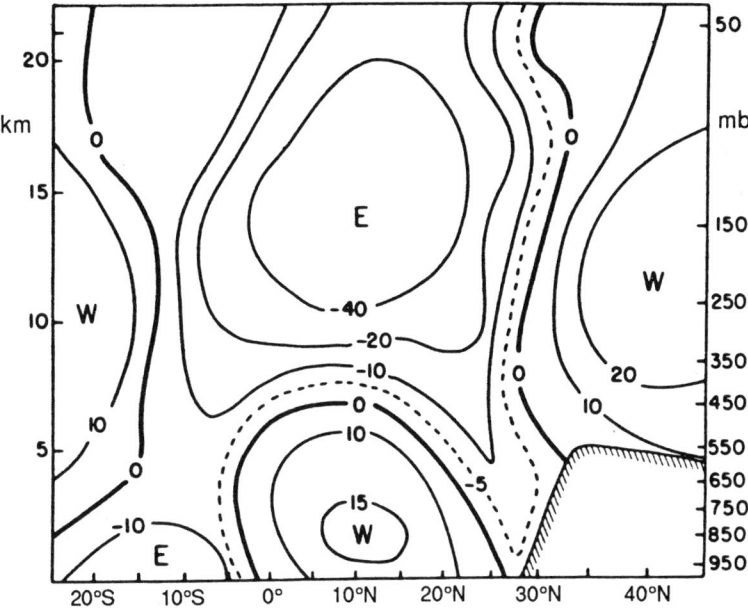

Fig. 16.14: Vertical cross section of mean-July zonal wind along 85° E. After Shukla, 1978. Unit: knots.

model to perform a 48-hr forecast of a well-developed monsoon depression over India during 4-8 August 1968. The energy analysis was carried out with the forecast simulation. The structure and energetics of monsoon depressions over the land may differ from those over the Bay of Bengal during the formation stage. Nevertheless, incorporating results of Godbole and Krishnamurti et al. with those of various aspects of monsoon depressions obtained by other studies using the MONEX data should offer a reasonable picture of the monsoon depression's structure and energetics.

The three-dimensional structure of the mature monsoon depression is portrayed by the composite vertical cross sections of meteorological variables over India in Fig. 16.15. For the zonal wind cross section along 83° E, the westerlies and easterlies appear south and north of the depression center, respectively. For the meridional wind cross-section along 23° N, the southerlies and northerlies are located to the east and west of the depression center, respectively. Both the maximum zonal and meridional winds of the depression occur in the lower half of the troposphere. A vertical transition from westerlies in the lower layer to easterlies in the upper layer south of the depression center takes place at about 400 mb, while that from southerlies to northerlies north of the center occurs at about 300 mb. It is clear that the monsoon depression consists of a cyclonic vortex confined mainly to the lower half of the troposphere with moderate wind speeds (less

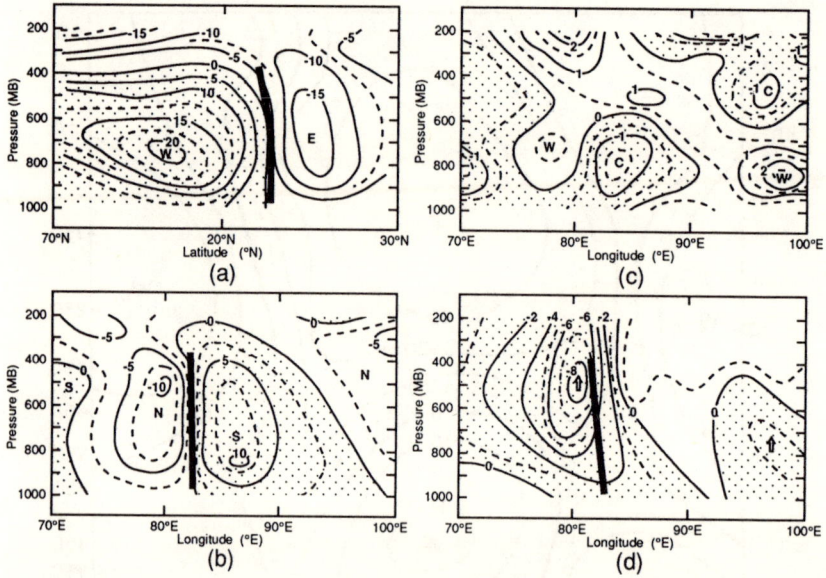

Fig. 16.15: Composite vertical cross sections of (a) the zonal wind (m s^{-1}) along 83° E, (b) the meridional wind (m s^{-1}) along 23° N, (c) temperature (°C) along 23° N and (d) vertical motion (10^{-5} mbs^{-1}) along 23° N. After Godbole, 1977.

than 20 m s^{-1}). The axis of the monsoon depression denoted by the thick solid line shows a slight southwest tilt. As revealed from numerous studies, there are as many depressions with eastward tilt as those with westward tilt.

It was shown in the previous two sections that a feature common to the thermal structure of synoptic-scale disturbances in the tropics is a cold core in the lower layer. The cold core of the monsoon depression with a maximum amplitude of -1.6°C at 800 mb appears in the thermal cross section along 23° N (Fig. 16.15c), and a reversal of the thermal field occurs at about 600 mb. A warm core shows up to the west of the depression center. Recall that all five monsoon depressions are at their mature stages over India. Godbole argued that this warm core results from the warm air advection from the desert in the western part of the Indian subcontinent. In fact, this warm core is also common to the monsoon depressions during the developing stage. However, the vertical motion field of the synoptic-scale disturbances treated so far is dominated by strong upward motion overlaying the troughs. The vertical motion cross section of the monsoon depression along 23° N (Fig. 16.15c) exhibits a different pattern, with a maximum upward motion at 500 mb to the west and a relatively weak subsidence to the east of the depression center.

ENERGETICS OF THE TROPICS

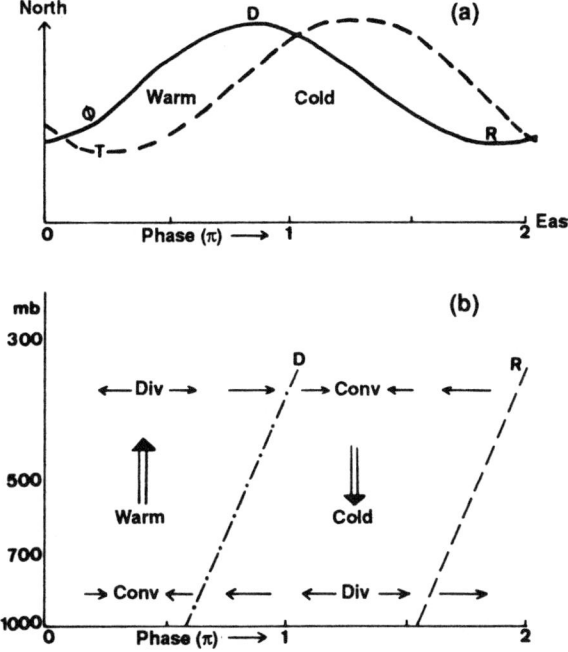

Fig. 16.16: Schematic diagrams of (a) the phase relations between geopotential (ϕ) and temperature (T) at 500 mb, and (b) the secondary divergent circulation of a developing monsoon depression. D and R denote the depression center and ridge, respectively. After Saha and Chang, 1983.

The correlation between the thermal structure and the vertical motion of the monsoon depressions suggests that warm air rises to the west and cold air sinks to the east of the depressions. Accordingly, it is inferred that there is strong convergence to the west and divergence to the east of the depression center in the lower troposphere, and a reversed situation occurs in the upper troposphere. Analyzing some developing monsoon depressions, Saha and Chang (1983) pointed out that the geopotential and temperature fields of the monsoon depressions have a phase lag such that the aforementioned secondary divergent circulation can be related to warm-air advection from the northwest sector of the depression and cold-air advection from the southeast sector. A schematic diagram summarizing this baroclinic process is shown in Fig. 16.16. The release of available potential energy by the thermally direct overturning is one of the important energetic processes maintaining the monsoon depressions.

To maintain the thermal structure of monsoon depressions, a supply of available potential energy through its generation by condensation heating may be one of the possible sources to counterbalance the release of available potential energy by the thermally direct overturning. Saha and Saha

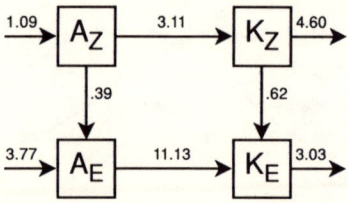

Fig. 16.17: Energy diagram of the forecasted monsoon depression during 4-6 August 1968 with a multi-level primitive equation. After Krishnamurti, 1976. Unit of energy conversions is W m^{-2}.

(1988) computed the diabatic heating associated with the MONEX monsoon depression (Fig. 16.12). They found that over the depression's entire life cycle, the diabatic heating occurs in the southwest quadrant where the temperature perturbation is positive. The correlation between diabatic heating and temperature of this MONEX monsoon depression indicates that its available potential energy may be generated there.

Finally, let us summarize the qualitative discussion above of the energetics of monsoon depressions. The energy diagram above (Fig. 16.17) was computed with a 48-hr forecast (4-6 August 1968) of a depression by Krishnamurti et al. (1976) and with a multi-level primitive equation model. The Lorenz approach, (16.3) and (16.4), was used. The eddy kinetic energy is maintained by both the barotropic, $C(K_E, K_Z)$, and baroclinic, $C(A_E, K_E)$, energy conversions, but primarily by the latter which was illustrated in Fig. 16.16. The eddy available potential energy is mainly furnished through its generation by condensation heating. The horizontal sensible heat advection is related to the secondary divergent circulation of the depression, but never makes the conversion between A_Z and A_E, an important energetic process. The large-scale environment over the Bay of Bengal is significantly different from that over the western Pacific. Surprisingly, the overall energy cycle of the monsoon depressions is similar to that of the Pacific equatorial waves.

17

ENERGETICS OF THE SOUTHERN HEMISPHERE

The atmospheric circulation derives its energy from the incoming solar radiation. Thus, the circulation of the two hemispheres could be assumed similar if symmetry about the equator existed in all respects. However, the atmospheric circulation is driven by the horizontal differential heating. Any cause inducing a spatial variation from this heating would result in the response of the atmospheric circulation. It is well known that the major mountain ranges and land-sea contrasts exist in the Northern Hemisphere, while the Southern Hemisphere contains smaller land masses and less extensive mountain ranges. Accordingly, the atmospheric diabatic heating is more uniform in the Southern hemisphere. Thus, the asymmetric components of the atmospheric circulation should be more pronounced in the Northern Hemisphere.

The *winter*-mean height fields at 200 mb in the two hemispheres (Fig. 17.1) show a distinct difference in the upper-air circulation. In the Northern Hemisphere, two deep troughs are located at the east coasts of the Asian and American continents, and a minor trough is present over the Mediterranean Sea. It was shown in Chapter 13 that these troughs contain the three subtropical jets. The upper-level circulation of the winter Southern Hemisphere is predominantly zonal, although a double-jet structure is associated with the significant horizontal gradients of the 200-mb height at about 50° south at Africa and at about 30° S stretching from the eastern Indian Ocean through Australia to the western Pacific. Even during the *summer* season, the upper-level circulation of the Northern Hemisphere is wavier than that of the Southern Hemisphere. Thus, the planetary-scale stationary waves induced by the orography and land-sea contrast are much more pronounced in the Northern Hemisphere, particularly in winter. Additionally, it is also revealed from an eyeball comparison of the 200-mb height fields between the two extreme seasons that the annual variation of the upper-level circulation is more developed in the Northern Hemisphere.

In view of the basic differences between the upper-level circulation systems of the two hemispheres, an energy analysis should enable us to understand the dynamics behind these differences. It might well be because of the lack of observations over the vast southern oceans that only a few attempts were made to investigate the atmospheric energetics of the Southern Hemisphere until the FGGE data became available in the early 1980s. Since then, several interesting aspects of the Southern Hemispheric energetics

298 FUNDAMENTALS OF ATMOSPHERIC ENERGETICS

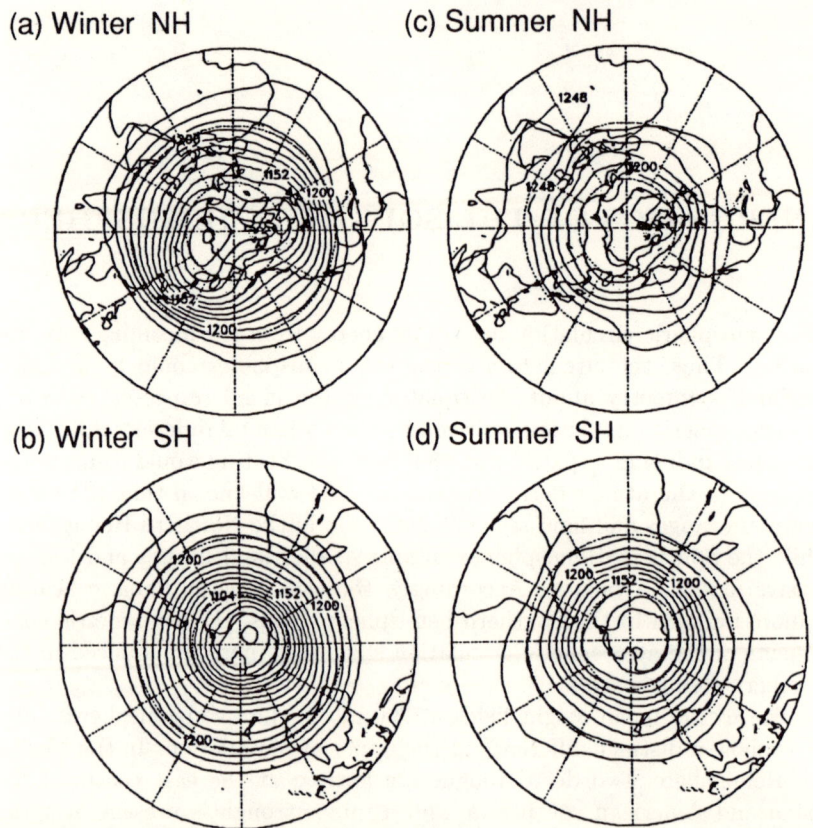

Fig. 17.1: The seasonal-mean 200-mb contours of both the Northern Hemisphere (NH) and Southern Hemisphere (SH) constructed with the height field generated by the Global Data Assimilation System of the European Centre for Medium Range Weather Forecasts. The northern (southern) winter is equivalent to the southern (northern) summer. We use December 1978 - Feburary 1979 and June 1979 - August 1979 as the two extreme seasons. The contour interval of the 200-mb height fields is 120 m.

have been examined. We shall concentrate in this chapter on four aspects. First, we would like to obtain an overview of the Southern Hemispheric energetics through a comparison of the annual variation of the Lorenz energy cycle between the two hemispheres. Second, it was shown that the upper-level circulation of the Southern Hemisphere is primarily zonal in the mean. One may question the function of atmospheric eddies in the general atmospheric circulation in this hemisphere. The spectral energy analysis may shed some light on this question. Third, the existence of the intermediate-wave (n=4-6) regime, especially wavenumber 5 with a period of 11 days, is a *ubiquitous* feature of the summer Southern Hemisphere. It was shown in Section 14.3.2 that the contribution of the intermediate-wave regime to the energy vacillation is just as important as that of the

ENERGETICS OF THE SOUTHERN HEMISPHERE

long-wave regime. Since the latter regime is much weaker in the Southern Hemisphere, does the intermediate-wave regime play a more vital role in the energy vacillation of the summer Southern Hemisphere? Finally, the single jet of the upper-level circulation in the summer Southern Hemisphere becomes a double jet in winter. These jets are not located over the east coasts of major continents. Can they be maintained by the same mechanism maintaining the winter subtropical jets of the Northern Hemisphere, illustrated in Section 13.3?

17.1 Comparison of the Annual Variations in the Atmospheric Energetics between the Southern and Northern Hemispheres

The structure of the atmospheric circulation can be described by the general circulation statistics, and the maintenance of the atmospheric circulation can be illustrated by means of an energy analysis, which involves some general circulation statistics. Although Fig. 17.1 provides a quick view of the upper-level circulation in the Southern Hemisphere, the three-dimensional structure of the upper-air circulation of this hemisphere may be supplemented with latitude-height cross sections of some general circulation statistics. To serve this purpose, some conventional general circulation statistics of the two extreme seasons over the entire globe are discussed. The annual variations in the atmospheric energetics of both hemispheres are presented next.

17.1.1 Sensible Heat and Momentum Transport

Fig. 17.2 shows the zonal-mean temperature $(\overline{T}_Z)$ (dashed lines) superimposed on the zonal-mean wind $\overline{u}_Z$ (solid lines) in which the easterlies are shaded. In the Northern Hemisphere, the winter westerly maximum in the upper troposphere at 30° N is weakened by a factor of 2 and moved to 30°N in summer. Surprisingly, the zonal-mean westerlies in the upper troposphere of the Southern Hemisphere exhibit double maxima at about 55° S and 25° S in winter. The magnitude of the westerly maximum is not reduced notably from winter to summer in this hemisphere, except for the existence of a single maximum at about 50° S. According to the thermal wind relationship, the significant weakening of the westerlies in the summer Northern Hemipshere results from the reduction of the north-south T_Z gradients in mid-latitudes. In contrast, the annual changes in the north-south T_Z gradients in the southern mid-latitudes are smaller compared to those in the northern mid-latitudes. The Antarctic continent is believed to be a major factor suppressing the annual variation in the high-latitude temperature of the Southern Hemisphere.

To maintain the thermal equilibrium of the atmospheric circulation, the

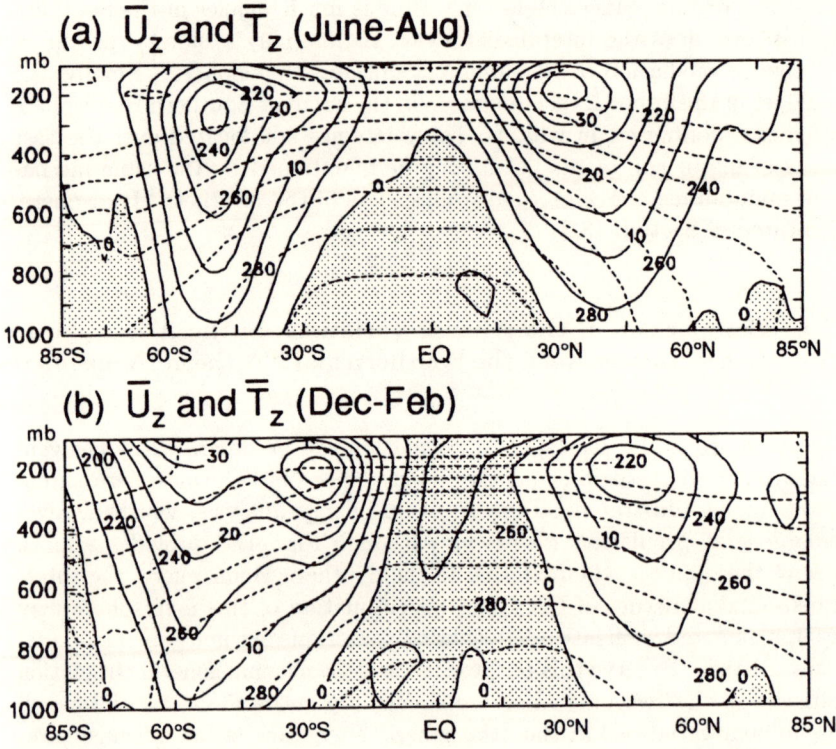

Fig. 17.2: The zonal-mean temperature (T_Z) (dashed line) and zonal wind (u_Z) (solid lines) of (a) June-August 1979 and (b) December 1978 - Feburary 1979. The contour intervals of T_Z and u_Z are 10 K and 5 m s^{-1}, respectively.

eddy sensible heat transport is an important process counterbalancing the north-south differential heating. The total and transient eddy sensible heat transports of the two extreme seasons over the entire globe are shown in Fig. 17.3. We shall highlight the following salient features in this figure:

1. The annual change in temperature is smaller in the high latitudes of the Southern Hemisphere. Since the poleward eddy sensible heat transport reduces the north-south temperature gradient, it is not surprising to see that the annual change in the eddy sensible transport is much less in the Southern Hemisphere than in the Northern Hemisphere.

2. As inferred from the difference between the total and transient eddy transports of sensible heat, stationary eddies transport as much sensible heat as transient eddies in the winter Northern Hemisphere. However, the fact that the eddy sensible heat transport in the win-

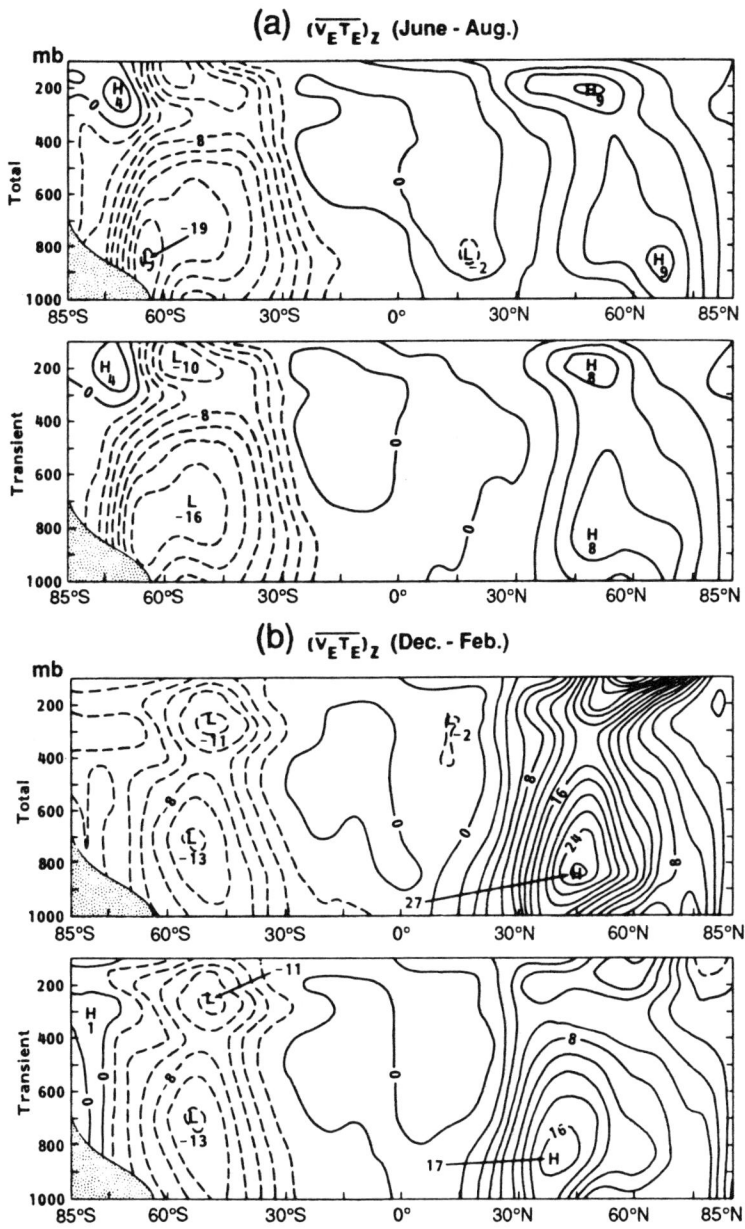

Fig. 17.3: The same as Fig. 16.2, but for total and transient eddy transports of sensible heat. The contour interval is 2 K m s^{-1}.

ter Southern Hemisphere is much less than in the wither Northern Hemisphere may be attributed to the insignificant contribution from stationary eddies.

3. During winter, the sensible heat transport by transient eddies is comparable in both hemispheres, but it is much weaker in the summer Northern Hemisphere. The weak sensible heat transport by transient eddies in this hemisphere results from the smaller north-south temperature difference.

4. The latitudinal distribution of transient eddy transport of sensible heat can be used as an indication of the baroclinic zone. The maximum centers of this transport process in both the winter and summer Southern Hemisphere and in the winter Northern Hemisphere are comparable in magnitude. The major difference between them is the locations of their centers. These centers are located at about 40° N in the winter Northern Hemisphere and at about 55° S in both the extreme seasons of the Southern Hemisphere. It is clear that the zone of maximum baroclinicity in the Southern Hemisphere is sustained by the Antarctic continent.

Let us next discuss the eddy momentum transport (Fig. 17.4):

1. A boundary between the poleward and equatorward eddy transport of momentum is located at the poleward side of the maximum westerlies in the Northern Hemisphere. Following the seasonal migration of the maximum westerlies, this boundary moves from about 45° N in winter to about 55° N in summer. The eddy momentum transport is mostly performed by transient eddies in the two extreme seasons, except for a minor part of the poleward transport and the equatorward transport in winter.

2. A single maximum of the westerlies exists in the summer Southern Hemisphere. The boundary between the poleward and equatorward transport of eddy momentum is also located on the poleward side of the maximum westerlies, as is its counterpart in the Northern Hemisphere. It seems possible that the dynamic relationship between the maximum westerlies and transient eddies in the summer Southern Hemisphere is the same as that in the Northern Hemisphere.

3. In the winter Southern Hemisphere, the boundary discussed here is located on the poleward side of the higher latitude westerly maximum, and the subtropical westerly maximum is located on the equatorward side of the center of the poleward eddy-momentum transport. The relationship between the westerly maxima and centers of eddy-momentum transport is more complicated in this season. These two westerly maxima may be maintained by different mechanisms.

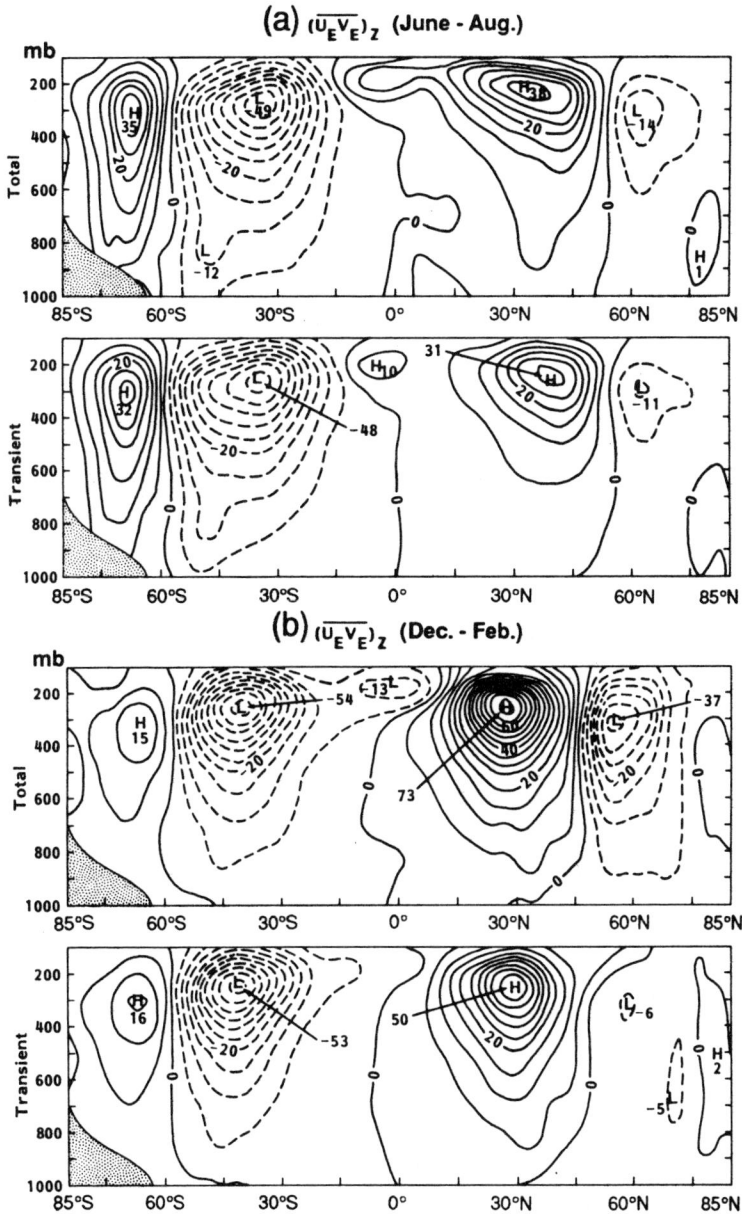

Fig. 17.4: The same as Fig. 16.3, but for the total and transient eddy transports of momentum. The contour interval is 5 m^2 s^{-2}.

4. A significant annual variation in the eddy-momentum transport exists in the Northern Hemisphere and south of 60° S. The poleward transport of eddy momentum in the Southern Hemisphere does not exhibit a marked annual change. This situation may be related to the small annual change in the maximum westerlies in mid-latitudes of the Southern Hemisphere.

17.1.2 Annual Variation of Energetics

With the general circulation statistics shown in the previous subsection, the annual variation of the Southern Hemisphere energetics will be illustrated by comparing it to that of the Northern Hemisphere energetics. Displayed in Fig. 17.5 are the monthly means of various energies (A_Z, A_E, K_Z, K_E) derived from three periods: June 1972-May 1976 and December 1978-November 1979 (FGGE year) for the Southern Hemisphere and February 1963-January 1964 for the Northern Hemisphere. The main differences in energies between the two hemispheres which emerge from this figure are:

1. The maximum westerlies of the two extreme seasons are comparable in magnitude in the Southern Hemisphere. In contrast, the maximum westerly of the Northern Hemisphere is reduced by about one half in magnitude from winter to summer. According to the thermal wind relation, the north-south T_Z gradients in the latter hemisphere decrease significantly from winter to summer as well. The stationary planetary-scale waves, which make up a large portion of eddy energies in the winter Northern Hemisphere, are very weak in the Southern Hemisphere and possess a smaller annual variation.

2. The Southern Hemisphere circulation is predominantly zonal (Fig. 17.1). This strong zonality makes the proportion of energy content in the zonal component larger in the Southern Hemisphere: the annual-averaged $A_Z/A \sim 90\%$ (80%) and $K_Z/K \sim 60\%$ (45%) in the Southern (Northern) Hemisphere.

3. Regardless of the less pronounced annual variation of the total available energy, $A + K$, in the Southern Hemisphere (Fig. 17.6), the annual mean of $A + K$ for this hemisphere is larger due mainly to larger A_Z.

4. Compared to the Southern Hemisphere, the ratios of monthly mean K to monthly mean A, K/A, over a year (Fig. 17.7) indicate that a larger portion of $A+K$ belongs to K in the Northern Hemisphere over the greater part of the year; K is less than 40% of A in magnitude during summer, but more than half of it in winter. In the Southern Hemisphere, the ratio K/A varies in a smaller range, between 30%

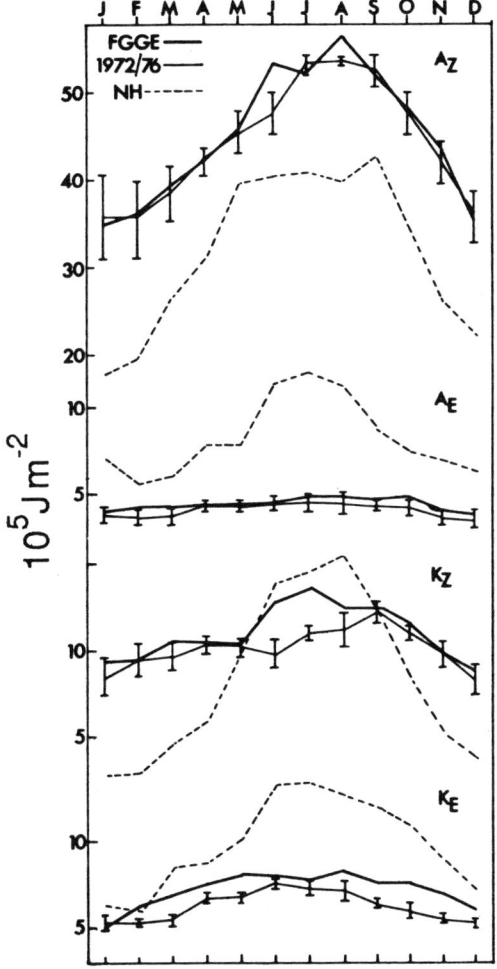

Fig. 17.5: The time variations of monthly mean $A_Z, A_E, K_Z,$ and K_E for the Southern Hemisphere during the FGGE year and 1972-76 (with one standard deviation error bar) (Price and Nicholls, 1982) and for the Northern Hemisphere during 1963-64 (Wiin-Nielsen, 1967). Unit: kJ m^{-2}.

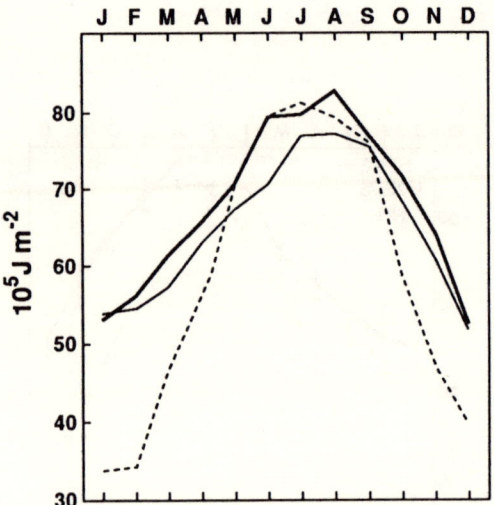

Fig. 17.6: The time variations of monthly mean $A + K$ for the Southern Hemisphere during the FGGE year (thick solid line) and 1972-76 (thin solid line) (Price and Nicholls, 1982) and for the Northern Hemisphere during 1963-64 (dashed line) (Wiin-Nielsen, 1967). Unit: kJ m^{-2}.

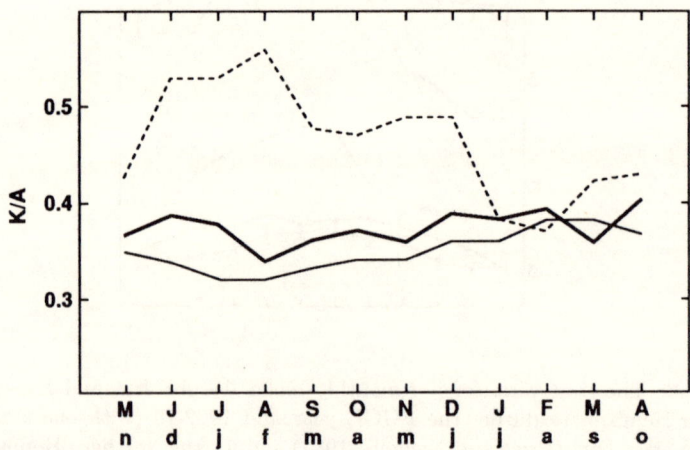

Fig. 17.7: The time variations of monthly mean K/A for the Southern Hemisphere during the FGGE year (thick solid line) and 1972-76 (thin solid line) (Price and Nicholls, 1982) and for the Northern Hemisphere during 1963-64 (dashed line) (Wiin-Nielsen, 1967). Unit: kJ m^{-2}.

Table 17.1: Amplitudes and phases of Fourier components of energies and energy conversions in the Southern Hemisphere during the FGGE year (December 1978-November 1979). Units of energy and energy conversions are 10^5 J m^{-2} and W m^{-2}, respectively.

n	0	1		2	
	F_0	F_1	r_1	F_2	r_2
A_Z	43.23	12.89	9/2	3.02	2/10
A_E	3.44	0.47	8/19	0.11	2/26
K_Z	8.75	3.14	7/28	0.16	1/25
K_E	6.41	0.88	7/26	0.06	1/26
$C(A_Z, K_Z)$	0.52	0.71	1/20	0.11	1/31
$C(A_Z, A_E)$	1.62	0.73	7/31	0.14	3/28
$C(A_E, K_E)$	1.12	0.17	8/1	0.10	1/22
$C(K_E, K_Z)$	0.37	0.10	2/5	0.04	2/28

and 40%. The annual variation of the ratio K/A is essentially in phase with those of K and A in the Northern Hemisphere, but remains relatively constant in the Southern Hemisphere over the year. The larger ratio K/A in the winter Northern Hemisphere suggests that the atmospheric circulation is more efficient in releasing its available potential energy.

To gain a more quantitative picture of the annual variation of the atmospheric energetics in the Southern Hemisphere, the Fourier analysis of the monthly mean energy variables with a one-year basic period (Section 14.1) is reperformed here. The Fourier analysis of all energy variables in time (outlined in Section 14.1) was done by Chen and Buja (1983) by evaluating the annual variations of energy variables contained in the Lorenz energy cycle. Shown in Table 17.1 are amplitudes and phases of the annual-mean (F_0), annual (F_1) and semiannual (F_2) components of energy variables in the Southern Hemisphere. For a comparison with those of the Northern Hemisphere, we refer to Table 14.1.

A comparison of the annual variation of atmospheric energies between the two hemispheres was given above. However, the quantitative comparison in magnitude and time can be obtained by contrasting the F_ns shown in Tables 14.1 and 17.1. We shall not repeat here the same detailed discussion as done in Section 14.3, but rather highlight some interesting results from the Fourier analysis. The date of the maximum value of each energy variable – that is, the phase – of each Fourier component is represented by r_n. In general, the time lag in the annual cycle of the Southern Hemisphere is somewhat larger than that in the Northern Hemisphere. This delay suggests that the response of the Southern Hemisphere circulation and energetics to the north-south differential heating is somewhat slower.

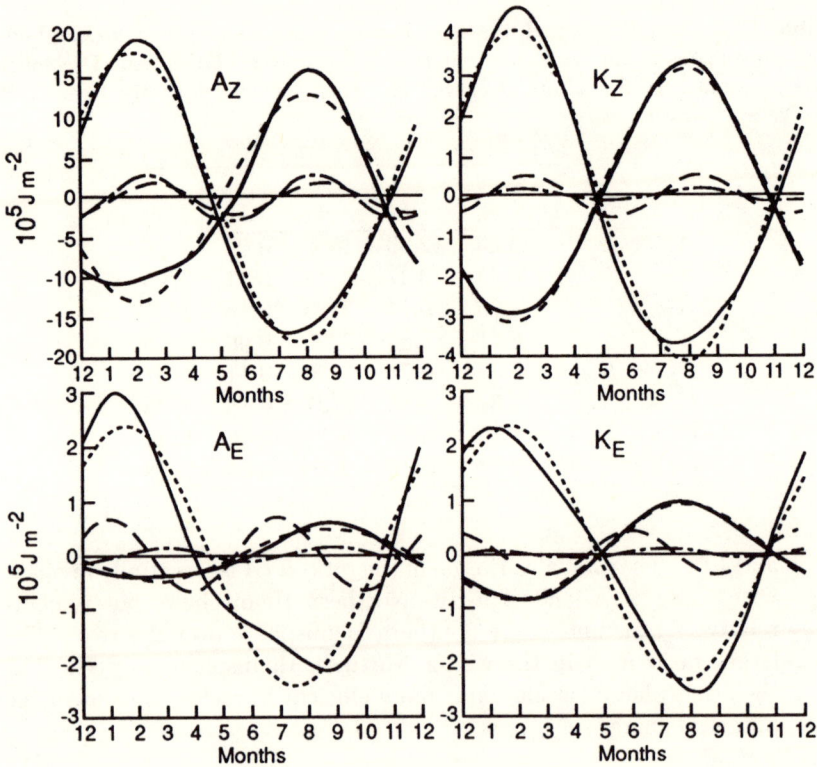

Fig. 17.8: The first (annual-cycle; dashed line) and second (semiannual-cycle; dashed-dotted line) Fourier components and the combination of them (solid line) for various energies. Energies of the Southern (Northern) Hemisphere are denoted by thick (thin) lines. Unit: 10^5 J m^{-2}.

To have a better view of the time evolution, the annual and semiannual components of various energies are shown in Fig. 17.8; the annual variation of energies is primarily determined by F_1. It was pointed out in Section 14.1 that energy variables have a slow decline in the spring and a faster buildup in the fall. This interesting feature of the annual variation in energies may occur when the semiannual-cycle component (F_1) reaches its maximum value with a phase somewhat ahead of F_1. The only case seen here is A_E of the Northern Hemisphere. In both hemispheres, F_2's of A_Z and K_Z are not negligible, but they peak more coincidentally in phase with both the maximum and minimum of F_1.

In addition to the statistical argument, we may also mention that the summer circulation in the Northern Hemisphere is so weak that it is only sporadically unstable. When the south-north temperature gradient becomes larger in the fall, the atmosphere passes into a stage of permanent

ENERGETICS OF THE SOUTHERN HEMISPHERE

baroclinic instability, which accounts for the rapid buildup of the energy. On the other hand, in the Northern Hemisphere spring, the decline of the circulation intensity is determined by the dissipation time and is thus more gradual. A similar argument cannot be applied to the Southern Hemisphere where the summer circulation is much stronger.

Comparing Fourier components of energy conversions shown in Tables 14.1 and 17.1, we find that the annual-mean values (F_0) of both hemispheres are relatively close, except $C(A_Z, K_Z)$. In contrast, most of the annual variations of energy conversions, represented essentially by the annual-cycle components, are much more pronounced in the Northern Hemisphere. A possible explanation of the difference in the annual variations in the evaluation of $C(A_Z, K_Z)$ and $C(K_E, K_Z)$ is that they may include some uncertainty in their phases of the annual-cycle component. In addition to these two energy conversions, $C(A_Z, A_E)$ and $C(A_E, K_E)$ have phases of their annual-cycle component behind the winter solstice larger than those in the Northern Hemisphere. This phase difference of energy conversions between the two hemispheres is consistent with that of the energies discussed previously. The semiannual-cycle components of the energy conversions are generally smaller in magnitude than the annual-cycle components. However, as shown in Fig. 17.9, the combination of the annual-cycle and semiannual-cycle components of $C(A_Z, K_Z)$, $C(A_E, K_E)$ and $C(K_E, K_Z)$ in the Southern Hemisphere results in a more pronounced maximum and a less pronounced minimum regardless of season.

17.2 Spectral Energetics

For the sake of discussion and comparison, the spectral energies analysis of the Northern Hemisphere atmosphere is summarized as follows:

1. The major content and the annual variation of eddy energies exist in the long-wave regime. The energy content of both A_n and K_n spectra decrease monotonically as wavenumber increases. A(-3) power law appears in the short-wave (n=8-15) regime.

2. A_Z is converted to A_n over the entire wave spectrum, particularly in the long-wave regime.

3. A_n is redistributed by the nonlinear cascade from the long-wave to the short-wave regime, and is also released by thermally direct overturning to maintain K_n with a maximum in the intermediate-wave regime.

4. K_n is redistributed by the nonlinear cascade from the intermediate-wave regime to the long-wave and short-wave regimes, and is also converted to maintain K_Z.

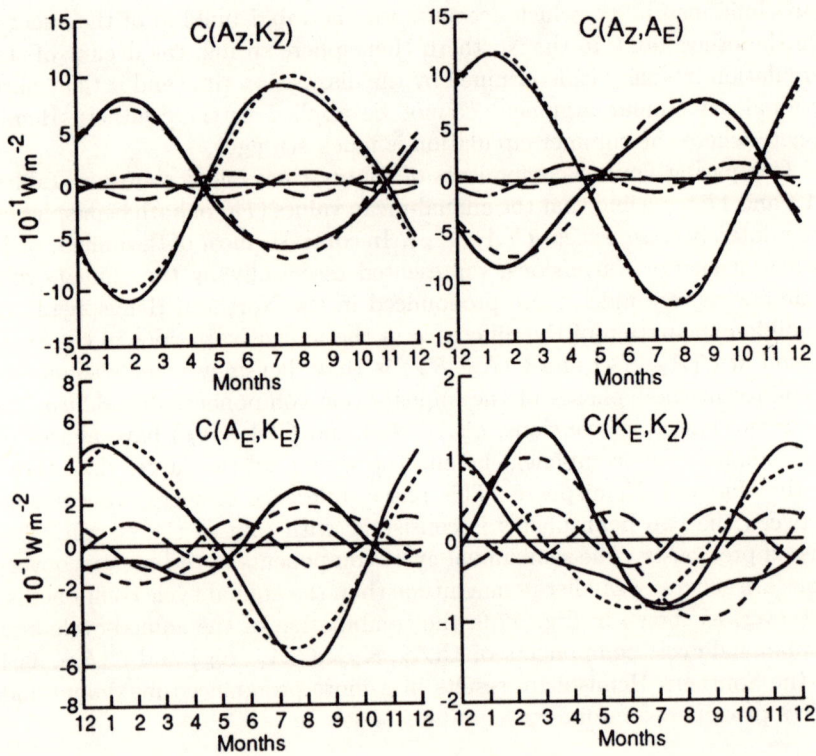

Fig. 17.9: The first (annual-cycle; dashed line) and second (semiannual-cycle; dashed-dotted line) Fourier components and the combination of them (solid line) for various energy conversions. Energy conversions of the Southern (Northern) Hemisphere are denoted by thick (thin) lines. Unit: 10^{-1} W m^{-2}

The spectral energetics of the Southern Hemisphere were analyzed with the multi-level data generated by the FGGE III-b analysis of three operational centers: the European Centre for Medium Range Weather Forecasts, the Geophysical Fluid Dynamics Laboratory, and the Goddard Laboratory for Atmospheric Sciences. The spectral distributions of energies, energy conversions and nonlinear cascades of energies are shown in Figs. 17.10-17.12. Compared to their counterparts in the Northern Hemisphere, long waves are much weaker in the Southern Hemisphere due to the lack of large mountain ranges. Consequently, the transient intermediate waves play a more vital role in the atmospheric energetics of the Southern Hemisphere. As in the A_n and K_n spectra of the Northern Hemisphere, a $A(-3)$ power law exists in the short-wave (N=8-15) regime. In contrast, neither the A_n nor K_n spectra of the Southern Hemisphere decrease monotonically as wavenumber 5. As inferred from the K_n spectra of the summer season, wavenumber 5 is a ubiquitous feature of the Southern Hemisphere circula-

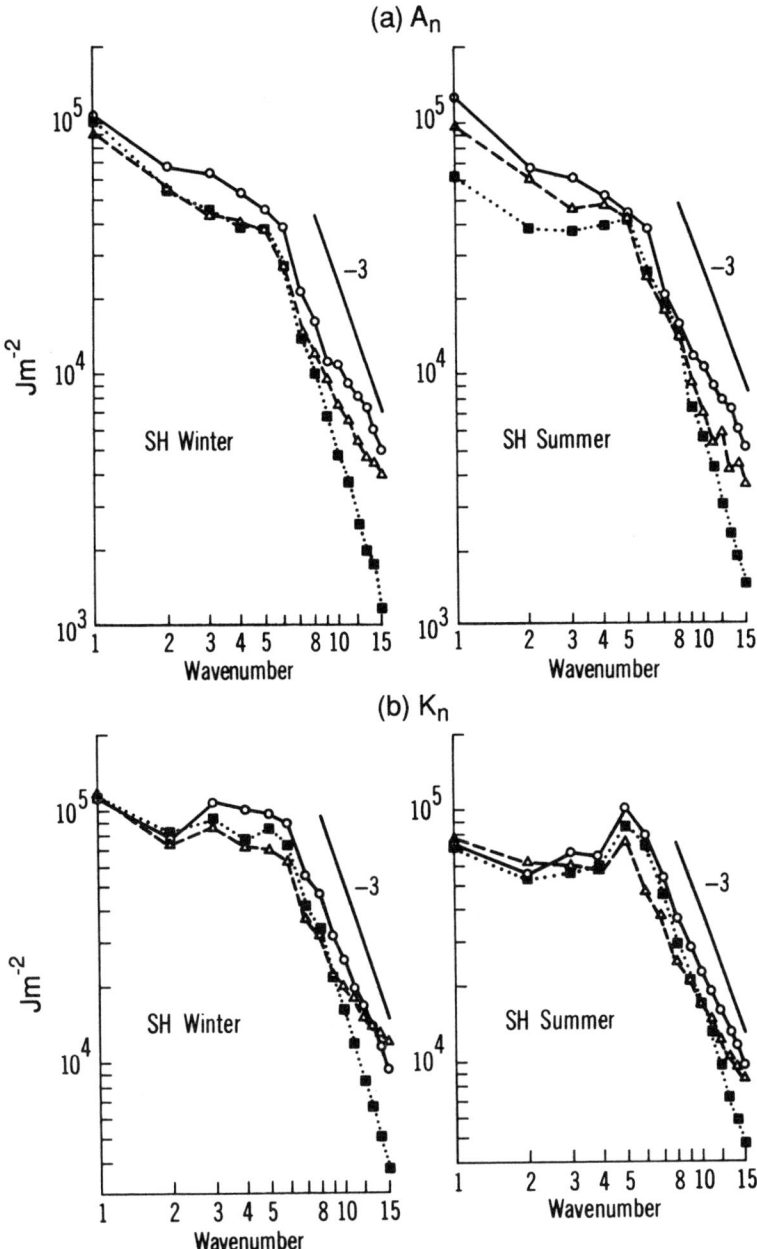

Fig. 17.10: Spectral distribution of (a) A_n and (b) K_n for the Southern Hemisphere summer (December 1978-February 1979) and winter (June-August 1979). After Chen and Lee, (1985).

tion in this season. It was pointed out in the previous section that both A_E and K_E have a much less pronounced annual variation in the Southern Hemisphere. Since the major contents of both A_E and K_E derive from the wavenumber regime N≤5, one expects that the A_n and K_n spectra in this wave regime do not have a significant annual variation. In fact, this is the case shown in Fig. 17.10, although the spectral distribution of K_n in the N≤5 wavenumber regime differs remarkably in the two extreme seasons.

Following the spectral energy scheme of Saltzman, the spectra of various energy conversions are shown in Fig. 17.11. In the winter Northern Hemisphere, the longwaves are responsible for a significant amount of sensible heat transport and, hence, contribute the major part of the conversion from A_Z to A_E. However, the wave dynamics in the Southern Hemisphere differ from those in the Northern Hemisphere. The sensible heat transport by intermediate waves becomes essential to the Southern Hemisphere's general circulation. It is not surprising to see in Fig. 17.11a that the conversion from A_Z to A_n in the Southern Hemisphere is largely derived from the intermediate-wave regime.

The large-scale motions in mid-latitudes are characterized as slanted convection. Therefore, the vertical sensible heat transport of large-scale motions is generally well correlated with the north-south sensible heat transport. Since the release of available potential energy is evaluated in terms of the covariance between vertical motion and temperature, it is expected that $C(A_n, K_n)$ should spatially and temporally correspond to $C(A_Z, A_n)$. Consequently, we may also expect that the spectral distribution of $C(A_n, K_n)$ follows that of $C(A_Z, A_n)$. As a matter of fact, this argument is confirmed by Fig. 17.11b.

So far, we have discussed the two baroclinic energy conversions. Let us now deal with the energy exchange between the zonal flow and the waves. Recall that the north-south zonal wind shear changes its sign across the maximum zonal-mean westerly in mid-latitudes, and that the westerly momentum is transported by eddies toward the upper troposphere north of the maximum zonal-mean westerlies. Because of these two factors, the contributions to $C(K_n, K_Z)$ alternate their signs latitudinally. The latitudinal integration of this energy conversion results in a small numerical value by cancellation between positive and negative values of $C(K_n, K_Z)$. Apparently, this energy conversion is not as informative as the two baroclinic energy conversions. Regardless of this special nature, $C(K_n, K_Z)$ behaves very differently in the two extreme seasons. Although its numerical values are smaller, the spectral distribution of $C(K_n, K_Z)$ in the summer Northern Hemisphere is similar to the two baroclinic energy conversions that correspond with the appearance of a maximum value in the intermediate-wave regime. Surprisingly, the maximum energy conversion is switched in the winter Southern Hemisphere to the long-wave regime, particularly to wavenumber 2. Comparing $C(A_2, K_2)$ and $C(K_2, K_Z)$ in the winter Southern Hemisphere, it is obvious that wavenumber 2 is an important wave component maintaining the zonal-mean westerlies.

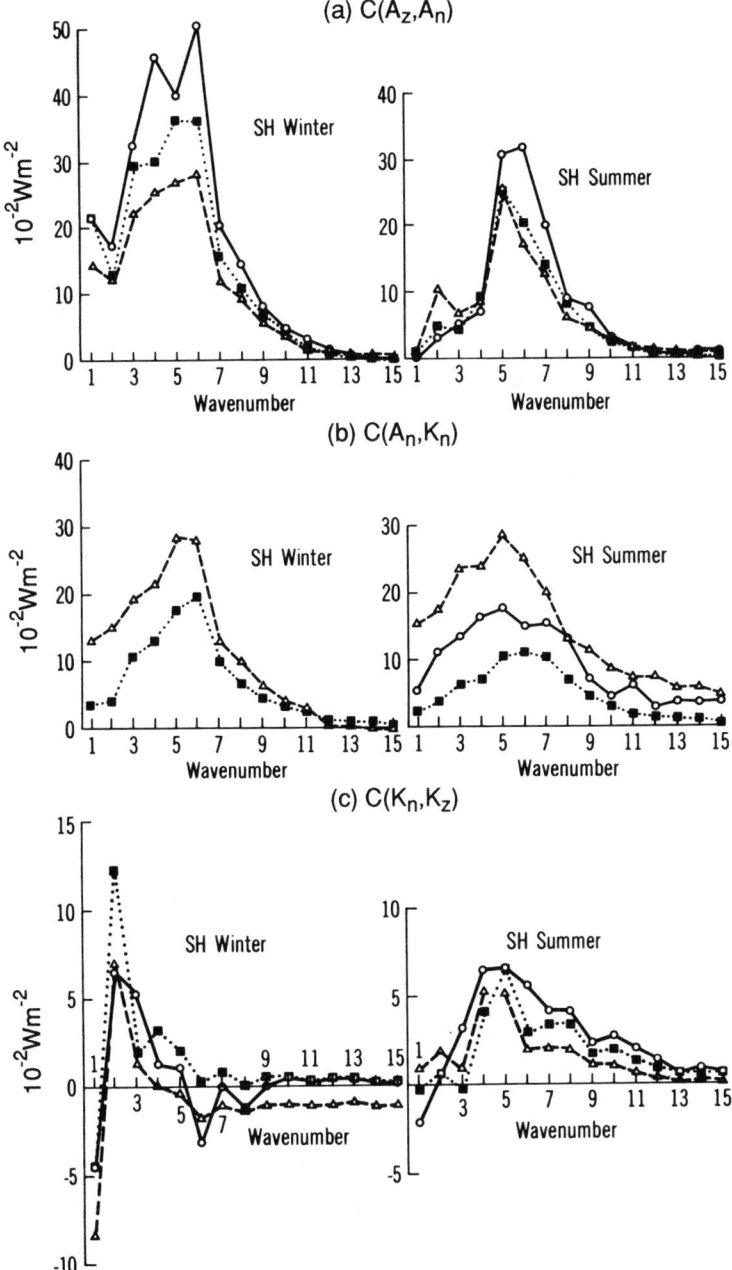

Fig. 17.11: Spectral distribution of (a) $C(A_Z, A_N)$, (b) $C(A_n, K_n)$, and (c) $C(K_n, K_Z)$ for the Southern–Hemisphere summer (June–August 1979) and winter (December 1978 – Feburary 1979). After Chen and Lee (1985).

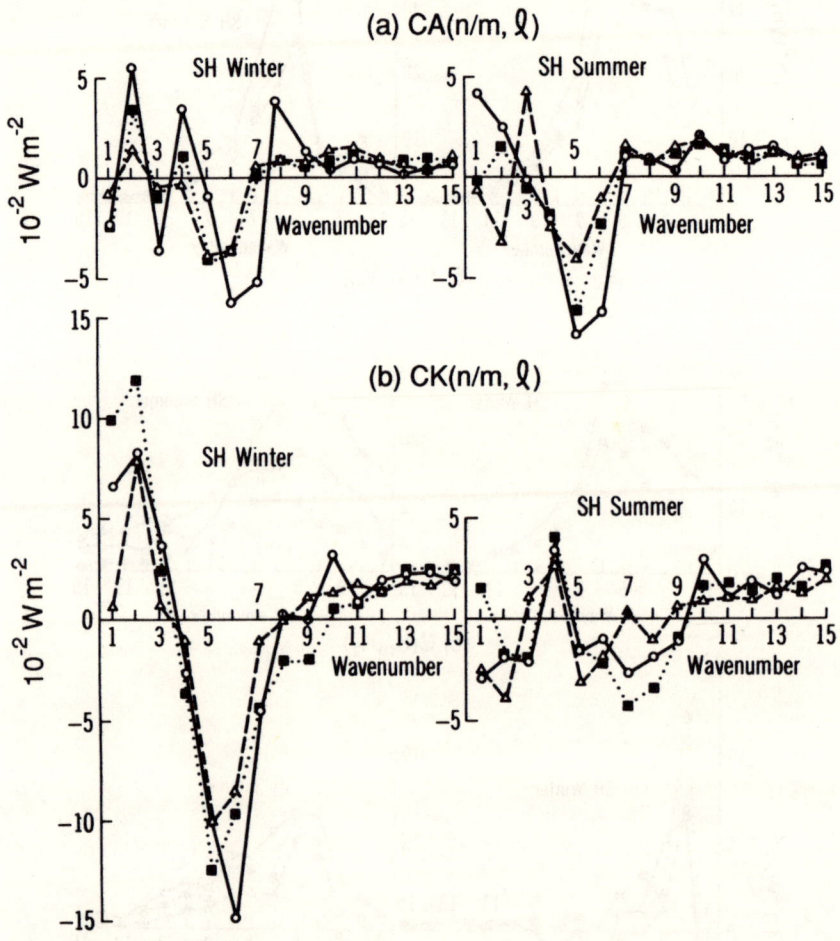

Fig. 17.12: Spectral distribution of (a) $CA(n/m, l)$, (b) $CK(n/m, l)$, and (c) $C(K_n, K_Z)$ for the Southern–Hemisphere summer (June–August 1979) and winter (December 1978 – Feburary 1979). After Chen and Lee (1985).

The wave motions in the Southern Hemisphere are also highly nonlinear. The nonlinear cascade provides a means for redistribution of the atmospheric energy through triplet interactions among waves. First, let us discuss the nonlinear cascade of available potential energy. In the Northern Hemisphere, $C(A_Z, A_n)$ takes place mainly in the long-wave regime. In other words, the major source of A_n exists in the long-wave regime. It was revealed from the energy spectra and energy conversions that the intermediate-wave regime is the most energetically active in the Southern Hemisphere. The conversion from A_Z to A_n occurs mainly in this wave regime. Interestingly, A_r is nonlinearly cascaded from the intermediate-wave regime to the long-wave and short-wave regimes (Fig. 17.11). The main supply to K_n through the release of available potential energy by thermally direct overturning exists in the intermediate-wave regime. It seems natural that K_n is cascaded from this wave regime to the long-wave and short-wave regime since it occurs in the Northern Hemisphere. In fact, this is true in the winter Southern Hemisphere, although K_n in the long-wave regime of the summer Southern Hemisphere is also cascaded to shorter waves.

Up to this point, it is clear from the spectral energy analysis that the intermediate-wave regime becomes essential to the atmospheric energetics of the Southern Hemisphere general circulation. This is attributed to the weakening of the long-wave regime in this hemisphere.

17.3 Vacillation of the Southern Hemisphere Atmospheric Energetics

It was shown in Section 14.3 that the short-wave regime, including intermediate waves, is as important as the long-wave regime to the energy vacillation of the Northern Hemisphere circulation. As we have learned in Section 17.1, one of the distinct differences between the atmospheric circulations of the two hemispheres is that the long-wave regime is much weaker in the Southern Hemisphere. Additionally, the spectral energy analysis discussed in Section 17.2 reveals that the atmospheric energetics of the Southern Hemisphere circulation are dominated by the intermediate-wave regime, particularly in summer. It seems conceivable that the energy vacillation of the Southern Hemisphere circulation in summer is primarily attributable to the intermediate waves. The investigation of this issue done by Chen et al. (1987) with the FGGE III-b data of the European Centre for Medium Range Weather Forecasts illustrated the role of the intermediate-wave regime in the energy vacillation of the summer Southern Hemisphere.

Following the Lorenz energy cycle, the summer-mean values and standard deviations of all energies and energy conversions averaged over the latitudinal zone between 30° S and 65° S are shown in Tables 17.2 and

Table 17.2: 1979 summer-mean energies of the Southern Hemisphere between 30° and 65° S and their standard deviations. Unit: 10^5 J m^{-2}

	Time mean	Standard deviation
A_Z	20.61	1.72
A_E	4.86	0.84
A_M	1.90	0.61
K_E	7.91	1.37
K_M	3.46	1.09
K_Z	11.48	1.12

Table 17.3: 1979 Summer-mean energy conversions of the Southern Hemisphere between 30° and 65° S and their standard deviations. Unit: W m^{-2}

	Time mean	Standard deviation
$C(A_Z, K_Z)$	-0.68	0.52
$C(A_Z, A_E)$	2.04	0.90
$C(A_Z, A_M)$	1.14	0.79
$C(A_E, K_E)$	1.22	0.53
$C(A_M, K_M)$	0.53	0.35
$C(K_E, K_Z)$	0.42	0.66
$C(K_M, K_Z)$	0.21	0.47

17.3, respectively. Most of the energies and the energy conversions in the intermediate-wave regime constitute only a portion, about 50% or less, of the total eddy contribution. Surprisingly, standard deviations of all energy quantities in the intermediate-wave regime are comparable to those of the total eddy. Certainly, the intermediate-wave regime usually explains variations of all of the energy quantities. Presumably, a vacillation exists in the atmospheric circulation of the summer Southern Hemisphere; the intermediate-wave regime should be essential to it.

Shown in Fig. 17.13 are time series of zonal and intermediate-wave energies, denoted by ( )$_M$. A quasi-periodic variation of these energy quantities emerges in these time series and the zonal and intermediate-wave energies vary oppositely in time. In order to gain a synoptic view of the vacillation of the Southern Hemisphere circulation in summer, four 500-mb contour maps are shown in Fig. 17.14. Two of these maps were selected when K_M reached its minimum values on 31 December 1978 (Day 30) and 11 January 1979 (Day 43). The other two maps were chosen when K_M peaked on 7 January (Day 38) and 20 January (Day 51), 1979. The difference between these two sets of synoptic maps shows clearly that the atmospheric circulation can vary from the highly undulatory condition, such as 31 December 1978 and 11 January 1979, to the relatively zonal condition, such as 31 December 1978 and 11 January 1979.

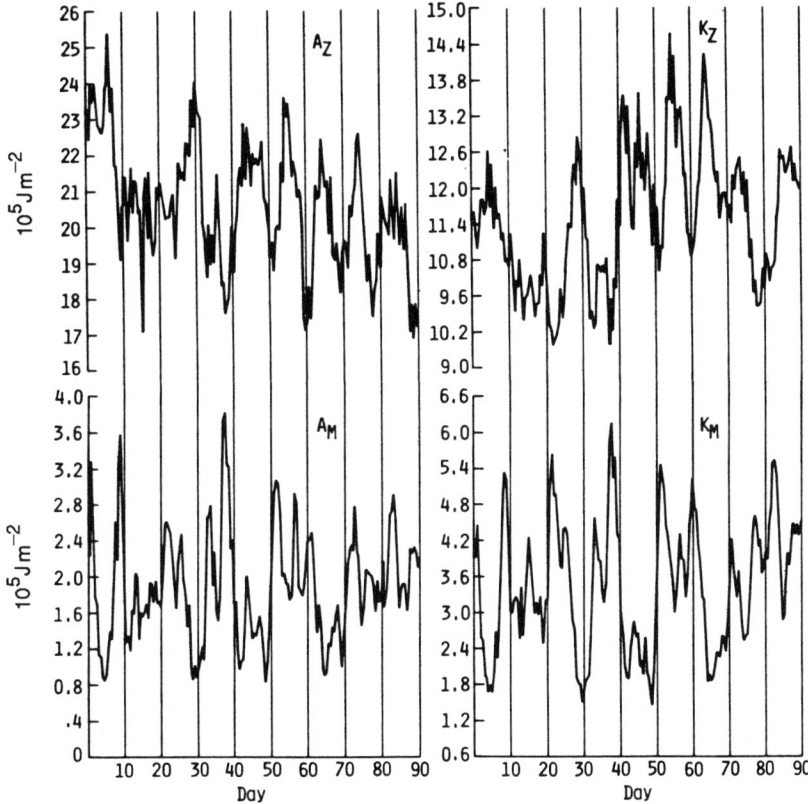

Fig. 17.13: Time series of energies (indicated on each time series) in the 30° S–65° S latitude zone during the 1978/79 winter. Day 1 is December 1, 1978 while Day 90 is Feburary 28, 1979.

To quantitatively assess the periodicity of the alternation between the maximum and minimum values of zonal and intermediate-wave energies, the statistical time-series analysis presented in Section 14.3 is adopted here. The cross spectra (solid lines) and coherencies (dashed lines) between energies related through the Lorenz energy cycle are shown in Fig. 17.15. The cross spectra between two energies peak at a period of about two weeks. The coherencies of these cross spectra around this particular period possess large values, which imply that the cross spectra of energies around this period are statistically significant. Coherencies and phases of cross spectra between energies are displayed in Table 17.4. The phases, either between zonal energies or between intermediate-wave energies, are insignificant. In contrast, the phases between energies of zonal flow and intermediate waves are about a half of the vacillation period. The phases shown in Table 17.4 confirm our subjective assessment based upon the comparison between the

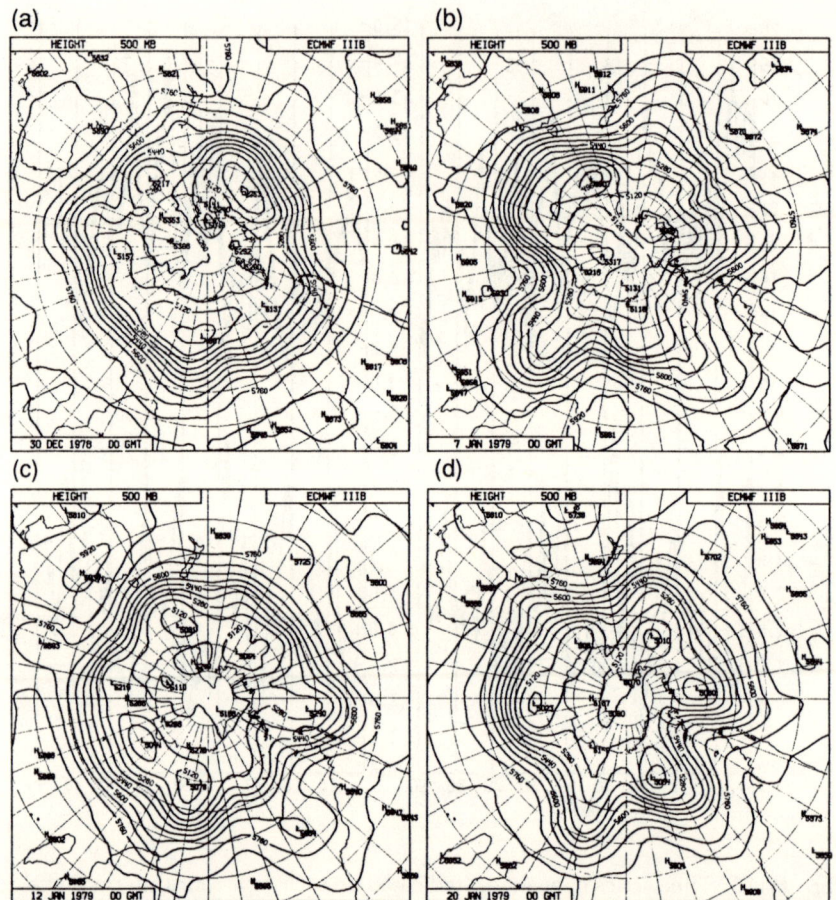

Fig. 17.14: The 500-mb contour charts of 30 December 1978 and 7, 14, and 20 January 1979. The first (Day 30) and third (Day 43) days correspond to K_M minima and K_Z maxima, while the second (Day 38) and fourth (Day 51) correspond to K_M maxima and K_Z minima.

time series of energies (Fig. 17.12) that A_M and K_M or A_Z and K_Z vacillate in phase and in time, but zonal energies and energies of intermediate wave vacillate out of phase.

According to the Lorenz energy cycle, an energy conversion may act as a source or sink to a connected energy reservoir within this energy cycle. Thus, we may expect that time variations of those energies and energy conversions connected with each other through the energy cycle should be correlated in a systematic way. Time series of various energy conversions averaged over the latitudinal zone between 30° S and 65° S are displayed in Fig. 17.16. A systematic relation may be subjectively revealed from the comparison of the time series of this figure and time series of related energies in Fig. 17.13. The systematic relation can be found by means of the

ENERGETICS OF THE SOUTHERN HEMISPHERE 319

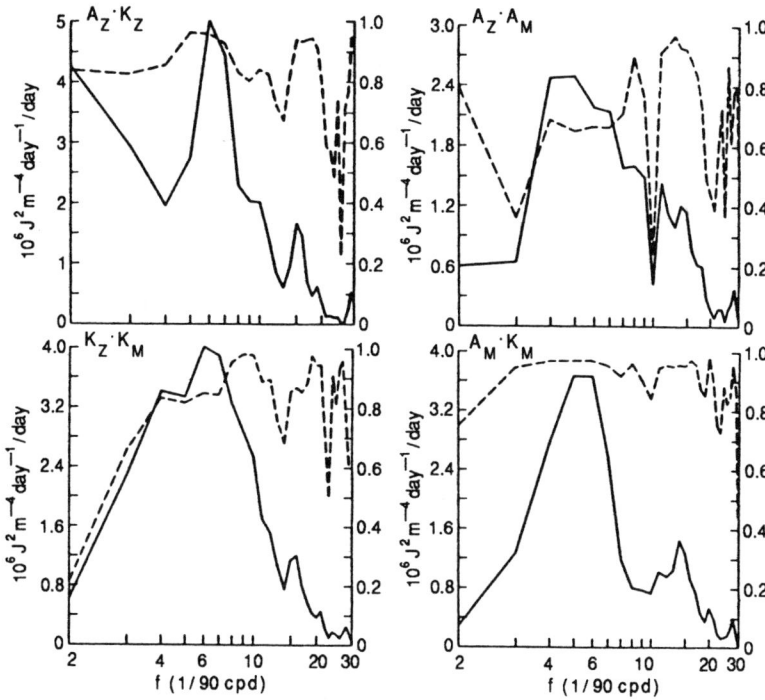

Fig. 17.15: Cross spectra (solid lines) and coherencies (dashed lines) between two energies shown in the top left corner of each figure.

cross spectra between the time series of energies and energy conversions, shown in Fig. 17.17. A spectral peak around a period of about two weeks appears in most of the cross spectra. Coherencies associated with the peak cross spectra are generally high. The energy conversions obviously vacillate coherently with related energies for a period of two weeks.

Next, to assess the way in which time variations of energy conversions and their related energies are correlated, the phases between energies and energy conversions around the two-week period are displayed in Table 17.5. One can easily find that phases between intermediate-wave energies and energy conversions are negligible and that those between zonal energies and energy conversions are about a half of the vacillation period. Based upon these phases, we are certain that intermediate-wave energies vary more or less in phase with the various energy conversions connected to them within the Lorenz energy cycle, but zonal energies vary almost oppositely in phase.

In the winter Northern Hemisphere, both the long-wave and short-wave regimes are important to the vacillation of atmospheric energetics with a period of two weeks. Interestingly, a clear indication of a two-week vacillation in the atmospheric energetics of the summer Southern Hemisphere circulation also stands out from the statistical time series of both the zonal and intermediate-wave energy variables. In addition to the significance of the intermediate-wave regime emerging from the time series analysis of the Southern Hemisphere energy variables in summer, we also found that standard deviations of the energies and energy conversions of the total and intermediate-wave regimes are comparable in this hemisphere. Based upon these two findings, we conclude that the intermediate-wave regime dominates the energy vacillation of the summer Southern Hemisphere circulation.

Table 17.4: Coherency and phase difference of cross spectra between energies.

Quantity	Frequency (cpd)	Coherency	Phase (days)
$A_Z \cdot K_Z$	5/90 (18 days)	0.86	1.61
	6/90 (15 days)	0.90	1.90
$A_Z \cdot A_M$	5/90 (18 days)	0.65	8.96
	6/90 (15 days)	0.66	-6.95
$A_M \cdot K_M$	5/90 (18 days)	0.97	-0.08
	6/90 (15 days)	0.97	-0.05
$K_Z \cdot K_M$	5/90 (18 days)	0.81	-8.48
	6/90 (15 days)	0.85	-6.50
$A_M \cdot A_E$	5/90 (18 days)	0.63	-0.83
	6/90 (15 days)	0.80	-0.70
$K_M \cdot K_E$	5/90 (18 days)	0.86	-0.23
	6/90 (15 days)	0.92	0.11

17.4 Jet Streams

It was shown in Section 13.2 that the winter subtropical jet streams of the Northern Hemisphere are located over the east coasts of Asia and North America and over North Africa, where there are pronounced baroclinic zones. As stressed earlier, the long-wave regime is much weaker in the Southern Hemisphere, and the upper air circulation of this hemisphere is more *zonal*. With the upper-air data collected during the International Geophysical Year (IGY), van Loon (1972) was able to describe the jet streams of the Southern Hemisphere in the two extreme seasons. During summer, a single jet with its axis at about 45° S extends from the South Atlantic to the south Indian Ocean. This single summer jet changes to a double jet structure in winter: a higher latitude jet at 50° S in the Indian Ocean and a subtropical jet at 25°-30° S over Australia and the western Pacific. In Section 13.2, we learned that theories were proposed and efforts were made in the past several decades to explore the maintenance of the winter subtropical jet streams in the Northern Hemisphere. Since

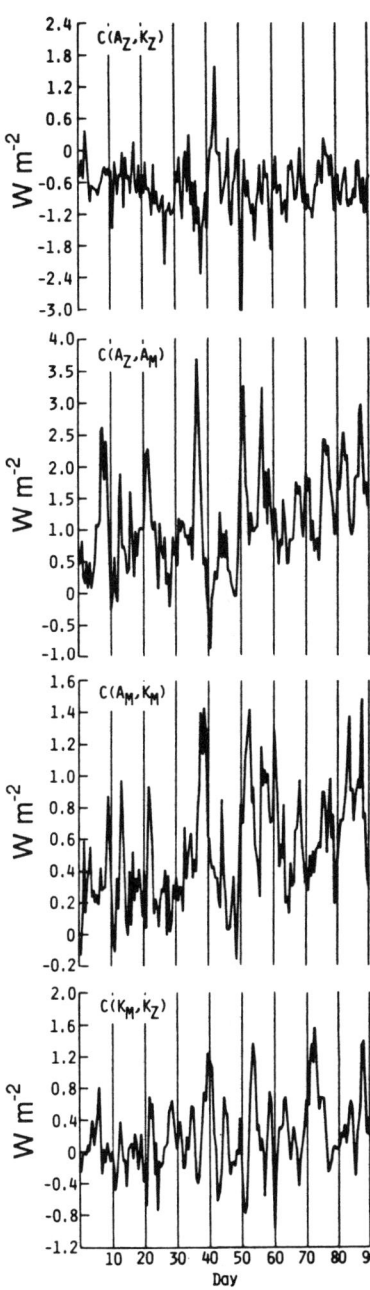

Fig. 17.16: Time series of energy conversions (given in the top left corner of each time series) in the $30°S$–$65°S$ latitude zone during the 1978/79 winter. Day 1 is December 1, 1978 while Day 90 is Feburary 28, 1979.

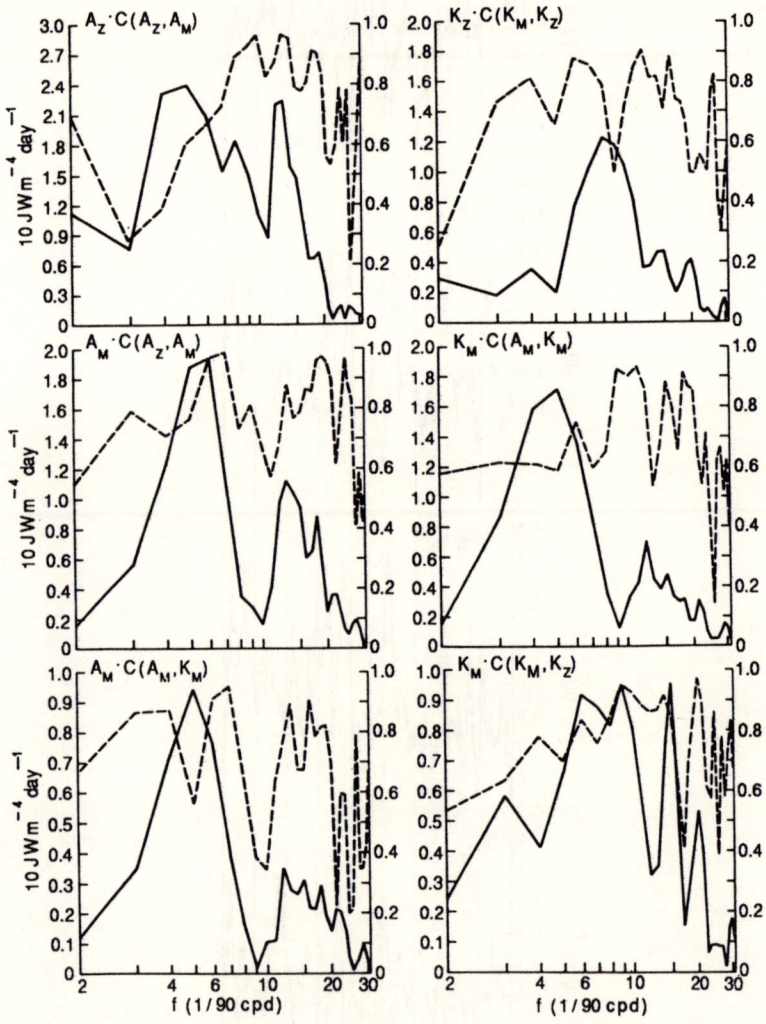

Fig. 17.17: Cross spectra (solid lines) and coherencies (dashed lines) between energies and energy conversions denoted in the top left corner of each figure.

ENERGETICS OF THE SOUTHERN HEMISPHERE

Table 17.5: Coherency and phase difference of cross spectra between energy conversions.

Quantity	Frequency (cpd)	Coherency	Phase (days)
$A_Z \cdot C(A_Z, K_Z)$	5/90 (18 days)	0.18	-9.39
	6/90 (15 days)	0.66	7.49
$A_Z \cdot C(A_Z, A_M)$	5/90 (18 days)	0.65	7.60
	6/90 (15 days)	0.66	7.17
$A_M \cdot C(A_Z, A_M)$	5/90 (18 days)	0.97	-1.83
	6/90 (15 days)	0.96	-1.04
$A_M \cdot C(A_M, K_M)$	6/90 (18 days)	0.96	-0.03
	6/90 (15 days)	0.93	0.35
$K_M \cdot C(A_M, K_M)$	5/90 (18 days)	0.95	-0.61
	6/90 (15 days)	0.92	0.52
$K_M \cdot C(K_M, K_Z)$	5/90 (18 days)	0.37	0.18
	6/90 (15 days)	0.52	0.67
$K_Z \cdot C(K_M, K_Z)$	5/90 (18 days)	0.19	7.69
	6/90 (15 days)	0.62	7.05
$K_Z \cdot C(A_Z, K_Z)$	5/90 (18 days)	0.13	9.28
	6/90 (15 days)	0.81	6.95

the upper-air circulation of the Southern Hemisphere differs from that of the Northern Hemisphere, one may question whether the jet streams of the former hemisphere are maintained by the same mechanisms maintaining those of the latter. We shall devote this section to discussing this matter.

17.4.1 Summer Australian Jet

Based upon the rainfall and appearance of the low-level monsoon westerlies, the onset of the 1979 Australian monsoon occurred between 21 December and 31 December 1978. The geographical distribution of the Australian jet stream can be inferred from the region of large k_R (200 mb) shown in Fig. 17.18a. To illustrate the time evolution of this jet stream associated with the Australian monsoon, the u (200 mb) time series of two key locations over the first two months of the 1978/1979 southern winter are shown in Fig. 17.18. The two key locations are selected to be as close as possible to both the jet cores before and after the monsoon onset and to observation stations. A comparison between these two u (200 mb) time series indicates that the poleward shift of the Australian jet stream took place within 10 days of the monsoon onset.

The latent heat released by cumulus convection constitutes the major part of the tropical diabatic heating and maintains divergent circulation. The interaction between condensation heating and divergent circulation can

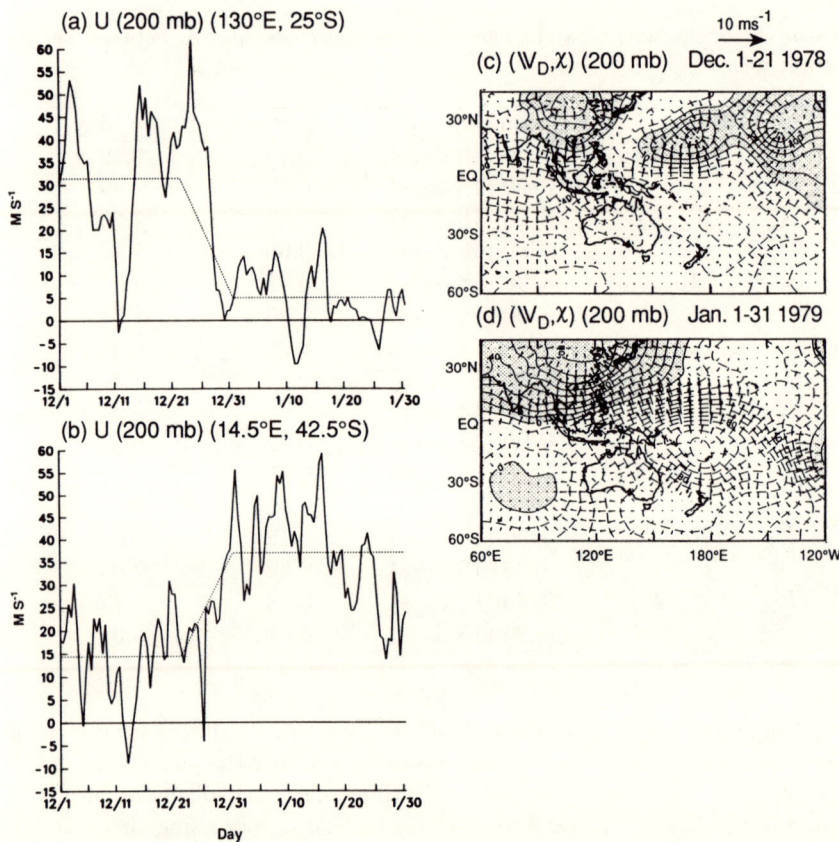

Fig. 17.18: Time series of the 200 mb zonal wind at (a) (130° E, 25° S) and (b) (147.5° E, 42.5° S) and divergent wind vectors (V_D) superimposed on the velocity potential (χ) at 200 mb averaged over (c) 1-21 December 1978 and (d) 1-31 January 1979. Positive values of χ are shaded and the contour interval of χ is 10^6 m^2 s^{-1}. After Chen et al. (1989).

be seen through the correlation between the enhancement of the monsoon rainfall and the intensification of the large-scale monsoon divergent circulation after the monsoon onset. The 200 mb divergent circulation over the Australian monsoon region (Fig. 17.18c) was not developed and organized before the monsoon onset, but became well defined and intensified with a divergent center near the Date Line after the monsoon onset. The rapid poleward shift of the Australian jet stream after the monsoon onset was first observed by Rodak and Grant (1957). The cause of this rapid poleward shift remained unclear for quite some time. Notice that the poleward shift of the Australian jet and the intensification of the monsoon divergent circulation took place simultaneously. A relation between them should exist. The Australian jet stream can be well depicted by the rotational flow.

The energy analysis of divergent and rotational flow may shed some light on this relation. To serve this purpose, let us rewrite (13.10) and (13.11) in symbolic forms and neglect insignificant terms:

$$\frac{\partial k_3}{\partial t} \sim -C(k_3, k_2) + G(k_3), \tag{17.1}$$

$$\frac{\partial k_2}{\partial t} \sim C(k_3, k_2) - FR(k) + G(k_2). \tag{17.2}$$

Shown in Fig. 17.19 is the energy budget analyis at 200 mb using (17.1) and (17.2) over the Australian monsoon region. An elongated region of large k_2, crossing the Australian continent before the monsoon onset (Fig. 17.19a), moves to the south of this continent after the monsoon onset (Fig. 17.19b). As inferred from the time evolution of the two u (200 mb) time series shown in Fig. 17.18a, this southward migration of the large-k_2 region reflects the poleward shift of the Australian jet stream. A question raised here is how the region of large k_2 associated with the Australian jet stream is maintained before and after the monsoon onset. Let us start with the k_3 budget. Recall that the planetary-scale divergent circulation over the Australian monsoon region is not well organized before the monsoon onset; $G(k_3)$ is significant only on the downstream side of the jet (Fig. 17.19b). In contrast, after the monsoon onset, $G(k_3)$ is not only enhanced over this region, but also over the south of Australia where the jet is located. Generally, $C(k_3, k_2)$ possesses a pattern similar to $G(k_3)$. The existence of significant $G(k_3)$ south of the Australian continent after the monsoon onset indicates that the interaction between divergent and rotational flows supports the poleward shifted jet in this region.

It was illustrated in Section 13.1.2 that the subtropical jet streams are maintained in such a way that their kinetic energy is generated on the upstream side and destroyed on the downstream side by an ageostrophic effect. The source and sink of a jet's kinetic energy are counterbalanced by the divergence and convergence of the kinetic energy flux on the upstream and downstream sides of the jets, respectively. As revealed from (17.2), $C(k_3, k_2)$ provides some input to maintain the Australian jet. However, the comparison between k_2 (Fig. 17.18a) and $G(k_2)$ (Fig. 17.18c) and between k_2 and $FR(k)$ (Fig. 17.18d) showed that the major energy processes maintaining the Australian jet stream are the same as those maintaining the subtropical jet streams of the Northern Hemisphere.

17.4.2 Winter Jets

It was shown in Section 13.1.1 by using the kinetic energy budget analysis of the time-mean flow that interactions betweeen transient eddies and the time-mean flow in the winter Northern Hemisphere are secondary in maintaining the latter flow component. Conversely, we have learned from

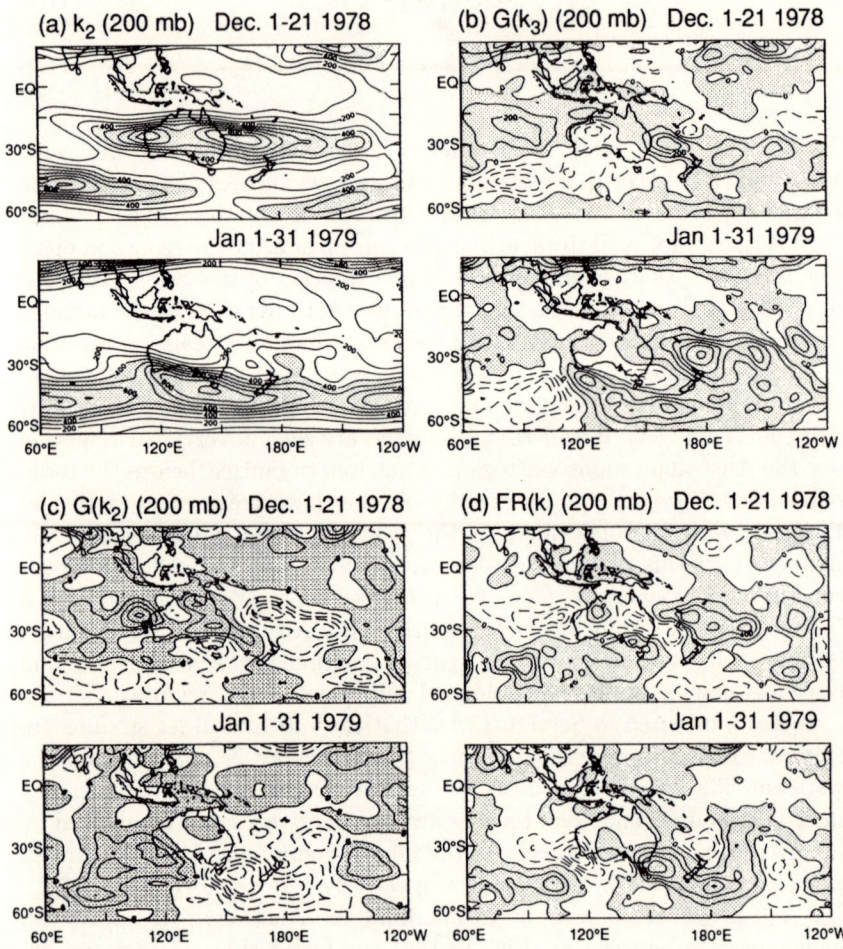

Fig. 17.19: (a) k_R (200 mb), (b) $G(k_D)$ (200 mb), (c) $G(k_R)$ (200 mb) and (d) FR (200 mb) averaged over 1-21 December 1978 and 1-31 January 1979. The contour interval of (a) is 10^2 m^2 s^{-2}, of (b) 10^{-3} m^2 s^{-3} and of (c) and (d) 2×10^{-3} m^2 s^{-3}. After Chen et al. (1989).

the spectral energy analysis of the Southern Hemisphere general circulation (Section 17.2) that the intermediate wave and short-wave regimes, which form the major part of transient eddies, play a much more active role in the atmospheric energetics of this hemisphere. It is conceivable that transient eddies may contribute to the maintenance of the Southern Hemisphere time-mean flow, in which the jet streams are the most conspicuous elements.

The maintenance of the winter subtropical jet streams in the Northern Hemisphere by the local ageostrophic circulation mechanisms were illustrated with the momentum equation (13.12). However, the effect of transient eddies is implicitly contained in the establishment of the cross-jet circulation in the downstream sides of the subtropical jet streams, rather than being expressed explicitly in a mathematical form. Since transient eddies are so vital to the Southern Hemisphere circulation, it is necessary to have this effect shown explicitly in a mathematical form. For this purpose, we may write the time-mean zonal momentum equation in the following form:

$$\frac{\bar{u}}{a\cos\varphi}\frac{\partial \bar{u}}{\partial \lambda} + \frac{\bar{v}}{a\cos\varphi}\frac{\partial}{\partial \varphi}(\bar{v}\cos\varphi) + \bar{\omega}\frac{\partial \bar{u}}{\partial p} \qquad (17.3)$$
$$= f\bar{v}_a - \left(\frac{1}{a\cos\varphi}\frac{\partial}{\partial \lambda}\overline{(u'u')} + \frac{1}{a\cos^2\varphi}\frac{\partial}{\partial \varphi}\overline{(v'u'\cos^2\varphi)}\right.$$
$$\left. + \frac{\partial}{\partial p}\overline{(\omega'u')}\right) - \overline{F}_\lambda.$$

The left-hand side of (17.3) represents the advection of zonal momentum by the time-mean circulation. Terms on the right-hand side of this equation are regarded as forcings of the time-mean flow. These forcings include the Coriolis force associated with the time-mean meridional ageostrophic flow and the convergence of momentum flux due to transient eddies. The latter forcing does not exist in (13.12). The frictional effect $\overline{F}_\lambda$ is negligible.

In the southern winter (December 1978-Feburary 1979), the salient feature of the upper-air circulation in the Southern Hemisphere revealed from the $\bar{u}$ (300 mb) field (Fig. 17.20a) is a double jet structure. A subtropical jet stretches from the Indian Ocean through Australia to the eastern Pacific with its axis located between 25° S and 30° S, and a higher latitude jet extends from the western Atlantic to the southern Indian Ocean centered at about 50° S. The core speeds of these two jets can reach 45 m s^{-1} and 35 m s^{-1}, respectively. It was revealed from the general circulation statistics of the Northern Hemisphere that the locations of the most active transient eddies, namely the storm tracks, are related to the subtropical jets in a special way. The variance of the 2.5-6 day bandpass filtered meridional wind (Fig. 17.20b) may be used as an indicator of the storm track. Two storm tracks can be easily identified by significant values of this variance; the storm track of the eastern Pacific center around (40° S, 105° W) lies in the downstream and poleward side of the subtropical jet, and the storm

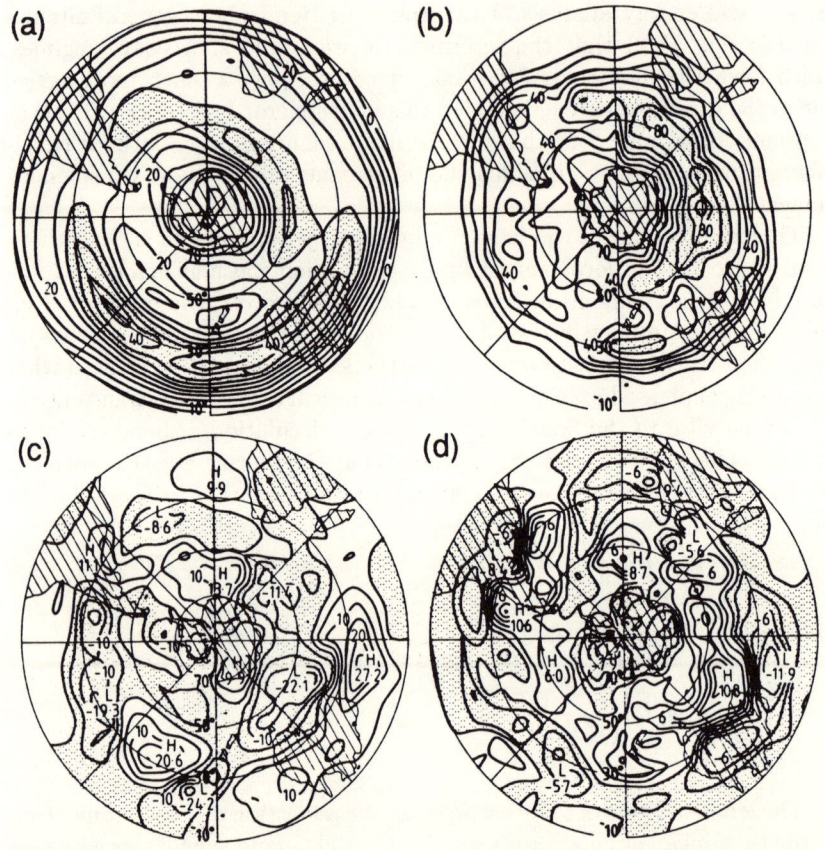

Fig. 17.20: (a) $\bar{u}$ (300 mb) average over the 1979 winter, (b) variance of 2.5-6 day bandpass filtered v(300mb), (c) Coriolis acceleration induced by $\bar{v}_a$ (300 mb) and (d) horizontal convergence of transient eddy flux of westerly momentum at 300 mb. Contour intervals of (a), (b), (c) and (d) are 5 m s^{-1}, 10m^2 s^{-2}, 5×10^{-5} m s^{-2} and 2×10^{-5} m^{-5} m s^{-2}, respectively. Shaded areas in (a) indicate $\bar{u}$ (300 mb) > 30 m s^{-1}, in (b) variance > 60 m^2 s^{-2}, in (c) and (d) are negative values. After Physick (1981).

track of the southern Indian Ocean coincides with the higher latitude jet. It is our intention in this section to determine the forcing on the time-mean flow exerted by these transient eddies.

According to (17.3), the primary forcing of the time-mean flow is due to the Coriolis force and the transient eddy effects. As shown in Figs. 17.20c and 17.20d, these two forcings exhibit noticeable longitudinal variations. The centers of the two forcings appear at the same locations but have opposite signs. Obviously, they counterbalance each other. Nevertheless, the Coriolis forcing dominates the transient eddy forcings north of 50° S. Based upon the momentum budget analysis presented here, one can see that the subtropical jet is accelerated by the poleward ageostrophic flow in the entrance region over the eastern Indian Ocean and the region between two jet cores in the Pacific Ocean. In the downstream region of this jet along the eastern Pacific storm track, the transient eddy forcing is surprisingly small. The subtropical jet in this region is primarily decelerated by the Coriolis forcing. A deceleration of the zonal flow also occurs south of Australia.

Along the higher latitude storm track in the southern Indian Ocean, both the Coriolis and transient eddy effects are comparable in their magnitudes. Obviously, the higher latitude jet is speeded up by the latter forcing and slowed down by the former forcing. An exception occurs in the region south of Africa where these two forcings work in concert to accelerate the zonal flow in the upstream region of the higher latitude jet. In view of the momentum budget analysis of time-mean flows in the winter Southern Hemisphere, mechanisms maintaining the subtropical jet and the higher latitude jet differ. The Coriolis forcing dominates the maintenance of the subtropical jet, but both the Coriolis and transient eddy forcings are equally important to the higher latitude jet. It is made clear in this section that the transient eddy forcing is not negligible, particularly in maintaining the higher latitude jet.

PROBLEMS

The purpose of this appendix is to formulate an exercise that is far from trivial. The goal has been to give a number of subproblems that are part of a general problem, and that make use of much of the material in the first 10 chapters of the textbook.

To this end it is necessary to simplify the chosen example, but, on the other hand, to stay so general that most of the processes entering atmospheric energetics are represented albeit in a rudimentary way. Simplification is obtained in two ways: we select to use a two-level, quasi-geostrophic model and we use a low-order representation, which permits nonlinear processes.

The quasi-geostrophic, two-level model will be very similar to the one formulated by Charney (1959). The heating is formulated in a Newtonian form:

$$H = -c_p \gamma (T - T_E), \qquad (.1)$$

where H is the heating per unit mass, c_p the specific heat for constant pressure, γ a constant of dimension t^{-1}, T the temperature, and T_E a reference temperature. T_E can be computed in various ways as shown by Wiin-Nielsen (1972) who shows how γ can be computed as well.

The stress is given partly as a surface stress

$$\vec{\tau}_0 = -c_d \rho V \vec{v} \qquad (.2)$$

and partly as a shearing stress applied at the middle level:

$$\vec{\tau} = -\mu \frac{\partial \vec{v}}{\partial z}. \qquad (.3)$$

With these assumptions we may write the equations for the two-level, quasi-nondivergent model in the following form:

$$\frac{\partial \zeta_*}{\partial t} + \vec{v}_* \cdot (\zeta_* + f) + \vec{v}_T \cdot \nabla \zeta_T = -\varepsilon \zeta_4, \qquad (.4)$$

$$\frac{\partial \zeta_T}{\partial t} + \vec{v}_* \cdot \nabla \zeta_T + \vec{v}_T \cdot \nabla (\zeta_* + f) = \frac{f_0}{P} \omega - \varepsilon_T \zeta_T + \varepsilon \zeta_4.$$

The numerical values are $\varepsilon = 3 \times 10^{-6}$ s^{-1} and $\varepsilon_T = 1.2 \times 10^{-6}$ s^{-1}. We recall that in this model ζ is the vorticity, $\vec{v}$ the horizontal nondivergent wind, ω the vertical velocity and f the Coriolis parameter. the subscripts indicate the pressure levels with 1 for 25 kPa, 2 for 50 kPa, 3 for 75 kPa, and 4 for 100 kPa. The subscripts * and T are defined by the expressions

$$(\ )_* = \frac{1}{2}[(\ )_1 + (\ )_3], \qquad (\ )_T = \frac{1}{2}[(\ )_1 - (\ )_3]. \tag{.5}$$

We shall furthermore use the value $P = 50$ kPa. In addition to (.5) we shall need the thermodynamic equation for this model. It is

$$\frac{\partial \psi_T}{\partial t} + \vec{v}_* \cdot \nabla \psi_T - \frac{f_0}{P}\frac{1}{\lambda^2}\omega = \frac{R}{c_p}\frac{1}{2f_0}H. \tag{.6}$$

One may elimate ω between the second equation in (.4) and (.6). Introducing at the same time (.1) for H, we may write the equations for the model in the form

$$\frac{\partial \zeta_*}{\partial t} + \vec{v}_* \cdot \nabla(\zeta + f) + \vec{v}_T \cdot \nabla \zeta_T$$
$$= -\varepsilon \zeta_4 \frac{\partial}{\partial t}\left(\zeta_T - \lambda^2 \psi_T\right) + \vec{v}_* \cdot \nabla\left(\zeta_T - \lambda^2 \psi_T\right) + \vec{v}_T \cdot \nabla(\zeta_* + f)$$
$$= -\varepsilon_T \zeta_T + \varepsilon \zeta_4 + \lambda^2 \gamma\left(\psi_T - \psi_E\right). \tag{.7}$$

We assume finally that

$$\zeta_4 = \frac{1}{2}\zeta_3 = \frac{1}{2}\left(\zeta_* - \zeta_T\right). \tag{.8}$$

Problem 1

The system (.7) shall be written in nondimensional form by setting

$$\psi = a^2 \Omega z, \qquad t = \Omega^{-1}\tau,$$

where the first relation holds for both subscripts. Show that the resulting equations are

$$\frac{\partial \nabla^2 z_*}{\partial \tau} = J\left(\nabla^2 z_*, z_*\right) + J\left(\nabla^2 z_T, z_T\right) - 2\frac{\partial z_*}{\partial \lambda} - E\left(\nabla^2 z_* - \nabla^2 z_T\right)$$
$$\frac{\partial}{\partial \tau}\left[\nabla^2 z_T - \Lambda z_T\right] = J\left(\nabla^2 z_T - \Lambda z_T, z_*\right) + J\left(\nabla^2 z_*, z_T\right) - 2\frac{\partial z_T}{\partial \lambda}$$
$$-E_T \nabla^2 z_T + E_0\left(\nabla^2 z_* - \nabla^2 z_T\right) + \Lambda\frac{\gamma}{\Omega}\left(z_T - z_E\right) \tag{.9}$$

in which
$$\Lambda = \lambda^2 a^2, \qquad E_0 = \frac{\varepsilon}{2\Omega}, \qquad E_T = \frac{\varepsilon_T}{\Omega}. \qquad (.10)$$

a is the radius of the Earth and ω the angular velocity of the Earth.

The system (.9)-(.10) will be used with a low-order representation of the fields. To be exact we shall permit only three components: a zonal component $\alpha = (0, n)$ and two-wave components $\beta = (\ell, n)$ and $\gamma = (\ell, n_2)$, and we shall select n as an even number. Recall, that in addition to $\beta = (\ell, n_1)$ and $\gamma = (\ell, n_2)$, we must also include $\bar{\beta} = (-\ell, n_1)$ and $\bar{\gamma} = (-\ell, n_2)$, and that the general formula for the interaction coefficient is (10.17).

Problem 2

Consider all nonzero interaction coefficients in the low-order system and show that they can be expressed in terms of one interaction coefficient, namely,

$$K(\gamma, \beta, \alpha) = \frac{\ell}{2} \int_{-1}^{+1} P_\gamma P_\beta \frac{dP_\alpha}{d\mu} \, d\mu.$$

Show, in particular, that

$$K(\bar{\gamma}, \bar{\beta}, \alpha) = -K(\gamma, \beta, \alpha),$$
$$K(\alpha, \beta, \bar{\gamma}) = -K(\alpha, \bar{\beta}, \gamma).$$

(Hint: Since n is given as an even number, it follows from the selection rules that n_1 and n_2 cannot both be even numbers. It is also impossible that both n_1 and n_2 are odd numbers. Thus one of them is even, and the other is odd.)

Let us now denote the amplitudes of the two zonal components by z^* and z^T. We shall also introduce the notations

$$z_\beta^* = \frac{1}{2}(x_1^* - iy_1^*), \qquad z_\beta^T = \frac{1}{2}(x_1^T - iy_1^T), \qquad (.11)$$

$$z_\gamma^* = \frac{1}{2}(x_2^* - iy_2^*), \qquad z_\gamma^T = \frac{1}{2}(x_2^T - iy_2^T). \qquad (.12)$$

It is then clear that the total system can be described by 10 real variables or by two real variables and four complex variables. The derivation of these equations can be made in a way analogous to the development from Eq. (10.10) to Eq. (10.19), by starting in this case from (.9) and (.10).

Problem 3

Show that the general spectral equations for the two-level model governed by Eqs. (.9) and (.10) are as follows:

$$\frac{dz_\gamma^*}{d\tau} = i\sum_\beta\sum_\alpha \frac{c_\beta - c_\alpha}{c_\gamma} z_\beta^* z_\alpha^* K + i\sum_\beta\sum_\alpha \frac{c_\beta - c_\alpha}{c_\gamma} z_\beta^T z_\alpha^T K$$
$$+ i\frac{2m_\gamma}{c_\gamma} z_\gamma^* - E\left(z_\gamma^* - z_\gamma^T\right), \tag{.13}$$

$$\frac{dz_\gamma^T}{d\tau} = i\sum_\beta\sum_\alpha \frac{d_\beta - c_\alpha}{d_\gamma} z_\beta^T z_\alpha^* K + i\sum_\beta\sum_\alpha \frac{c_\beta - d_\alpha}{d_\gamma} z_\beta^* z_\alpha^T K$$
$$+ i\frac{2m_\gamma}{d_\gamma} z_\gamma^T - \frac{\Lambda}{d_\gamma}\frac{\gamma}{\Omega}\left(z_\gamma^T - z_\gamma^E\right)$$
$$- E_T \frac{c_\gamma}{d_\gamma} z_\gamma^T + E\frac{c_\gamma}{d_\gamma}\left(z_\gamma^* - z_\gamma^T\right), \tag{.14}$$

in which

$$K = K(\gamma, \beta, \alpha) = \frac{1}{2}\int_{-1}^{+1} P_\gamma \left(m_\beta P_\beta \frac{dP_\alpha}{d\mu} - m_\alpha P_\alpha \frac{dP_\beta}{d\mu}\right) d\mu,$$
$$m_\gamma = m_\beta + m_\alpha,$$
$$d_\gamma = c_\gamma + \Lambda.$$

The next point in the problem is to reduce the general equations (.13), (.14) to the three-component, low-order system. The reduction is rather tedious, but not difficult. It requires essentially good bookkeeping. We shall therefore give the resulting 10 equations, but the reader is encouraged to derive at least one of the equations to know the general procedure. For convenience we have also introduced a number of constants listed after the equations. The equations are:

$$\frac{dz^*}{d\tau} = \frac{1}{2}g_1\left(x_1^* y_2^* - x_2^* y_1^* + x_1^T y_2^T - x_2^T y_1^T\right) - E_0\left(z^* - z^T\right),$$
$$\frac{dx_1^*}{d\tau} = g_2\left(z^* y_2^* + z^T y_2^T\right) + F_1 y_1^* - E_0\left(x_1^* - x_1^T\right),$$
$$\frac{dy_1^*}{d\tau} = -g_2\left(z^* x_2^* + z^T x_2^T\right) - F_1 x_1^* - E_0\left(y_1^* - y_1^T\right),$$
$$\frac{dx_2^*}{d\tau} = -g_3\left(z^* y_1^* + z^T y_1^T\right) + F_2 y_2^* - E_0\left(x_2^* - x_2^T\right),$$
$$\frac{dy_2^*}{d\tau} = g_3\left(z^* x_1^* + z^T x_1^T\right) - F_2 x_2^* - E_0\left(y_2^* - y_2^T\right),$$
$$\frac{dz^T}{d\tau} = \frac{1}{2}d_1\left(x_1^T y_2^* - y_1^T x_2^*\right) + \frac{1}{2}e_1\left(y_2^T x_1^* - x_2^T y_1^*\right) + Ez^* - \Gamma z^T + Nz^E,$$

$$\frac{dx_1^T}{d\tau} = d_2 z^* y_2^T + e_2 z^T y_2^* + +G_1 y_1^T + E_1 x_1^* - \Gamma_1 x_1^T + N_1 x_1^E,$$

$$\frac{dy_1^T}{d\tau} = -d_2 z^* x_2^T - e_2 z^T x_2^* + G_1 x_1^T + E_1 y_1^* - \Gamma_1 y_1^T + N_1 y_1^E,$$

$$\frac{dx_2^T}{d\tau} = -e_3 z^* y_1^T - d_3 z^T y_1^* + G_2 y_2^T + E_2 x_2^* - \Gamma_2 x_2^T + N_2 x_2^E,$$

$$\frac{dy_2^T}{d\tau} = e_3 z^* x_1^T + d_3 z^T x_1^* + G_2 x_2^T + E_2 y_2^* - \Gamma_2 y_2^T + N_2 y_2^E. \tag{.15}$$

The equations in (.15) have been ordered in such a way that the nonlinear interaction terms appear first on the right hand sides followed by the Coriolis terms. Thereafter come the terms related to forcing and dissipation. Note that all three components interact with each other.

The constants are defined as follows:

$$g_1 = \frac{c_\beta - c_\gamma}{c_\alpha} K, \qquad g_2 = \frac{c_\gamma - c_\alpha}{c_\beta} K, \qquad g_3 = \frac{c_\alpha - c_\beta}{c_\gamma},$$

$$e_1 = \frac{d_\beta - d_\gamma}{d_\alpha} K, \qquad e_2 = \frac{c_\gamma - d_\alpha}{d_\beta} K, \qquad e_3 = \frac{c_\alpha - d_\beta}{d_\gamma} K,$$

$$d_1 = \frac{d_\beta - c_\gamma}{d_\alpha} K, \qquad d_2 = \frac{d_\gamma - c_\alpha}{d_\beta} K, \qquad d_3 = \frac{d_\alpha - c_\beta}{d_\gamma} K,$$

$$F_1 = \frac{2\ell}{c_\beta}, \qquad F_2 = \frac{2\ell}{c_\gamma}, \qquad G_1 = \frac{2\ell}{d_\beta}, \qquad G_2 = \frac{2\ell}{d_\gamma},$$

$$E = E_0 \frac{c_\alpha}{d_\alpha}, \qquad E_1 = E_0 \frac{c_\beta}{d_\beta}, \qquad E_2 = E_0 \frac{c_\gamma}{d_\gamma},$$

$$\Gamma = \frac{\Lambda}{d_\alpha} \frac{\gamma}{\Omega} + \frac{c_\alpha}{d_\alpha}(E_0 + E), \qquad \Gamma_1 = \frac{\Lambda}{d_\beta} \frac{\gamma}{\Omega} + \frac{c_\beta}{d_\beta}(E_0 + E_T),$$

$$\Gamma_2 = \frac{\Lambda}{d_\gamma} \frac{\gamma}{\Omega} + \frac{c_\gamma}{d_\gamma}(E_0 + E_T),$$

$$N = \frac{\Lambda}{d_\alpha} \frac{\gamma}{\Omega}, \qquad N_1 = \frac{\Lambda}{d_\beta} \frac{\gamma}{\Omega}, \qquad N_2 = \frac{\Lambda}{d_\gamma} \frac{\gamma}{\Omega}. \tag{.16}$$

For some of the energy conversions we need the vertical velocity. We scale this parameter by the relation

$$\omega = P\Omega W. \tag{.17}$$

In the two-level, quasi-geostrophic model we can obtain ω as the solution of a diagnostic equation [see Eq. (3.75)]. The vertical velocity should naturally be expressed as a sum of the components included in the system. We may write

$$W = \quad r P_m(\mu) + r_1 P_{n_1}^\ell(\mu) \cos(\ell\lambda) + s_1 P_{n_1}^\ell(\mu) \sin(\ell\lambda)$$
$$+ \quad r_2 P_{n_2}^\ell(\mu) \cos(\ell\lambda) + s_2 P_{n_2}^\ell(\mu) \sin(\ell\lambda). \tag{.18}$$

The five coefficients in (.18) are determined from the ω-equation. This procedure is also cumbersome, although rather straightforward. We give the values of the coefficients below. The notation $c_\alpha = c$, $c_\beta = c_1$, and $c_\alpha = c_2$ has been introduced.

$$
\begin{aligned}
r &= \left(1+\frac{c}{\Lambda}\right)\left[\begin{array}{l}(c_1 - c_2 + c)^{-1}\frac{K}{2}\left(x_1^* y_2^T - y_1^* x_2^T\right) + (c_1 - c_2 - c)\frac{K}{2}\left(x_1^T y_2^* y_2^*\right) \\ -y_1^T x_2^*) + c\left\{\frac{\gamma}{\Omega}\left(z_T - z_E\right) - E_T z^T + E_0\left(z^* - z^T\right)\right\}\end{array}\right] \\
r_1 &= \left(1+\frac{c_1}{\Lambda}\right)^{-1}\left[\begin{array}{l}(c_2 - c + c_1) Kz^T y_2^* + (c_2 - c - c_1) Kz^* y_2^T + 2\ell y_1^T \\ + c_1\left\{\frac{\gamma}{\Omega}\left(x_1^T - x_1^E\right) - E_T x_1^T + E_0\left(x_1^* - x_1^T\right)\right\}\end{array}\right] \\
s_1 &= \left(1+\frac{c_1}{\Lambda}\right)^{-1}\left[\begin{array}{l}-(c_2 - c + c_1) Kz^T x_2^* - (c_2 - c - c_1) Kz^* x_2^T + 2\ell x_1^T \\ + c_1\left\{\frac{\gamma}{\Omega}\left(y_1^T - y_1^E\right) - E_T y_1^T + E_0\left(y_1^* - y_1^T\right)\right\}\end{array}\right] \\
r_2 &= \left(1+\frac{c_2}{\Lambda}\right)^{-1}\left[\begin{array}{l}-(c - c_1 + c_2) Kz^* y_1^T - (c - c_1 - c_2) Kz^T y_1^T + 2\ell y_2^T \\ + c_2\left\{\frac{\gamma}{\Omega}\left(x_2^T - x_2^E\right) - E_T x_2^T + E_0\left(x_2^* - x_2^T\right)\right\}\end{array}\right] \\
s_2 &= \left(1+\frac{c_2}{\Lambda}\right) T-1\left[\begin{array}{l}(c - c_1 + c_2) Kz^* x_1^T + (c - c_1 - c_2) Kz^T x_1^* + 2\ell x_2^T \\ + c_2\left\{\frac{\gamma}{\Omega}\left(y_2^T - y_2^E\right) - E_T y_2^T + E_0\left(y_2^* - y_2^T\right)\right\}\end{array}\right]
\end{aligned}
\tag{.19}
$$

The first lines in each expression give the contribution from the nonlinear interactions and the Coriolis terms, while the second lines contain the contribution from the external forcing and the dissipative forces.

We are now ready to start on the energetics of the low-order model. To start the calculations it is necessary to adapt the general formulas to the variables that are used in the model. For example, the specific volume α appears in some of the integrals. It is adapted to the model as follows:

$$
\alpha_2 = -\left(\frac{\partial \phi}{\partial p}\right)_2 = -f_0\left(\frac{\partial \psi}{\partial p}\right)_2 = -f_0 \frac{\psi_3 - \psi_T}{P} = \frac{2f_0}{P}\psi_T. \tag{.20}
$$

If the variable is T_2 we get

$$
T_2 = \frac{P}{R}\alpha_2 = \frac{2f_0}{R}\psi_T. \tag{.21}
$$

It should also be recalled that the variables are nondimensionalized by the relations appearing in Problem 1 and by (.17).

Problem 4

Show that the available potential energy as expressed by

$$A = \frac{1}{gs} \int_0^{p_0} \int_S \frac{1}{2\bar{\sigma}} \alpha'^2 \, dS \, dp$$

in the low-order model becomes

$$A = \frac{Pa^2\Omega^2}{g} \Lambda \frac{1}{4\pi} \int_{-1}^{+1} \int_0^{2\pi} z_T^2 \, d\lambda \, d\mu,$$

and that

$$A_z = C_0 \Lambda \left(z^T\right)^2,$$
$$A_E = C_0 \Lambda \left[\left(x_1^T\right)^2 + \left(y_1^T\right)^2 + \left(x_2^T\right)^2 + \left(y_2^T\right)^2\right],$$

with

$$C_0 = \frac{Pa^2\Omega^2}{g}.$$

Problem 5

Show that the kinetic energy of the vertical mean flow (here identified with the quantities with subscript $*$) is

$$K_m = C_0 \frac{1}{4\pi} \int_{-1}^{+1} \int_0^{2\pi} (-z_*) \nabla^2 z_* \, d\lambda \, d\mu$$

and that the low-order model gives

$$K_{Mz} = C_0 c \left(z^*\right)^2$$

and

$$K_{ME} = C_0 \left[c_1 \left\{\left(x_1^*\right)^2 + \left(y_1^*\right)^2\right\} + c_2 \left\{\left(x_2^*\right)^2 + \left(y_2^*\right)^2\right\}\right].$$

It follows naturally by analogy that the shear flow kinetic enegy is expressed as the mean flow kinetic energy, when M is replaced by S and $*$ by T.

Our next concern is to evaluate the generations, conversions, and dissipations.

Problem 6

Show that the generation of available potential energy may be written in the form

$$G(A) = \Gamma_0 2\Lambda \frac{\gamma}{\Omega} \frac{1}{4\pi} \int_{-1}^{+1} \int_0^{2\pi} z_T \left(z_E - z_T \right) d\lambda \, d\mu$$

with

$$\Gamma_0 = \frac{p_0}{g} a^2 \Omega^3.$$

Show next that

$$\begin{aligned}G(A) = \quad & \Gamma_0 2\Lambda \frac{\gamma}{\Omega} z^T \left(z_E - z^T \right) \\ + \; & \Gamma_0 2\Lambda \frac{\gamma}{\Omega} \left[x_1^T \left(x_1^E - x_1^T \right) + y_1^T \left(y_1^E - y_1^T \right) \right. \\ & \left. + x_2^T \left(x_2^E - x_2^T \right) + y_2^T \left(y_2^E - y_2^T \right) \right].\end{aligned}$$

Problem 7

Show that

$$\begin{aligned}C(A, K) & = -\Gamma_0 2 \frac{f_0}{\Omega} \frac{1}{4\pi} \int_{-1}^{+1} \int_0^{2\pi} 2\pi W z_T \, d\lambda \, d\mu \\ & = \Gamma_0 2 \frac{f_0}{\Omega} \left[-r z^T - \left(r_1 x_1^T + s_1 y_1^T + r_2 x_2^T + s_2 y_2^T \right) \right].\end{aligned}$$

We are now going to turn our attention to the dissipation. This quantity as well as others directly concerned with the kinetic energy should be evaluated recalling that the streamfunction, from which the velocities and the vorticities are obtained, is available at levels 1 and 3. In calculating the integrals for the dissipation it is then the safest procedure to calculate the contributions from the upper and lower halves of the atmosphere separately. This procedure is recommended for the next problem dealing with the dissipation of kinetic energy.

Problem 8

Show that

$$\begin{aligned}D(K) = \; & -\Gamma_z \left(E_T \frac{1}{4\pi} \int_{-1}^{+1} \int_0^{2\pi} z_T \nabla^2 z_T \, d\lambda \, d\mu \right. \\ & \left. + E_0 \frac{1}{4\pi} \int_{-1}^{+1} \int_0^{2\pi} \left(z_* - z_T \right) \nabla^2 \left(z_* - z_T \right) d\lambda \, d\mu \right)\end{aligned}$$

$$= \Gamma_0 \left(E_T \left\{ c\left(z^T\right)^2 + c_1 \left[\left(x_1^T\right)^2 + \left(y_1^T\right)^2 \right] + c_2 \left[\left(x_2^T\right)^2 + \left(y_2^T\right)^2 \right] \right\} \right.$$
$$+ E_0 \left\{ c\left(z^* - z^T\right)^2 + c_1 \left[(x_1^*) + \left(y_1^* - y_1^T\right)^2 \right] \right.$$
$$\left. \left. + c_2 \left[\left(x_2^* - x_2^T\right)^2 + \left(y_2^* - y_2^T\right)^2 \right] \right\} \right).$$

Problem 8, which given the total dissipation, indicates also how the dissipation can be interpreted. Similar statements can be made for the results obtained in Problems 6 and 7. The terms in the first line make up the dissipation $D(K_T) = D(K_{Tz}) + D(K_{TE})$, while those in the next two lines with the coefficient E_0 can be written $D(K_M) = D(K_{Mz}) + D(K_{ME})$.

The information gathered in the solutions of Problems 4 to 8 is sufficient to calculate the most simple energy diagrams. This leads us to the next problem.

Problem 9

We consider a triplet (0,2), (6,6), and (6,7). The following information is given:

$z^E = -0.005$	$z^T = -0.003$	$z^* = -0.0125$
$x_1^E = -0.003$	$x_1^T = -0.001$	$x_1^* = -0.003064$
$y_1^E = 0$	$y_1^T = 0.001732$	$y_1^* = 0.002571$
$x_2^E = -0.003$	$x_2^T = -0.001414$	$x_2^* = -0.003759$
$y_2^E = 0$	$y_2^T = 0.001414$	$y_2^* = 0.001368$

Calculate:

1. the interaction coefficient,
2. the amount of A,
3. the amount of K,
4. the generation $G(A)$,
5. the conversion $C(A, K)$,
6. the dissipation $D(K)$.

Energy amounts should be expressed in k Jm^{-2}. Generations, conversions and dissipations in W m^{-2}.

The next goal is to be able to divide the state into the zonal components and the eddy components. In addition to the quantities already calculated,

PROBLEMS 339

we should also obtain $C(A_z, A_E)$ and $C(K_E, K_z)$. These integrals are more complicated, and we shall therefore show in some detail how we can get $C(A_z, A_E)$ from the general quasi-geostrophic formula. Leaving out the primes we get from (7.29)

$$C(A_z, A_E) = \frac{p_0}{g 4\pi a^2} \int_{-\frac{\pi}{2}}^{\frac{\pi}{2}} \int_0^{2\pi} \frac{\alpha_z}{\overline{\sigma} a \cos \varphi} \frac{\partial}{\partial \varphi} \left[\cos \varphi \, (\alpha_E v_E)_z\right] a^2 \cos \varphi \, d\lambda \, d\varphi. \quad (.22)$$

Converting to the variables of the two-level model we find

$$C(A_z, A_E) = \frac{p_0}{ga} \frac{4f_0^2}{\overline{\sigma} P^2} \frac{1}{4\pi} \int_{-1}^{+1} \int_0^{2\pi} \psi_{Tz} \frac{d}{d\mu} \left[\cos \varphi \, (\psi_{TE} v_E)_z\right] d\lambda \, d\mu, \quad (.23)$$

or, using nondimensional variables,

$$C(A_z, A_E) = \frac{p_0}{g} a^2 \Omega^3 \cdot 2 \cdot \frac{1}{4\pi} \int_{-1}^{+1} z_{Tz} \frac{d}{d\mu} \left(\frac{1}{2\pi} \int_0^{2\pi} z_{TE} \frac{\partial z_{*E}}{\partial \lambda} d\lambda \right) d\lambda \, d\mu. \quad (.24)$$

We calculate as a matter of convenience the heat transport separately, and we find

$$\frac{1}{2\pi} \int_0^{2\pi} z_{TE} \frac{\partial z_{*E}}{\partial \lambda} d\lambda = \frac{l}{2} \left(x_1^T y_1^* - y_1^T x_1^*\right) P_{n_1}^l(\mu) P_{n_1}^l(\mu)$$
$$+ \frac{l}{2} \left(x_1^T y_1^* - y_1^T x_1^*\right) P_{n_1}^l(\mu) P_{n_2}^l(\mu) \quad (.25)$$
$$+ \frac{l}{2} \left(x_1^T y_2^* - y_1^T x_2^* + x_2^T y_1^* - y_2^T x_1^*\right) P_{n_1}^l(\mu) P_{n_2}^l(\mu).$$

Equation (.25) is substituted back in (.24). After rearrangement we find

$$C(A_z, A_E) =$$
$$2\Gamma_0 \Lambda l \left[z^T \left(x_1^T y_1^* - y_1^T x_1^*\right) \frac{1}{2} \int_{-1}^{+1} P_n(\mu) \frac{d \left[P_{n_1}^l(\mu)\right]^2}{d\mu} d\mu \right.$$
$$+ z^T \left(x_2^T y_2^* - y_2^* - y_2^T x_2^*\right) \frac{1}{2} \int_{-1}^{+1} P_n(\mu) \frac{d \left[P_{n_2}^l(\mu)\right]^2}{d\mu} d\mu \quad (.26)$$
$$\left. + z^T \left(x_1^T y_2^* - y_1^T x_2^* + x_2^T y_1^* - y_2^T x_1^*\right) \frac{1}{2} \int_{-1}^{+1} P_n(\mu) \frac{d \left[P_{n_1}^l P_{n_2}^l(\mu)\right]}{d\mu} d\mu \right].$$

At this point we recall that we have restricted n to be even, and we found that n_1 and n_2 have one even and one odd number. From this it follows that the first two integrals in (.26) vanish because the square of a function is even, its derivative odd, and the integrand is thus odd and integrates to zero. The final result is therefore

$$C(A_z, A_E) = 2\Gamma_0 \Lambda z^T \left(x_1^* y_2^T - y_1^* x_2^T + x_2^* y_1^T - y_2^* x_1^T \right) K, \qquad (.27)$$

where K is the interaction coefficient.

$C(K_E, K_z)$ is the only remaining conversion necessary to calculate all quantities in the diagram with the reservoirs A_z, A_E, K_E, K_z. It is evaluated in the following problem.

Problem 10

Show that $C(K_E, K_z)$ can be calculated from the following formula

$$\begin{aligned} C(K_E, K_z) &= \Gamma_0 K (c_1 - c_2) \left[z^* \left(x_1^* y_2^* - y_1^* x_2^* + x_1^T y_2^T - y_1^T x_2^T \right) \right. \\ &\quad \left. + z^T \left(x_1^* y_2^T - y_1^* x_2^T + x_1^T y_x^* - y_1^T x_2^* \right) \right]. \end{aligned}$$

Hint: It is an advantage to start from the most basic form of the quasi-geostrophic equation by writing

$$C(K_E, K_z) = \frac{P}{g 4\pi a^2} \int_S \left(\frac{\psi_{1z}}{a \cos \varphi} \frac{\partial (\zeta_{1E} v_{1E})_z \cos \varphi}{\partial \varphi} \right. \\ \left. + \frac{\psi_{3z}}{a \cos \varphi} \frac{\partial (\zeta_{3E} v_{3E})_z \cos \varphi}{\partial \varphi} \right) dS.$$

Problem 11

With the data given in Problem 9 the following quantities should be calculated:

1. A_z, A_E, K_E, K_z in k Jm^{-2},
2. $G(A_z)$, $G(A_E)$ in W m^{-2},
3. $C(A_z, A_E)$ in W m^{-2},
4. $C(A_z, K_z)$, $C(A_E, K_E)$ in W m^{-2},
5. $C(K_E, K_z)$ in W m^{-2},
6. $D(K_z)$, $D(K_E)$ in W m^{-2}.

The simple low-order model may also be used to calculate the energy exchange between the baroclinic and barotropic flows as discussed in Chap-

PROBLEMS

ter 5. Since the model is quasi-geostrophic, we can calculate only

$$C_{ND}(K_s, K_M) = \frac{p_0}{gS} \int_S \vec{k} \cdot \left(\vec{v}_* \times \vec{v}_T \right) \zeta_T \, dS. \qquad (.28)$$

Equation (.28) is more cumbersome to evaluate than most other energy conversions because it contains three quantities in the integrand, but the simplicity of the low-order model, where we have one longitudinal wavenumber only, eases the calculations. Note that (.28) already has been adapted to the two-level model.

Problem 12

Show that the energy conversion from the baroclinic to the barotropic component of the flow in the low-order model is

$$\begin{aligned} C(K_s, K_M) &= \Gamma_0 K \left[(c_2 - c_1) z^* \left(x_2^T y_1^T - y_2^T x_1^T \right) \right. \\ &\quad + (c_1 - c) z^T \left(x_2^* y_1^T - y_2^* x_1^T \right) \\ &\quad + \left. (c - c_2) z^T \left(x_2^T y_1^* - y_2^T x_1^* \right) \right]. \end{aligned}$$

Hint: Show first that any term containing three trigonometric factors gives a vanishing contribution to the energy conversion.

Problem 13

Using the data given in Problem 9, calculate the following quantities:

1. A, K_s, K_M in kJ m^{-2},
2. $G(A)$, $C(A, K_s)$, $C(K_s, K_M)$ in W m^{-2},
3. $D(K_s)$, $D(K_M)$ in W m^{-2}.

EXERCISES

No. 1

In a vertical column of unit cross section the wind speed varies as follows:

$$V = \begin{cases} V_m p/p_m, & 0 \leq p \leq p_m \\ V_m \dfrac{p_0 - p}{p_0 - p_m}, & p_m \leq p \leq p_0. \end{cases}$$

Calculate:

1. the kinetic energy per unit area,
2. the kinetic energy per unit area of the vertical mean flow,
3. the kinetic energy per unit area of the vertical shear flow.

No. 2

Consider an atmosphere with a constant lapse-rate. To be exact

$$T = T_0 - \gamma z \qquad (T_0 = 283 \ K, \qquad \gamma = 6.5 \times 10^{-3} \ K \ m^{-1})$$

1. Determine the internal energy per unit area and calculate the amount in the unit k Jm^{-2} ($p_0 = 100$ kPa).
2. Calculate the potential energy per unit area.

No. 3

Consider an atmosphere, where the static stability varies as inversely proportional to the square of the pressure – that is,

$$\sigma = -\alpha \frac{d \ln \theta}{dp} = \frac{\sigma_0 p_0^2}{p^2} \qquad (\sigma_0 = 1 \ m^4 \ s^2 \ t^{-2}, \qquad p_0 = 100 \ k \ Pa).$$

Determine the temperature as a function of pressure in this atmosphere, and calculate the "total' potential energy (i.e., internal plus potential energy) in a vertical column with unit cross section.

No. 4

The ω-equation for the two-level, quasi-geostrophic model may be written in the following form:

$$\nabla^2\omega - q^2\omega = \frac{2}{\sigma P}\left(\nabla^2(V_* \cdot \nabla\phi_T) - V_* \cdot \nabla\left(\nabla^2\phi_T\right)\right)$$
$$-V_T \cdot \nabla\left(\nabla^2\phi_*\right) - \beta\frac{\partial\phi_T}{\partial x}\,),$$

where ϕ_* and ϕ_T are the geopotentials for the mean flow and the thermal flow, respectively; V_* and V_T are the horizontal winds for the same two fields, while σ is the static stability parameter, $P = 50$ k Pa, $\beta = df/dy =$ constant and $q^2 = 2f_0^2/(\sigma P^2)$.

Let ϕ_* and ϕ_T be given by

$$\phi_* = -f_0 U_* y + A_* \cos(kx),$$
$$\phi_T = -f_0 U_T y + A_T \cos[k(x + x_*)]. \tag{.1}$$

1. Solve the ω-equation using the specification (.1) and show that the solution may be written in the form

$$\omega = \omega_\beta \sin[k(x + x_*)] + \omega_A \sin(kx).$$

where the specific expression for ω_β and ω_A should be given.

2. Use the solution for ω to discuss the vertical velocity in a baroclinic wave.

3. The energy conversion $C(A, K)$ can for this model be written in the form

$$C(A, K) = -\frac{4}{g}\frac{1}{L}\int_0^L \omega\phi_T\, dx.$$

Use the solution for ω to evaluate $C(A, K)$ and discuss when the conversion is positive, negative, and zero.

4. Calculate $C(A, K)$ in the unit W m^{-2}, when the wavelength is 4000 km, $q^2 = 2.5\times 10^{-12}$ m^{-2}, $v_{*max} = kA_*/f_0 = 5$ m s^{-1}, $v_{Tmax} = kA_t/f_0 = 5$ m s^{-1}, $U_T = 10$ m's^{-1}, $p_0 = 100$ kPa, $g = 9.8$ m s^{-2}, $kx_* = \pi/4$.

No. 5

It is assumed that the barotropic model applies at 500 kPa. A channel with width D in the south-north direction and length L in the west-east direction is considered. The eddy streamfunction is given by

$$\psi_E(x,y) = \quad x_1 \sin(\mu y) \cos(kx) + y_1 \sin(\mu y) \sin(kx)$$
$$+ \quad x_3 \sin(3\mu y) \cos(kx) + y_3 \sin(3\mu y) \sin(kx).$$

1. Show that the momentum transport is

$$(u_E v_E)_z = \frac{1}{2}\mu k \left(x_1 y_3 - x_3 y_1\right) \left[\sin(4\mu y) - 2\sin(2\mu y)\right].$$

Note that $\mu = \pi/D$ and $k = 2\pi/L$. Let the zonal wind be given by

$$u_z = \frac{1}{2} U_m (1 - \cos 2\mu y).$$

2. Show that the energy conversion from eddy to zonal kinetic energy is

$$C(K_E, K_z) = \frac{1}{2}\frac{p_0}{g} U_m \mu^2 k \left(x_1 y_3 - x_3 y_1\right).$$

3. Discuss and illustrate the momentum transport, its convergence, and the wave structure necessary to have $C(K_E, K_z)$ positive.

   Hint: To answer question 3 it may be an advantage to introduce an amplitude and a phase angle as, for example, $x = R\cos\delta$, $y = R\sin\delta$.

No. 6

1. Show that the conversion from zonal to eddy available potential energy in a two-level, quasi-geostrophic model can be written in the form

$$C(A_z, A_E) = \frac{1}{g}\frac{1}{D}\int_0^{p_0}\int_0^D \frac{\alpha z}{\sigma}\frac{\partial (\alpha_E v_E)_z}{\partial y} dy\, dp$$
$$= 2\frac{p_0}{g}q^2\frac{1}{D}\int_0^D \psi_{Tz}\frac{\partial (\psi_T v_*)_z}{\partial y} dy.$$

2. Let

$$\psi_{*E} = x_* \sin(\mu y)\cos(kx) + y_* \sin(\mu y)\sin(kx),$$
$$\psi_{TE} = x_T \sin(\mu y)\cos(kx) + y_T \sin(\mu y)\sin(kx),$$

where $\mu = \pi/D$, where D is the width of the channel, and $k = \pi/L$, where L is the wavelength. Show that the "heat transport" is

$$(\psi_T v_*)_z = \frac{1}{2}k\left(x_T y_* - y_T x_*\right)\sin^2(\mu y).$$

EXERCISES

3. Let
$$\psi_{Tz} = \Psi_m \cos(\mu y).$$

Show that
$$C(A_z, A_E) = \frac{4}{3\pi} \frac{p_0}{g} q^2 \mu k \left(x_T y_* - y_T x_* \right) \Psi_m.$$

4. Introducing the amplitudes R_* and R_T and the phase angles δ_* and δ_T (see the hint in Problem 5), show that
$$C(A_z, A_E) = \frac{4}{3\pi} \frac{p_0}{g} q^2 \mu k R_* R_T \sin(\delta_* - \delta_T) \Psi_m.$$

5. Calculate $C(A_z, A_E)$ in the unit Wm^{-2} when $p_0 = 100$ kPa, $g = 9.8$ m s^{-2}, $q^2 = 2.5 \times 10^{-12}$ m^{-2}, $D = 10^7$ m, $kR_* = 10$ m s^{-1}, $R_T = 7.2 \times 10^6$ m^2 s^{-1}, $\Psi_m = 4.3 \times 10^7$ m^2 s^{-1}, $\delta = 40°$ and $\delta_T = 20°$.

No. 7

The ω-equation for quasi-geostrophic flow may be used to make simple estimates of the mean meridional circulation. In this problem we shall look at the direct effect of the heating. As a first approximation we may assume that
$$H_z = H_m \cos(\mu y), \qquad \mu = \frac{\pi}{D}, \tag{.2}$$

where D is the width of the channel. The zonally averaged ω-equation including only the heating as forcing is in the two-level case:
$$\frac{d^2 \omega_z}{dy^2} - \lambda^2 \omega_z = -\frac{\kappa}{\sigma P} \frac{d^2 H_z}{dy^2}, \tag{.3}$$

where $\lambda^2 = 2.5 \times 10^{-12}$ m^{-2}, $\kappa = 0.286$, $\sigma = 3.2$ m^4 s^{2}t^{-2} and $P = 50$ kPa.

If $H_m = 0.1$ kJ t^{-1} s^{-1} we want a solution of (.3) for ω_z. Furthermore, knowing ω_z calculate the velocity potential in the upper and lower layers. Calculate finally the meridional, zonally-averaged wind component.

Illustrate finally the derived mean meridional circulation.

No. 8

In this problem we shall estimate the mean meridional circulation created by the transport of sensible heat. The ω-equation is for this purpose:
$$\frac{d^2 \omega_z}{dy^2} - \lambda^2 \omega_z = \frac{p}{f_0} \lambda^2 \frac{R}{2f_0} \frac{d^3 (T_E v_E)_z}{dy^3}. \tag{.4}$$

Determine $w_z = w_z(y)$ when $(T_E v_E)_z = T_H \sin(\mu y)$, where $T_H = 20$ K m s^{-1}, $R = 287$ m^2 s^{-2} K^{-1}, $f_0 = 10^{-4}$ s^{-1}, $\lambda^2 = 2.5 \times 10^{-12}$ m^{-2} and $p = 50$ dPa.

Discuss your result, make a figure of this mean-meridional circulation and compare this circulation with the one obtained in Probem No. 7.

No. 9

$C(K_E, K_z)$ should be calculated in this problem. For the two-level, quasi-geostrophic model we have

$$C(K_E, K_z) = -\frac{P}{g}\frac{1}{D}\int_0^D \left(u_{z1}\frac{\partial M_1}{\partial y} + u_{z3}\frac{\partial M_3}{\partial y} \right) dy. \quad (.5)$$

It is known that the momentum transport has its maximum in the upper part of the troposphere. In this exercise we shall assume that the contribution from level three can be neglected. We have thus

$$C(K_E, K_z) = -\frac{P}{g}\frac{1}{D}\int_0^1 u_z \frac{\partial M}{\partial \eta} d\eta, \quad (.6)$$

where we have dropped the subscript 1, $M = (u_E v_E)_z$, and $\eta = y/D$. In the first part of the problem we shall specify u_z and M as follows:

$$u_z = 4U_m \eta(1 - \eta), \qquad 0 \leq \eta \leq 1$$

and

$$M = M_s = 16 M_m \eta \left(\frac{1}{2} - \eta\right), \qquad 0 \leq \eta \leq \frac{1}{2}$$

$$M = M_N = -16 M_m \left(\eta - \frac{1}{2}\right)(1 - \eta)(5 - 12\eta + 8\eta^2), \qquad \frac{1}{2} \leq \eta \leq 1.$$

From the above information calculate $C(K_E, K_z)$ as given in (.6).

Discuss the distribution of the momentum transport and its divergence based on the definition of M given above and compare the schematic distribution to the observed distribution as calculated by Chen (1982). The values $U_m = 30$ m s^{-1} and $M_m = 45$ m^2 s^{-2} should be used in the calculations.

Repeat the calculation of $C(K_E, K_z)$ when the momentum transport remains the same, but u_z is replaced by

$$u_z = \frac{27}{4} U_m \eta (1 - \eta)^2.$$

EXERCISES

No. 10

All quantities entering atmospheric energetics have an annual variation. A major part of the variation can be described by the mean value for the year and the first Fourier component in time taking a year as the basic period. Let the annual variation of the available potential energy, the kinetic energy, and the generation of available potential energy be given by

$$A = 3980 + 1910 \cos \nu(t - 16),$$
$$K = 1850 + 1400 \cos \nu(t - 18),$$
$$G = 4.2 + 1.2 \cos \nu(t - 354).$$

where ν is the frequency corresponding to the period of one year (i.e., $\nu = 2 \times 10^{-7}$ s^{-1}), and where energy amounts are given in the unit kJ m^{-2}, while the generation use the unit W m^{-2}. Finally, the phase of the Fourier component is given in days.

Using the basic two-box diagram calculate the annual variation, expressed in the same form and in the same unit as G above, of the energy conversion $C(A, K)$ and the dissipation $D(K)$.

No. 11

With reference to Problem 10 let it be assumed in this problem that

$$A = 3980 + 1475 \cos \nu(t - 20),$$
$$K = 1850 + 980 \cos \nu(t - 24).$$

We assume furthermore that the frictional dissipation is proportional to the kinetic energy, say

$$D(K) = \alpha K,$$

with $\alpha = 9 \times 10^{-7}$ s^{-1}.

Calculate the annual variation of $G(A)$ and $C(A, K)$ under these assumptions.

No. 12

The temperature distribution is given by

$$T = T(\lambda, \varphi, p) = T_0 + (T_1 - T_0)\left(\frac{p}{p_0}\right)^{R/c_p}$$
$$- kL(\varphi) + h \sin(2\varphi) \sin(n\lambda).$$
$$L = \frac{1}{4}(1 - 3\cos 2\varphi).$$

T_0, T_1, p_0, k, and n are constants, while λ is longitude, φ is latitude, and p is pressure.

Calculate:

1. The area averaged temperature $(\overline{T})$ over the whole globe.

2. The static stability $\overline{\sigma}$ corresponding to $\overline{T}$.

3. Zonal and eddy potential energy considering the whole globe. Use the quasi-geostrophic expressions.

4. The total potential energy per unit area.

No. 13

The zonal heating at 50 kPa is

$$H'_z = H_m \cos(\mu y), \qquad \mu = \frac{\pi}{D},$$
$$D = 10^7 \text{ m}, \qquad H_m = 2.5 \times 10^{-3} \text{ kJ t}^{-1} \text{ s}^{-1}.$$

The temperature at the same level is given by

$$T'_z = T_m \cos(\mu y), \qquad T_m = 20° \text{ C}.$$

Assuming no horizontal motion in the atmosphere, calculate:

1. The zonally averaged vertical velocity $w_z = w_z(y)$

2. The generation of zonal available potential energy.

3. The conversion of zonal available potential energy to zonal kinetic energy

State the assumptions you make to solve these problems. $f_0 = 10^{-4} \text{ s}^{-1}$, $\overline{\sigma} = 2 \text{ m}^4 \text{ s}^2 \text{ t}^{-2}$.

Hint: See Exercise No. 7.

No. 14

Suppose that equipment to measure upper air winds had never been invented, but that surface winds and winds in the boundary layer could be measured. This situation simulates conditions in the early part of the twentieth century.

EXERCISES

1. What can one say about the meridional transport of momentum if the profile $u_z = u_z(\varphi)$ is known? Hint: One will want to include a relation between the surface stress and the surface wind.

2. Using a methodology developed under A, say as much as possible about the momentum transport if

$$u_z(\varphi) = -U_0 \cos(4\varphi), \qquad -\frac{\pi}{2} < \varphi < \frac{\pi}{2},$$

where φ is latitudes and $U_0 = 5$ m s^{-1}.

No. 15

The divergence of the vertically averaged momentum transport is given by

$$\frac{1}{a\cos^2\varphi} \frac{\partial (u_E v_E)_{zM} \cos^2\varphi}{\partial \varphi} = D_M \cos(4\varphi).$$

Suppose furthermore that the vertical mean of the zonal wind is

$$U_{zM} = 16 U_m \mu^2 (1-\mu)^2, \qquad \mu = \sin\varphi.$$

Finally, we assume that the vertical variation of both of the above quantities can be described by a function of pressure only. This means for example that

$$u_z(\mu, p) = u_{zM} A(p).$$

1. Calculate the energy conversion $C(K_E, K_z)$ under these assumptions, when $D_M = 1.5 \times 10^{-6}$ m s^{-2} and $U_m = 20$ m s^{-1}. $A(p)$ is give as follows:

$$A(p) = \begin{cases} 8p/p_0 & 0 \le p \le 25 \text{ kPa} \\ \frac{8}{3}(1 - p/p_0) & 25 \le p \le 100 \text{ kPa}. \end{cases}$$

2. One may test the sensitivity of the result to the zonal wind profile by using

$$u_{zM} = 4 U_m \mu (1-\mu).$$

No. 16

The mean-zonal vertical velocity is approximated by

$$\omega_z(\mu, p) = kB(p)P(\mu),$$

where $k = 10^{-5}$ kPa s^{-1} and

$$B(p) = 4\frac{p}{p_0}\left(1 - \frac{p}{p_0}\right), \qquad p_0 = 100 \text{ kPa},$$

while

$$P(\mu) = \frac{1}{16}\left(231\mu^6 - 315\mu^4 + 105\mu^2 - 5\right), \qquad \mu = \sin\varphi.$$

1. Determine the mean meridional vertical velocity v_z.

2. Estimate the change in the zonal mean of the temperature at 45° N and at 50 kPa in one day as caused by the mean meridional circulation.

3. Estimate the change in the mean zonal wind at 45° N and at 25 kPa in one day as caused by the mean meridional circulation. You may use

$$\sigma = 2 \text{ m}^4 \text{ s}^2 \text{ t}^{-2} \qquad (\text{at } 50 \text{ kPa})$$
$$\Omega = 7.29 \times 10^{-5} \text{ s}^{-1}$$
$$R = 287 \text{ m}^2 \text{ s}^{-2} \text{ K}^{-1}.$$

ANSWERS

No. 1

1. $V_m^2 p_0/(6g)$
2. $V_m^2 p_0/(8g)$
3. $V_m^2 p_0/(24g)$

No. 2

$$I = \frac{c_v}{g}\frac{T_0 p_0}{1+\frac{\gamma R}{g}} = 1.74 \times 10^6 \text{ kJ m}^{-2}$$

$$P = \frac{R}{c_v}I = 7 \times 10^5 \text{ kJ m}^{-2}$$

No. 3

$$T = \frac{c_p \sigma_p p_0^2}{R^2} + \left(T_0 - \frac{c_p \sigma_0 p_0^2}{R^2}\right)\left(\frac{p}{p_0}\right)^{R/c_p}$$

$$I + P = c_p \int_0^{p_0} T\, dp = 2.5 \times 10^7 \text{ kJ m}^{-2}$$

No. 4

1. $\omega_\beta = -\frac{2}{\sigma P}\frac{C_R}{1+q^2/k^2}kA_T \qquad \omega_A = \frac{2}{\sigma P}\frac{2U_T}{1+q^2/k^2}kA_*$

2. Discussion
3.
$$C(A,K) = \frac{2p_0}{g}kv_{T,max}v_{*,max}U_T\frac{q^2}{k^2+q^2}\sin(kx_*)$$

4. 2.85 W m^{-2}

No. 5

Derivations with given results.

No. 6

A-D: Derivations with given results.

E. $C(A_z, A_E) = 3.6$ W m^{-2}.

No. 7

$$\omega_z = -\frac{R/c_p}{\sigma P}\frac{\mu^2}{\mu^2+\lambda}H_m\cos(\mu y)$$
$$\chi_{1z} = -\chi_{3z} = -1.38\times 10^6\cos(\mu y)\text{ m}^2\text{ s}^{-1}$$
$$v_{1z} = -v_{3z} = 0.4\sin(\mu y)\text{ m s}^{-1}$$

No. 8

$$\omega_z = \frac{R}{\sigma T}T_H\mu\frac{\mu^2}{\mu^2+\lambda^2}\cos(\mu y)$$
$$\omega_z = -6.4\times 10^{-7}\text{ kPa}^{-1}\ (\approx W_z \approx 1\text{ mm s}^{-1})\cos\mu y$$

No. 9

$$C(K_E, K_z) = 0.18\text{ W'm}^{-2}$$
$$C(K_E, K_z) = 0.76\text{ W m}^{-2}$$

No. 10

$$C = 4.2 + 1.08\cos\nu(t-8)$$
$$D = 4.2 + 1.07\cos\nu(t-23)$$

No. 11

$$C = 1.66 + 1.01\cos\nu(t-10)$$
$$G = 1.66 + 1.10\cos\nu(t-360)$$

No. 12

1. $\overline{T} = T_0 + (T_1 - T_0)\left(\frac{p}{p_0}\right)^{R/c_p}$

2. $\sigma = \frac{R^2}{c_p}T_0\frac{1}{p^2}$

3. $A_z = \frac{1}{10}\frac{p_0}{g}c_p k^2 \qquad A_E = \frac{2}{15}\frac{p_0}{g}\frac{c_p}{T_0}h^2$

ANSWERS

4. $P + I = \frac{c_p}{g} \frac{p_0}{R+c_p} T_0 R \left(1 + \frac{T_1}{T_0} \frac{c_p}{R}\right)$

No. 13

1. $\omega_z = -\frac{R/c_p}{\bar{\sigma} P} H_m \frac{\pi^2}{\pi^2 + \lambda^2 D^2} \cos\left(\pi \frac{y}{D}\right)$

2. $G(A_z) = \frac{1}{2} \frac{p_0}{g} \frac{1}{c_p} \frac{R^2}{\bar{\sigma} P^2} H_m T_m = 4.2$ W m^{-2}

3. $C(A_z, K_z) = \frac{R}{g} \frac{R/c_p}{\bar{\sigma} P} \frac{\pi^2}{\pi^2 + \lambda^2 D^2} H_m T_m = 0.1$ W m^{-2}

No. 14

$$(uv)_{zM} = -\frac{g}{p_0} k_F a \frac{1}{\cos^2 \varphi} \int_{-\pi/2}^{\varphi} u_z(p_0) \cos^2 \varphi \, dp$$

with $k_F \approx c_d \rho_0 V_0 \approx 3.7 \times 10^{-5}$

$$(uv)_{zM} = \frac{1}{4} \frac{g}{p_0} k_F a U_0 \frac{1}{1 + \cos 2\varphi} \left[\sin(2\varphi) + \sin(4\varphi) + \frac{1}{3}\sin(6\varphi)\right]$$

No. 15

$$C(K_E, K_z) = \frac{16 \times 8}{45} \frac{p_0}{g} D_M \overline{A^2} U_m = 1.14 \text{ W m}^{-2} \qquad \left(\overline{A^2} = \frac{4}{3}\right)$$

$$C(K_E, K_z) = -\frac{13 \times 4}{105} \frac{p_0}{g} D_M \overline{A^2} U_m = -0.71 \text{ W m}^{-2}$$

No. 16

1. $v_z = a k \frac{dA}{dp} \frac{1}{16} \sqrt{1-\mu^2} \left(33\mu^4 - 30\mu^2 + 5\right) \mu$

2. $\Delta T_z \approx -0.05 \circ C$ day^{-1}

3. $\Delta u_z \approx -0.6$ m s^{-1} day^{-1}

BIBLIOGRAPHY

Alpert, J. C., 1981: An analysus if the kinetic energy budget for live extratropical cyclones: The vertically averaged flow and the vertical shear flow. *Mon. Wea. Rev.* 109 , 1219-32.

Aspliden, C. I., G. A. Dean, and H. Landers, 1966: Satellite study, Tropical North Atlantic, 1963. Florida State University, Tallahassee, Tech. Rept., Grant WBG58.

Austin, J. F., 1980: The blocking of middle-latitude westerly winds by planetary waves. *Quart. J. Roy. Meteor. Soc.* 106, 327-50.

Barnes, S. L., 1964: A technique for maximizing detail in numerical map analysis. *J. Appl. Meteor.* 3, 396-409.

Barros, V. R., and A. Wiin-Nielsen, 1974: On quasi-geostrophic turbulence: A numerical experiment. *J. Atmos. Sci.* 31, 609-21.

Bendat, J. S., and A. G. Piersol, 1971: *Random Data: analysis and measurement procedures.* Wiley-interscience, John Wiley and Sons, Inc., New York, 407pp.

Berggreon, R., B. Bolin, and C.-G. Rossby, 1949: An aerological study of zonal motion, its perturbation and beakdown. *Tellus*, 1, 14-37.

Blackmon, M. L., J. M. Wallace, N. G. Lau, and S. L. Mullen, 1977: An observational study of the Northern Hemisphere winter circulation. *J. Atmos. Sci.* 34, 1040-53.

Boer, G. J., 1976: Reply to comments by J. Egger on "Zonal and eddy forms of the available potential energy equtions in pressure coordinates." *Tellus*, 28, 379.

Bolin, B., 1950: On the influence of the earth's orography on the general character of the westerlies. *Tellus*, 2, 184-95.

Boyle, J. S., and L. F. Bosart, 1986: Cyclone / anticyclone couplets over North America. Part II: Analysis of a major cyclone event over the eastern United States. *Mon. Wea. Rev.* 114, 2432-65.

Brown, J. A., Jr., 1964: A diagnostic study of tropospheric, diabatic heating, and the generation of available potential energy. *Tellus,* 16, 371-88.

Buch, H. S., 1954: *Hemispheric wind conditions during the year 1950.*, M.I.T. Final Report, No.AF 19-122-153, 126 pp.

Buechler, D. E., and H. E. Fuelberg, 1986: Budget of divergent and rotational kinetic energy during two periods of intense convection. *Mon. Wea. Rev.* 114, 95-114.

Buechler, D. E., and H. E. Fuelberg, 1989: Energy analysis of convectively induced wind perturbations. *Mon. Wea. Rev.* 117, 745-64.

Burpee, R. W., 1972: The origin and structure of easterly waves in the lower troposphere of North Africa. *J. Atmos. Sci* 29, 77-90.

Charney, J. G., 1959: On the theory of the general circulation of the atmosphere. in *The Atmosphere and Sea in Motion.* Rockefeller Inst. Press, Editor: B. Bolin. 178-93.

Charney, K. B., and A. Eliassen, 1949: A numerical method for predicting the perturbations on the middle-latitude westerlies. *Tellus* 1, 38-54.

Chen, T.-C., 1980: On the energy exchange between the divergent and rotational components of atmospheric flow over the tropics and subtropics at 200 mb during two Northern Hemisphere summers. *Mon. Wea. Rev.* 108, 896-912.

Chen, T.-C., 1982a: A further study of spectral energetics in the winter atmosphere. *Mon. Wea. Rev.* 110, 947-61.

Chen, T.-C., 1982b: On the kinetic energy budget of summertime-mean flow at 200 mb in the tropics. *Tellus* 34, 55-62.

Chen, T.-C., 1983: On the energy exchange between the baroclinic and barotropic components of atmospheric flow in the tropics during the FGGE summer. *Mon. Wea. Rev.* 111, 1389-1396.

Chen, T.-C., 1985: On the maintenance of enstrophy in the tropics during the 1979 northern summer. *Mon. Wea. Rev.* 113, 624-40.

Chen, T.-C., 1987: 30-50 day oscillation of 200 mb temperature and 850 mb height during the 1979 northern summer. *Mon. Wea. Rev.* 115, 1589-1605.

Chen, T.-C., and A. Wiin-Nielsen, 1976: On the kinetic energy of the divergent and nondivergent flows in the atmosphere. *Tellus* 18, 486-98.

Chen, T.-C., and A. Wiin-Nielsen, 1978: On nonlinear cascades of atmospheric energy and enstrophy in a two-dimensional index. *Tellus* 30, 313 and 332.

Chen, T.-C., and Yen-Huei Lee, 1982: A note on the maintenance of atmospheric kinetic energy. *PAGEOPH* 120, 642-47.

Chen, T.-C., and Yen-Huei Lee, 1983: A study of kinetic energy generation with general circulation models. *J. Meteor. Soc. Japan* 61, 439-48.

Chen, T.-C., and Lawrence E. Buja, 1983: A comparsion study for the time variation of the atmospheric energetics between two hemispheres during the FGGE year: Annual variation and vacillation. in *the First International Conference on Southern Hemisphere Meteorology*. San Jose dos Campos, Brazil, 31 July -6 August 1983, 21-24.

Chen, T.-C., and J. Shukla, 1983: Diagnostic analysis and energetics of blocking event generated in the GLAS climate model. *Mon. Wea. Rev.* 111, 3-22.

Chen, T.-C., and Hal G. Marshal, 1984: Time variation of atmospheric energetics during the FGGE winter. *Tellus* 36A, 251-68.

Chen, T.-C., and Yen-Huei Lee, 1985: A comparison study of spectral energetics analysis using various FGGE IIIb data. in *Proceedings of the First National Workshop on the Global Weather,* , Woods Hole, Mass., 9-20 July, 1984. Washington, D. C.: National Academy Press, 247-66.

Chen, T.-C., and H. van Loon, 1987: On the interannual variation of the tropical easterly jet. *Mon. Wea. Rev.* 115, 1739-59.

Chen, T.-C., and Ming-Cheng Yen, 1989: The effect of the divergent circulation on some aspects of the 1978/79 Southern Hemisphere Monsoon. *J. Climate*, 2, 1270-88.

Chen, T.-C., and Jau-Ming Chen, 1990: A note on the maintenance of stationary eddies with the streamfunction budget analysis. *J. Atmos. Sci.* 47, 2818-24.

Chen, T.-C., and Ming-Cheng Yen, 1991a: A study for the diabatic heating associated with the planetary-scale Madden-Julian oscillation. *J. Geophys. Res.* 96, 13,163-13,177.

Chen, T.-C., and Ming-Cheng Yen. 1991b. Intraseasonal variations of the tropical easterly jet during 1979 northern summer. *Tellus* 43A, 213-25.

Chen, T.-C., and Ming-Cheng Yen. 1991c: Interaction between intraseasonal oscillations of midlatitude flow and tropical convection during 1979 northern summer: Pacific Ocean. *J. Climate* 4, 653-71.

Chen, T.-C., A. R. Hansen, and J. J. Tribbia, 1981: A note on the release of available potential energy. *J. Atmos. Sci.* 37, 1157-76.

Chen, T.-C., M.-C. Yen, and D. D. Nune, 1987: Dynamic aspects of the Southern Hemisphere medium-scale waves during the southern summer season. *J. Meteor. Sci. Japan* 65, 401-21.

Chen, T.-C., M.-C. Yen, and H. van Loon, 1989: The effect of the divergetn circulation on some aspects of the 1978/79 Southern Hemisphere monsoon. *J. Climate* 2, 1270-1288.

Chen, T.-C., Jordan C. Alpert, and Thomas W. Schlatter, 1978: The effect of divergent and nondivergent winds on the kinetic energy budget of midlatitude cyclone system: a case study. *Mon. Wea. Rev.* 106, 458-68.

Chen, T.-J., and L. F. Bosart, 1977: Quasi-Lagrangian kinetic energy budgets of composite cyclone-anticyclone couplets. *J. Atmos. Sci.* 34, 452-64.

Defant, A., 1921: Die Zirkulation der Atmosphare in den gemassigten Breiten der Erde. *Geogr. Ann.* 3, 209-66.

Depradine, C. A., 1980: Energetics of large-scale motion in the tropics during GATE at 250 mb. *Mon. Wea. Rev.* 108, 886-95.

DiMego, G. J., and L. F. Bosart, 1982: The transormation of tropical storm Agnus into an extratropical cyclone. Part II: moisture, vorticity and kinetic energy budget. *Mon. Wea. Rev.* 110, 412-33.

Dutton, J. A., and D. R. Johnson, 1967: The theory of available potential energy and a variational approach to atmospheric energetics. *Adv. in Geophys.* 12, 334-443.

Endlich, T. M., 1967: An iterative method for altering the kinetic porperties of wind fields. *J. Appl. Meteor.* 6, 837-44.

Eliasen, E., 1958: A study of the long atmospheric waves on the basis of zonal harmonic analysis. *Tellus* 10, 106-215.

Fischer, G., 1984: Spectral energetics analyses of blocking events in a general circulation model. *Contribution to Atmospheric Physics*, 57, 183-200.

Fjørtoft, R., 1953: On the changes in the spectral distribution of kinetic energy for two-dimensional, nondivergent flow. *Tellus* 5, 225-30.

Flohn, H., 1968: Contributions to meteorology of the Tibetan Highlands. Colorado State University, Fort Collins, Rept. No. 130, 1-120.

Fuelberg, H. E., and P. A. Browning, 1983: Roles of divergent and rotational winds in the kinetic energy balance during intense convective activity. *Mon. Wea. Rev.* 111, 2176-93.

Godbole, R. V., 1977: The composite structure of the monsoon depression. *Tellus* 29, 25-40.

Haltiner, G. J., 1971: *Numerical Weather Prediction*, New York: John Wiley and Sons 317 pp.

Hansen, A. R., and T.-C. Chen, 1982: A spectral energetics analysis of atmospheric blocking. *Mon. Wea. Rev.* 110, 1146-65.

Heddinghaus, T. R., and E. C. Kung, 1980: An analysis of climatological patterns of the Northern Hamispheric circualtion. *Mon. Wea. Rev.*, 108, 1-17.

Hesselberg, Th., 1914: Die Reibung in der Atmosphare, Braunschweig, reprinted in *Norwegian Classical Meteorological Papers*. Oslo Univ. Press, 126-38.

Hollingsworth, A., K. Arpe, M. Tiedthe, M. Capaldo, and H. Savijärvi, 1980: The performance of a medium - range forecast model in winter-impact of physical parameterizations. *Mon. Wea. Rev.* 108, 1736-73.

Holopainen, E. O., 1970: An observational study of the energy balance of the stationary disturbances in the atmosphere. *Quart. J. Roy. Meteor. Soc.* 96, 626-44.

Holopainen, E. O., 1973: An attempt to determine the effects of turbulent friction in the upper troposphere from the balance requirements of the large-scale flow: a frustrating experiment. *Geophysica* 112, 151-76.

Holopainen, E. O., 1978: A diagnostic study of the kinetic energy of the long-term mean flow and the associated transient fluctuations in the atmosphere. *Geophysica* 15, 125-45.

Holopainen, E. O., and C. Fortelius, 1987: High-frequency transient eddies and blocking. *J. Atmos. Sci.* 44,1632-45.

Holopainen, E.O., and K. Eerola, 1979: A diagnostic study of the long-term balance of kinetic energy of atmospheric large scale motion over the British Isles. *Quart. J. Roy. Met. Soc.*, 105, 849-58.

Holton, J. R., 1992: *An Introduction to Dynamic Meteorology*, Academic Press, Inc., Third Edition, San Diego, California, 507 pp.

Hoskins, B. J., and D. J. Karaly, 1981: The Steady linear response of a spherical atmosphere in thermal and orographic forcing. *J. Atmos. Sci.*, 38, 1179-96.

Hunt, B. G., 1978: Atmospheric vacillation in a general circulation model (I). The large-scale energy cycle. *J. Atmos. Sci.* 35, 1133-43.

Illari, L., 1984: Diagnostic study of the potential study, of the potential vorticity in a warm blocking anticyclone. *J. Atmos. Sci.* 41, 3518-26.

Jeffries, J., 1926: On the dynamics of geostrophic winds. *Quart. J. Roy. Meteor. Soc.* 52, 85-104.

Johnson, D. R., R. D. Townsend, and M.-Y. Wei, 1985: The thermally coupled response of the planetary scale circulation to the global distribution of heat sources and sinks. *Tellus* 37A, 106-25.

Julian, P. R., 1971: Some aspects of variance spectra of synoptic scale tropospheric wind components in midlatitudes and in the tropics. *Mon. Wea. Rev.* 99, 954-65.

Kallen, E., 1981: The nonlinear effects of orographic and momentum forcing in a low-order barotropic model, *J. Atmos. Sci.* 38, 2150-63.

Kanamitsu, M., T. N. Krishnamurti, and C. A. Depradine, 1972: On scale interactions in the tropics during northern summer. *J. Atmos. Sci.*

29, 698-706.

Koteswaram, P., 1958: The easterly jet stream in the tropics. *Tellus*, 10, 43-57.

Kraus, E. B., and E. N. Lorenz, 1966: Numerical experiments with large-scale seasonal forcing. *J. Atmos. Sci.*, 23, 3-12.

Krishnamurti, T. N., 1961: The subtropical jet stream of winter. *J. Meteor.* 18, 172-91.

Krishnamurti, T. N., 1971a: Tropical east-west circulations during the northern summer. *J. Atmos. Sci.* 28, 1342-47.

Krishnamurti, T. N., 1971b: Observational study of the tropical upper tropospheric motion during the Northern Hemisphere summer. *J. Appl. Meteor.* 10, 1066-96.

Krishnamurti, T. N., 1979: Tropical metrorology. in *Compendium of Meteorology II* WMO no. 364 Edited by A. Wiin-Nielsen. Geneva: World Meteorological Organization., 428.

Krishnamurti, T. N., M. Kanamitsu, W. J. Ross, and J. D. Lee, 1973a: Tropical east-west circulations during the northern winter. *J. Atmos. Sci.* 30, 780-87.

Krishnamurti, T. N., J. Molinari, H.-L Pan, and V. Wong, 1977: Downstream amplification and formation of monsoon disturbances. *Mon. Wea. Rev.* 105, 1281-97.

Krishnamrti, T. N., M. Kanamitsu, R. Godbole, C. B. Chang, F. Carr, and J. H. Chow, 1976: Study of a monsoon depression. (II) Dynamic structure. *J. Meteor. Soc. Japan* 54, 208-25.

Kraichnan, R. H., 1967: Inertial subranges in two-dimensional turbulence. *Phys. Fluid* 10, 1417-23.

Krueger, A. F., J. S. Winston, and D. A. Haines, 1965: Computations of atmospheric energy and its transformation for the northern hemisphere for a recent five-year period. *Mon. Wea. Rev.* 93, 227-38.

Kung, E. C., 1966a: Kinetic energy generation and dissipation in the large-scale atmospheric circulation. *Mon. Wea. Rev.* 94, No.2, 67-82.

Kung, E. C., 1966b: Large-scale balance of kinetic energy in the atmosphere. *Mon. Wea. Rev.*, 94, 627-40.

Kung, E. C., 1967: Diurnal and long-term variation of the kinetic energy generation and dissipation for a five-year period. *Mon. Wea. Rev.* 95, 593-606.

Kung, E. C., 1969: Further study on the kinetic energy budget. *Mon. Wea. Rev.* 97, 573-81.

Kung, E. C., and S. Soong, 1969: Seasonal variation of kinetic energy in the atmosphere. *Quart. J. Roy. Meteor. Soc.* 95, 501-12.

Kung, E. C., and W. E. Baker, 1975: Energy transformations in the middlelatitude disturbances. *Quart. J. Roy. Meteor. Soc.* 101, 793-815.

Kung, E. C., and W. E. Baker, 1986: Comparative energetics of the observed and simulated Northern Hemisphere general circulation during blocking periods. *J. Atmos. Sci.* 43, 2792-2812.

Lau, N.-C., 1979: The structure and energetics of transient disturbances in the Northern Hemisphere winter circulation. *J. Atmos. Sci.* 36, 982-95.

Lau, N.-C., 1984a: *Circulation statistics based on FGGE Level IIIB analyses*, produced by GFDL, NOAA Data Report ERL, GFDL-5.

Lau, N.-C., 1984b: *A comparison of circulation statistics based on FGGE Level IIIB analyses produced by GFDL and ECMWF for the special observing periods*, NOAA Data Report ERL, GFDL-6.

Lawniczak, G. E., 1969: *On a multi-layer analysis of atmospheric diabatic processes and the generation of available potential energy*, Sci. Rept., University of Michigan, Ann Arbor: 08759-5-T, 111 pp.

Lee, Y.-H., 1983: A comparison study of the energetics of standing and transient eddies between two general circulation models. The GLAS Climate Model and the NCAR Community Climate Model. Iowa State University: Ph.D. dissertation, 185 pp.

Lee, Y.-H., and Tsing-Chang Chen, 1986: On the structure and maintenance of standing eddies in the NCAR community climate model and the GLAS climate model. *Mon. Wea. Rev.* 114, 2057-77.

Leith, C. E., 1968: Diffusion approximation for two-dimensional turbulence. *Phys. Fluids* 11, 671-73.

Lejenäs, H., 1977: On the breakdown of the westerlies. *Atmosphere* 15, 89-113.

Lorenz, E. N., 1955: Available potential energy and the maintenance of the general circulation. *Tellus* 7, 157-67.

Lorenz, E. N., 1963: The mechanics of vacillations. *J. Atmos. Sci.* 20, 448-64.

Lorenz, E. N., 1967: *The nature and theory of the general circulation of the atmosphere*, Geneva World Meteorological Organization, 161 pp.

Lorenz, E. N., 1987: Available energy and the maintenance of a moist circulation. *Tellus* 30, 15-31.

Madden, R. A. and P. R. Julian, 1972: Detection of 40-50 day oscillation in the zonal wind in the tropical Pacific. *J. Atmso. Sci.*, 28, 702-08.

Maddox, R. A., 1980: Mesoscale convective complexes. *Bull. Amer. Meteor. Soc.* 61, 1374-87.

Margules, M., 1904: Über die Energie der Sturme. *Jahrb. Zentralan. Meteor. Vienna*, 1-26.

McGuirk, J. P., and E. R. Reiter, 1976: A vacillation in atmospheric energy parameters. *J. Atmos. Sci.*, 33, 2079-93.

Miller, A. J., 1974: Periodic variation of the atmospheric circulation at 14-16 days. *J. Atmos. Sci.* 31, 720-26.

Miyakoda, K., 1963: Some characteristic features of winter circulation in the troposphere and the lower stratosphere. University of Chicago Tech. Rept. No. 14 Available from Photoduplication Dept., Library of the University of Chicago, Chicago, IL, 60637.

Murakami, T., 1963: On the maintenance of kinetic energy of the large-scale stationary disturbances in the atmosphere. *Final Rep. Planetary Circulation Project, MIT*, 120-61.

Namias, J., 1978: Multiple causes of the North American abnormal winter 1976-77. *Mon. Wea. Rev.* 106, 279-95.

Namias, J., and P. F. Clapp, 1949: Confluence theory of the high tropospheric jet stream. *J. Meteor.* 6, 330-36.

Newell, R. E., J. W. Kidson, D. G. Vincent, and G. J. Boer, 1972: The general circulation of the tropical atmosphere and interactions with extratropical latitudes. Vol.1. The M.I.T. Press, Cambridge, Mass., IJSA, 258 pp.

Nitta, T., 1970: A study of generation and conversion of eddy available potential energy in the tropics. *J. Meteor. Soc. Japan* 48, 524-28.

Nitta, T., 1972: Energy budget of wave disturbances over the Marshall islands during the years of 1956 and 1958. *J. Meteor. Soc. Japan* 50, 71-84.

Nitta, T., 1982: *Tropical Meteorology* (in Japanese), Tokyo, Japan, Tokyo Publication Co., 215 pp.

Norquist, D. C., E. E. Recker, and R. J. Reed, 1977: The energetics of African wave disturbances as observed during Phase III of GATE. *Mon. Wea. Rev.* 105, 334-42.

Obasi, G. O. P., 1965: On the maintenance of the kinetic energy of the mean flow in the Southern Hemisphere. *Tellus* 17, 95-105.

O'Brian, J. J., 1970: Alternative solutions to the classical vertical velocity problem. *J. Appl. Meteor.*, 9, 197-203.

O'Conner, J. F., 1963: The weather and circulation of January 1963. *Mon. Wea. Rev.* 19, 209-18.

Oort, A. H., 1983: *Global atmospheric circulation statistics*, 1958-73, NOAA Prof. Paper, No.14, 180 pp.

Oort, A. H., and J. P. Peixóto, 1974: The annual cycle of the energetics of the atmosphere on a planetary scale. *J. Geophys. Res.* 79, 2705-19.

Palmén, E. and C. Newton, 1969: *Atmospheric Circulation Systems*. New York: Academic Press, 603 pp.

Paulin, G., 1970: A study of the energetics of January 1959., *Mon. Wea. Rev.* 91, 209-18.

Pearce, R. P., 1974: The design and interpretaion of diagnostic studies of synoptic scale atmospheric systems. *Quart. J. Roy. Meteor. Soc.* 100, 265-85.

Petterssen, S., 1956: Convective clouds and weather. in *Weather Analysis and Forecasting*, 2nd ed., New York: McGraw-Hill, 133-95.

Petterssen, S., and S. J. Smebye, 1971: On the development of extratropical cyclones. *Quart. J. Roy. Meteor. Soc.* 97, No. 3, 457-82.

Pfeffer, R., G. Buzyna, and W. W. Fowlis, 1974: Synoptic features and energetics of wave amplitude vacillation in a rotating differentially heated fluid. *J. Atmos. Sci.* 31, 622-45.

Phillips, N. A., 1956: The general circulation of the atmosphere: a numerical experiment. *Quart. J. Roy. Met. Soc.* 82, 123-64.

Phillips N. A., 1963: Geostrophic motion. *Rev. Geophys.* 1, 123-76.

Physick, W. L., 1981: Winter depressiona tracks and climatological jet streams in the Southern Hemisphere during the FGGE year. *Quart. J. Roy. Meteor. Soc.* 107, 883-98.

Pitcher, E. J., R. C. Malone, V. Ramanathan, M. L. Blackmon, K. Puri, and W. Bourke, 1983: January and July simulations with a spectral general circulation model. *J. Atmos. Sci.* 40, 580-604.

Platzman, G. W., 1960: The spectal form of the vorticity equation. *J. Meteor.* 17, 635-44.

Price, P. G., and N. Nicholls, 1982: Southern Hemisphere energetics: a reappraisal of some earlier estimates. *Tellus*, 34, 406-08.

Rasmusson, E. M., and T. H. Carpenter, 1982: Variations in tropical sea surface temperature and surface wind fields associated with the southern oscillation/El Niño. *Mon. Wea. Rev.*, 110, 354-84.

Reed, R. J., and E. E. Recker, 1971: Structure and properties of synoptic-scale wave disturbances in the equatorial western Pacific. *J. Atmos. Sci.* 28, 1117-33.

Reed, R. J., D. C. Norquist, and E. E. Recker, 1977: The structure and properties of African wave disturbances as observed during phase III of GATE. *Mon. Wea. Rev.* 105, 317-33.

Rex, D. F., 1950: Blocking action in the middle of troposphere and its effect upon regional climate. Part I. *Tellus* 2, 196-221. Part II. *Tellus* 2, 275-301.

Riehl, H., 1954: *Tropical Meteorology*, New York: McGraw-Hill, 392 pp.

Rodak, H., and A. M. Grant, 1957: Variations in the high tropospheric mean flow over Australia and New Zealand. *J. Meteor.* 14, 141-49.

Rossby, C. G. and Collaborators, 1947: On the general circulation of the atmosphere in middle latitudes. *Bull. Amer. Meteor.* 28, 255-79.

Ryd, V. H., 1923: *Traveling cyclones*, Medd. Danish Met. Inst., No.5, 125 pp.

Ryd, V. H., 1927: *The energy of the winds*, Medd. Danish Met. Inst., No.7, 96 pp.

Saha, K., and C. P. Chang, 1983: The baroclinic processes of monsoon depression. *Mon. Wea. Rev.* 111, 1506-14.

Saha, K., and S. Saha, 1988: Thermal budget of a monsoon depression in the Bay of Bengal during FGGE-MONEX 1971. *Mon. Wea. Rev.* 116, 242-54.

Saha, K., F. Sanders, and J. Shukla, 1981: Westward propagating predecessors of monsoon depressions. *Mon. Wea. Rev.,* 109, 330-43.

Sadler, J. C., 1967: The tropical upper tropospheric trough as a secondary source of typhoons and a primary source of trade wind disturbances. Final Rept., AFCRL-67-0203, Hawaii Institute of Geophysics, University of Hawaii, 1-48.

Saltzman, B., 1957: Equations governing the energetics of larger scales of atmospheric turbulence in the domain of wave numbers. *J. of Meteor.* 14, 513-23.

Saltzman, B., and S. Teweles, 1964: Further statistics on the exchange of kinetic energy between harmonic components of the atmospheric flow. *Tellus* 16, 432-35.

Sheu, Y.-T. P., and P. J. Smith, 1981: A kinetic energy climatalogy of flow regimes associated with 500 mb waves over Norht America in winter and spring. *Mon. Wea. Rev.* 109, 1862-78.

Shukla, J., 1978: CISK-barotropic-baroclinic instability and the growth of monsoon depressions. *J. Atmos. Sci.* 35, 495-508.

Shukla, J., and B. Bangaru, 1979: Effect of a Pacific sea surface temperature anomaly on the circulation over North America. *Fourth*

National Aeronautics and Space Administration Weather and Climate Program Science Review, the NASA Goddard Space Flight Center, Greenbelt, MD. 20771, 171-176.

Shutts, G. J., 1983: The propagation of eddies in diffuent jetstream: Eddy vorticity forcing of bloching flow fields. *Quart. J. Roy. Meteor. Soc.* 109. 737-61.

Smagorinsky, J., 1953: The dynamical influence of large-scale heat sources and sinks on the quasi-stationary mean motions in the atmosphere. *Quart. J. Roy. Meteor. Soc.* 100, 342-36.

Smagorinsky, J., 1963: General circulation experiments with the primitive equations. I. The basic experiment. *Mon. Wea. Rev.* 91, 99-164.

Smith, P. J., 1980: The energetics of extratropical cyclones. *Rev. Geophys. Space Phys.* 18, 378-386.

Starr, V. P., and A. Oort, 1973: Five-year climatic trend for the Northern Hemisphere. *Nature* 242, 310-13.

Steinberg, L., A. Wiin-Nielsen, C.H. Yang, 1971: On nonlinear cascades in the large-scale flow. *J. Geoph. Res.* 76, 8629-40.

Thompson, R. M., S. W. Payne, E. E. Recker, and R. J. Reed, 1979: Structure and properties of synoptic-scale wave disturbances in the Intertropical Convergence Zone of the eastern Atlantic. *J. Atmos. Sci.* 36, 53-72.

Tibaldi, S., A. Buzzi and P. Malguzzi, 1980: induced cyclogenesis: Analysis of numerical experiments. *Mon. Wea. Rev.* 108, 1302-14.

van Loon, H., 1972: Winds in the Southern Hemisphere. *Meteor. Monogr.* 13, No. 35, 87-111.

van Mieghem, J., 1956: The energy available in the atmosphere for conversion into kinetic energy. *Beitr. Phys. Free Atmos.* 29, 129-42.

van Meighem, J., 1973: *Atmospheric Energetics*, Oxford Monographs in Meteorology, 306 pp.

Webster, P. J., and J. L. Keller, 1975: Atmospheric variations, vacillations and index cycles. *J. Atmos. Sci.* 32, 1283-1300.

Webster, P. J., 1983: Large-scale structure of the tropical atmosphere. in *Large-Scale Dynamical Processes in the Atmosphere.* Edited by B. J.

Hoskins and R. P. Pearce, New York Academic Press, 397 pp.

Wiin-Nielsen, A., 1961: On Short- and long-term variations in quasi-barotropic flow. *Mon. Wea. Rev.* 89, 461-76.

Wiin-Nielsen, A., 1962: On the transformation of kinetic energy between the vertical shear flow and the vertical mean flow in the atmosphere. *Mon. Wea. Rev.* 90, 79-92.

Wiin-Nielsen, A., 1967: On the annual variation and spectral distribution of atmospheric energy. *Tellus* 19, 540-59.

Wiin-Nielsen, A., 1972: Simulations of the annual variation of the zonally averaged state of the atmosphere. *Geofys. Publ,* 28, 1-45.

Wiin-Nielsen, A., 1973: *Dynamic Meteorology*, World Meteorological Organization. 334 pp.

Wiin-Nielsen, A., 1985: *The scientific problem of medium-range weather prediction*, in Proc. of Seminar, 10th anniversary of the European Centre for Medium-Range Weather Forecasts, 31-53.

Wiin-Nielsen, A., 1991: On resolution in atmospheric numerical weather prediction models and cascade processes. *Atmosfera* 4, 3-22.

Wiin-Nielsen, A., and J. A. Brown, Jr., 1962: On diagnostic computations of atmospheric heat sources and sinks and the generation of available potential energy. in *Proc. Int. Conf. on Num. Wea. Prediction*, Tokyo, 1960, 593-613.

Wiin-Nielsen, A., and M. Drake, 1965: On the energy exchange between the baroclinic and barotropic components of atmospheric flow. *Mon. Wea. Rev.* 39, 79-92.

Wiin-Nielsen, A., and M. Drake, 1966: An observational study of kinetic energy conversions in the atmosphere. *Mon. Wea. Rev.* 94, 221-30.

Wiin-Nielsen, A., J.A. Brown Jr., and M. Drake, 1963: On atmospheric energy conversions between the zonal flow and the eddies. *Tellus* 15, 261-79.

Wiin-Nielsen, A., J. A. Brown, Jr., and M. Drake, 1964: Further studies of energy exchange between the zonal flow and the eddies. *Tellus* 16, 168-80.

Winston, J. S., and A. F. Krueger, 1961: Some aspects of a cycle of available potential energy. *Mon. Wea. Rev.* 89, 307-18.

Yang, C.-H., 1967: *Nonlinear aspects of the large-scale motion in the atmosphere*, University of Michigan, Sci. Rept., 08759-1-T, 173 pp.

AUTHOR INDEX

Alpert, J. C., 143, 144
Aspliden, C. I., 227
Austin, J. F., 163

Baker, W. E., 132, 138, 164
Bangaru, B., 165
Barnes, S. L., 149
Barros, V. R., 121
Blackmon, M. L., 151, 152, 155, 156, 161, 187
Boer, G. J., 135
Bolin, B., 177
Bosart, L. F., 137, 148
Boyle, J. S., 148
Brown, J. A., 65, 79
Browning, P. A., 148
Buch, H. S., 65
Buechler, D. E., 148, 149
Buja, L. E., 194, 201–203, 209, 307
Burpee, R. W., 284

Carpenter, T. H., 194
Chang, C. P., 295
Charney, J. G., 52, 77, 110, 177
Chen, T.-C., 92, 108, 118, 123, 131, 145–150, 152, 154, 158, 165, 170, 178, 186–187, 190, 194, 199–203, 209–211, 231–232, 237–238, 246, 249, 252, 269–271, 307, 311, 313–315, 324
Chen, T.-J., 137, 138

Defant, A., 55
Depradine, C. A., 231
DiMego, G. J., 148
Drake, M., 54, 65, 144
Dutton, J. A., 3

Eerola, K., 210
Eliasen, E., 163

Endlich, T. M., 94

Fischer, G., 162, 164
Fjørtoft, R., 110
Flohn, H., 227
Fortelius, C., 164
Fuelberg, H. E., 148–149

Godbole, R. V., 292, 294
Grant, A. M., 324

Haltiner, G. J., 89, 91
Hansen, A. R., 162, 164–165, 170
Heddinghaus, T. R., 208
Hesselberg, Th., 16
Hollingsworth, A., 177
Holopainen, E. O., 151, 158, 164, 181, 186, 194, 210
Hoskins, B. J., 177
Hunt, B. G., 210, 213–214

Illari, L., 164

Jeffries, J., 55
Johnson, D. R., 3, 241
Julian, P. R., 212, 237

Kanamitsu, M., 231, 256–257
Koraly, D. J., 177
Keller, J. L., 210, 214
Koteswaram, P., 227, 268
Kraichnan, R. H., 120, 123
Kraus, E. B., 210
Krishnamurti, T. N., 152, 158, 225, 227, 229, 233, 235–236, 252, 256–257, 291–292, 296
Krueger, A. F., 162, 197, 209
Kung, E. C., 132, 138, 164, 194, 204, 206–208

Lau, N.-C., 65, 67, 151, 177, 181, 184, 193, 210

Lawniczak, G. E., 79
Lee, Y.-H., 154, 178, 181, 187, 199, 210, 271, 311, 313–314
Leith, C. E., 120, 123
Lejenäs, H., 162
Lorenz, E. N., 2, 31, 33–34, 40, 55, 65, 81, 210

Madden, R. A., 212, 237
Maddox, R. A., 148
Margules, M., 12, 14, 16
Marshall, H. G., 211, 238
McGuirk, J. P., 214
Miller, A. J., 213–214
Miyakoda, K., 162
Murakami, T., 193

Namias, J., 151, 156, 165
Newell, R. E., 226
Newton, C., 80, 131, 133, 152, 195, 224
Nicholls, N., 305–306
Nitta, T., 276, 279–282
Norquist, D. C., 287–289

O'Brian, J. J., 279
O'Conner, J. F., 82
Obasi, G. O. P., 100
Oort, A. H., 15, 65, 67, 81, 93–94, 194–195, 199, 202–203, 207–209

Palmén, E., 131, 133, 152, 195, 224
Paulin, G., 162
Pearce, R. P., 145–146
Peixóto, J. P., 65, 93–94, 194–195, 199, 202–203, 207–209
Petterssen, S., 142, 147
Pfeffer, R., 210
Phillips, N. A., 2, 89
Physick, W. L., 328
Pitcher, E. J., 181–182
Platzman, G. W., 110, 111

Price, P. G., 305–306

Rasmusson, E. M., 194
Recker, E. E., 276–277
Reed, R. J., 276–277, 283–285
Rex, D. F., 162–163
Riehl, H., 276
Rodak, H., 324
Rossby, C. G., 151
Ryd, V. H., 9, 14, 16

Saha, K., 291–292, 295
Saha, S., 291, 295
Saltzman, B., 2, 95, 117, 164, 231
Sheu, Y.-T. P., 144
Shukla, J., 162, 165, 292–293
Shutts, G. J., 164
Smagorinsky, J., 2, 95, 177
Smebye, S. J., 142, 147
Smith, P. J., 132, 135, 144
Soong, S., 194, 204, 206–207
Starr, V. P., 65
Steinberg, L., 121, 211

Teweles, S., 117
Thompson, R. M., 283, 286
Tibaldi, S., 148

van Loon, H., 229, 256, 269–270, 320
van Mieghem, J., 3

Webster, P. J., 210, 214, 228, 230
Wiin-Nielsen, A., 2, 22–23, 29, 45, 54, 64–65, 78–79, 92–93, 95, 118, 121, 123, 126–127, 129, 145–146, 150, 159, 194, 197, 199–201, 203, 209–210, 305–306
Winston, J. S., 162

Yang, C.-H., 111, 121
Yen, M.-C., 152, 178, 237

SUBJECT INDEX

abnormal energy conversion, 81-84
African Wave, 283-290
    structure, 284-287
    energetics, 287-290
    energy diagram, 287-288
angular momentum transport, 65
annual variation
    Available potential energy, 45
    comparison of energetics between the Southern and Northern Hemisphere, 304-310
    Kinetic energy budget over North America, 204-210
    Lorenz energy cycle, 195-204
approximation to vorticity equation, 91-92
available potential energy
    annual variation of, 45
    budget analysis of extratropical cyclone, 135-136
    elementary considerations, 4-8
    elementary derivations of, 43-45
    Eulerian equation in an open system, 138-140
    general formulation, 31-34
    Lagrangian equation in an open system, 138-140
    quasi-geostrophic formulation, 41-43

B-integral, 37
baroclinic
    conversion to barotropic flow, 48-54
    conversion to barotropic flow in the tropics, 247-251
    energy cascade, 118-122
    flow, 48
    flow in an open domain, 140-144
    flow in the tropics, 246-251
    predictability, 126-130
barotropic
    conversion to baroclinic flow, 48-54
    conversion to baroclinic flow in the tropics, 247-251
    energy cascade, 118-122
    flow, 41
    flow in an open domain, 140-144
    flow in the tropics, 246-251
    predictability, 118-122
blocking 82-84, 162-177
    constructive interference, 162-164
    nonlinear forcing, 170-177
    spectral energetics, 164-177
boundary layer friction, 52

cascade
    and available potential energy, 103-104
    and enstrophy, 106
    and kinetic energy, 105-106
    and wave interaction, 112-117
    Fourier representation of, 95-99
    Legendre representation of, 101-102
catalytic quantity, 26, 94
coherency, 211-216, 219-230, 319-320
confidence level, 211-215, 217-223
confluence theory, 151
continuity equation in isentropic coordinates, 35
continuity in pressure coordinates, 46

conversion from total potential to kinetic energy, 39-40
conversion, general, see energy conversions
cross spectral, 213-223, 318-323

definition of available potential energy, 31-34
dissipation, 18-19
    and hydrostatic conditions, 24-26
    in baroclinic flow, 48, 53-54
    in barotropic flow, 48, 53-59
    of zonal kinetic energy, 79
    eddy kinetic energy, 79
divergence of
    available potential energy flux, 135
    horizontal Eulerian kinetic energy flux, 137
    horizontal system kinetic energy flux, 137
    kinetic energy flux, 132, 145
    kinetic energy flux of tropical easterly jet, 268-270
divergent and non-divergent flows
    cyclone, 148
    global, 85-90
    in an open domain, 145-146
    mesoscale convective complex, 148-150
    tropical, 251-257

east-west circulation, 225, 227-228
easterly waves, 276
eddy available potential energy, 70
eddy kinetic energy, 70
eddy quantity, 55
efficiency factor, 39, 134
energy
    total amounts, 16
    hydrostatic, 22-27
    non-hydrostatic, 17-19
    stationary eddies, 186-187
energy conversions between
    available potential and kinetic energy, 38-39, 42
    available potential and kinetic energy of extratropical cyclones, 134-135
    baroclinic and barotropic flow, 49-51
    baroclinic and barotropic flows in an open domain, 140-144
    baroclinic and barotropic flows in the tropics, 253-257
    divergent and non-divergent flow, 88-94
    divergent and non-divergent flows in an open domain, 144-150
    divergent and non-divergent flows in the tropics, 253-257
    internal and kinetic energy, 18-19
    potential and kinetic energy, 18-19
    zonal and eddy energy, 74-75, 77-78
energy conversions in hydrostatic systems, 27-28
energy conversion in non-hydrostatic systems, 17-19
enstrophy, 106, 110, 257-268
    spectra in the tropics, 260, 263
    spectra of the budget in the tropics, 259-263, 264-268
equatorial waves, 276-283
    energetics, 279-283
    energy diagram, 282-283
    structure, 276-279
exchange of kinetic energy between low and middle latitudes, 271-274
Eulerian
    available potential energy equation in an open system, 133-136

SUBJECT INDEX 373

kinetic energy equation in an open system, 131-133
extratropical cyclone
    available potential energy budget, 135-136
    kinetic energy budget, 132

Fjørtofts theorem, 111
Fourier analysis
    amplitude, 201
    energy variables of the Northern Hemisphere, 200-204
    energy variables of the Southern Hemisphere, 307-309
    general, 95-99
    for conversion from eddy available to eddy kinetic energy, 107, 109
    for conversion from eddy to zonal kinetic energy, 109
    for conversion from zonal to eddy potential energy, 108
    for eddy available potential energy, 106
    for kinetic energy, 107
    phase, 201
fractional variance, 206-207

generation
    available potential energy, 36, 39, 42, 134-135, 139
    baroclinic kinetic energy, 141-142
    barotropic kinetic energy, 140, 142
    by frictional heating, 19
    divergent kinetic energy, 146-148, 161
    divergent kinetic energy of the summer Australian jet, 325
    kinetic energy, 132-133, 137, 139, 145, 208-210, 271-274
    kinetic energy of the tropical easterly jet, 268-270
    of available potential energy in quasi-geostrophic flow, 42
    of eddy available potential energy, 74-76
    of internal energy, 18-20
    of zonal available potential energy, 74-76
    rotational kinetic energy, 145, 158-162
    rotational kinetic energy of the summer Australian jet, 325
gradient wind vortex, 9-12

Hadley circulation, 226
heating
    frictional and non-frictional, 19
    calculation of, 79
heat transport
    by eddies, 57
    by mean meridional circulation, 57
    by vertical motion, 56
homogeneous fluid with free surface, 4-8
hydrostatic equation, 28

interaction
    between stationary and transient eddies, 190-193
    between zonal-mean state and stationary eddies, 187-190
    nonlinear, 219-223
influence of limited region, 99-100
internal energy, 15, 19
internal friction, 52-53
intertropical convergence zone, 227-228
isentropic coordinates, 33, 35-37

justification (in part) of quasi-geostrophic flow, 92-94

kinetic energy

budget of baroclinic and barotropic flows for cyclones, 142-144
budget of extratropical cyclone, 132, 137-138
budget of rotational and divergent flow for cyclones, 147-148
budget of rotational and divergent flow for mesoscale convective complex, 148-150
budget of subtropical jet streams, 152-162
budget of the summer Australian jet, 323-326
budget of the tropical easterly jet, 268-270
budget over North America, 204-210
equation of baroclinic and barotropic flows in an open domain, 140-144
equation of divergent and rotational flows in an open domain, 145-146
in barotropic and baroclinic flows, 47-48
in divergent and non-divergent flows, 85-86
in non-hydrostatic systems, 17-19
in simple systems, 9-12, 14
in vertical mean flows, 47
in vertical shear flows, 48
in wave number regimes, 105, 107
in wave-wave interactions, 114
in zonal and eddy flows, 71-74

Lagrangian, 136-140
available potential energy equation, 138-140
kinetic energy budget of extratropical cyclone and anticyclone, 137-138
kinetic energy equation, 136-137
Legendre function analysis
general, 101-102
available potential energy, 102-104
enstrophy, 106
kinetic energy, 105
longitudinal wave number, 95-96
Lorenz-diagram, 75
low-frequency variation of the tropical energetics, 237-246

Margules examples, 12-14
mean
vertical mean, 46
zonal mean, 70-71
momentum transport, 59-62
mid-oceanic troughs, 227-229
momentum transport by stationary eddies, 187
monsoon depression, 291-296
energy diagram, 295-296
structure, 293-295

non-divergent part of kinetic energy, 85
non-divergent part of conversion from shear and mean flow, 51
non-frictional heating, 21
non-hydrostatic flow, 17-19

omega equation, 79
Oort diagram, 82

parcel method, use of, 43-44
Parceval's identity, 98-99
potential energy
in non-hydrostatic system, 15-19
in elementary systems, 4-6, 7-9
potential temperature, 31
potential vorticity, 127
predictability

SUBJECT INDEX

    in barotropic systems, 124-126
    in quasi-geostrophic systems, 126-130
  proportionality between potential and internal energy, 22-23

quasi-geostrophic system, 41-43
quasi-geostrophic predictability, 126-130
quasi-periodic variation of atmospheric energetics, 194-223
    annual variation of Lorenz energy cycle, 195-204
    annual variation of the kinetic energy budget over North America, 204-210
    annual variation of the Southern Hemisphere energetics, 304-315
    semiannual variation of energy variables, 195-210
    vacillation of the Northern Hemisphere, 210-223
    vacillation of the Southern Hemisphere, 315-320

Richardson's equation, 23-24

sensible heat transport
    by stationary eddies, 187
    of the Southern Hemisphere, 299-302
Southern Hemisphere
    annual variation of energetics, 304-309
    momentum transport, 302-304
    sensible heat transport, 299-302
    spectral energetics, 309-315
    vacillation of energetics, 315-320
spectral energetics
    of baroclinic and barotropic flows in the tropics, 246-251
    of blocking, 162-177
    of divergent and nondivergent flow in the tropics, 251-257
    the Southern Hemisphere, 309-315
    the tropics, 231-237
stability factor, 41
stationary or standing waves, 65-66
    energetics analysis, 177-193
    interaction with the zonal-mean state, 187
    interaction with transient eddies, 190
    momentum transport by, 187
    release of available potential energy by, 190
    sensible heat transport by, 190
    three-dimensional structure, 181-184
streamfunction, 85
structure functions, vertical, 127-128
subtropical jet streams
    maintenance by hemispheric interaction mechanism, 158-160
    maintenance by regional ageostrophic mechanism, 152-158
    relation between the two maintenance mechanisms, 161-162
summer Australian jet, 323-325

thermodynamic equation, 35, 41
time mean, 65
time-mean zonal momentum equation, 327-329
time series analysis, 211
total potential energy, 27
transient waves, 65
transport

of momentum by stationary
  eddies, 187
of sensible heat, 55-58
of sensible heat by stationary
  eddies, 190
of momentum, 58-62
tropical easterly jet, 268-270

vacillation
  coherency, 211-223, 317-323
  confidence level, 211-223
  cross spectra, 211-223, 317-322
  power spectra, 211-223
  time series analysis of energy variables, 211
velocity potential, 85
vertical mean flow, 46
vertical mean momentum transport and frication, 60-61
vertical shear flow, 46
vertical structure functions, 127-128

wave interactions, 119-122
wave spectra
  for available potential energy, 115-120
  for kinetic energy, 112-115
wave structure and transport of heat, 63-64
wave structure and transport of momentum, 61-64

zonal available potential energy, 70
zonal kinetic energy, 70
zonal mean, 55